Trace elements in animal production systems

Organized and sponsored by:

Under the patronage of:

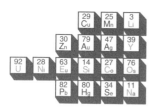

Trace elements

in
animal production systems

edited by:

P. Schlegel
S. Durosoy
A.W. Jongbloed

Wageningen Academic
P u b l i s h e r s

ISBN 978-90-8686-061-6

First published, 2008

Wageningen Academic Publishers
The Netherlands, 2008

Table of contents

Preface

Human population continues to increase worldwide at a high rate. Farm animal populations increase accordingly to compensate the demand of new consumers for animal products. It has therefore become urgent to verify how both populations can live together in a sustainable way. Increases in animal production do not only have major impact on food safety, but pose also new challenges or threats to the environment.

Globalisation and economical forces change, accelerate and intensify the international competition in animal production. Increasing intensive production systems lead to amplified concentrations of nutrients in manure which are threatening by their accumulations in soils, their run-off and leaching into ground and surface waters. Nitrogen and phosphorous have been the first nutrients of environmental concern which lead to new legal limitations in several countries. Trace minerals may in some areas become the next limiting factors for further development of animal farming. Not only should dietary levels of supplementation be carefully controlled, but also the treatment of animal wastes may make a considerable contribution to alleviation of the problem. Everyone (government, farmers, industry, scientists and the public) has their own responsibility and interest in creating a sustainable environment.

Given the above, we decided to organize the first Scientific International Symposium on Trace Elements in Animal Production Systems on 14th and 15th June 2007 in Geneva, Switzerland under the title OTEANE, meaning Organic Trace Elements for Animal Nutrition and the Environment. The first day was especially dedicated to environmental aspects of trace minerals and the second day to the nutritional aspects of trace elements. Under the term 'organic trace elements' are meant trace elements bound to organic compounds containing carbon and nitrogen, such as amino acids, resulting in superior bioavailability. As a result, mineral supply levels may be lowered, while maintaining adequate supply to the animals. This is the reason why Pancosma, as a manufacturer of bioavailable organic trace mineral sources (B-TRAXIM® range), was proud to be the major sponsor of the first edition of this Symposium.

This book contains 18 peer-reviewed proceedings, as well as, peer-reviewed abstracts from 25 presented short communications. This book will be a valuable resource for researchers and professionals in the live sciences of animal nutrition, soil and water quality, for actors in the feed industry and policy making.

The editors gratefully acknowledge the contributions by all authors, and especially those from the Members of the Scientific Committee, who generously gave their time and expertise to OTEANE's success and to this book. The editors also thank Mr. Frans Verstraete, from the E.U. Commission for his contribution in presenting the actual E.U. legislation to trace elements and heavy metals in animal nutrition. The patronage by the Institute of Trace Elements from the UNESCO was also highly appreciated.

A warm thank you is given to Mrs Esra Yilmaz, General Secretary of the first Scientific International Symposium on Trace Elements in Animal Production Systems, for her successful preparation and organisation of the event.

The editors

P. Schlegel
S. Durosoy
A.W. Jongbloed

Introduction

The creation in 1996 of Trace Element-Institute for UNESCO under the auspices of UNESCO, confirms the importance of trace elements as indicated in 1987 by Federico Mayor, Director-General of UNESCO:

> 'Trace elements are increasingly recognized as important factors in sustaining life… international cooperation in this field is of the essence.'

According to UNESCO's recommendations at the World Conference on Science, 1999, Budapest:

> 'Industrialized countries should cooperate with developing countries through jointly defined science projects that respond to the basic problems of the population in the latter.'

> 'All countries should share scientific knowledge and cooperate to reduce avoidable ill-health throughout the world.'

The supply of trace elements (which can be 'essential') to living cells, is linked to an ecosystem in a state of continual change from soils to man. The consequences of trace element deficiencies are well-documented in plants, in man and in animals. For this reason, the first Scientific International Symposium on Trace Elements in Animal Production Systems, OTEANE has provided valuable scientific data on trace elements for animal nutrition and health.

On behalf of UNESCO, we would like to congratulate the organizers and participants of the International Symposium on Trace Elements in Animal Production Systems, OTEANE.

G. Chazot
President of the Trace Element-Institute for UNESCO

Scientific committee

Jongbloed, A.W., Wageningen UR, The Netherlands
Hill, G.M., Michigan State University, U.S.A.
Lall, S.P., National Research Council, Canada
Menzi, H., Swiss College of Agriculture, Switzerland
Nicholson, F.A., ADAS, U.K.
Novak, J.M., United States Department of Agriculture, U.S.A.
Öborn, I., Swedish University of Agricultural Sciences, Sweden
Poulsen, H.D., University of Aarhus, Denmark
Römkens, P.F.A.M., Wageningen UR, The Netherlands
Schlegel, P., Pancosma, Switzerland
Spears, J.W., North Carolina State University, U.S.A.
Van Ryssen, J.B.J., University of Pretoria, South Africa
Windisch, W., University of Natural Resources and Applied Life Sciences, Austria

Environment

Worldwide growth of animal production and environmental consequences

P.J. Gerber and H. Steinfeld

Keywords: climate change, water resources, biodiversity, environmental policy, livestock

Introduction

Ensuring environmental sustainability is one of the Millennium Development Goals to which FAO and its Members are committed. The conservation, improvement and sustainable utilization of natural resources, including land, water, forests, fisheries and genetic resources for food and agriculture is one of the three global goals of FAO's Strategic Framework between 2000 and 2015. In this document, the issue of livestock's impact on the environment is presented.

Driven by growing populations and incomes, the increase in demand for animal products will be stronger than for most food items. Global production of meat is projected to more than double from 229 million tonnes in 1999/2001 to 465 million tonnes in 2050, and that of milk to increase from 580 to 1043 million tonnes (FAO, 2006). The bulk of the growth in meat and milk production will occur in developing countries, with China, India and Brazil representing two thirds of current meat production and India predicted to grow rapidly, albeit starting from a low base. Among the meat products, poultry will be the commodity of choice for reasons of acceptance across cultures and technical efficiency in relation to feed concentrates. It is expected that intensive systems will contribute to most of the increase in production, as they have done in the past three decades.

The livestock sector has a primary and growing role in the agricultural economy. It is a major provider of livelihoods for the larger part of the world's poor people. It is also an important determinant of human health and diet. Global demand for livestock products is projected to double by 2050 – despite this growth, per capita consumption levels in developing countries will be not more than half the level of developed countries. But already the livestock sector stresses many ecosystems and contributes to global environmental problems. Greenhouse gas emissions from livestock production and their waste, and from pasture expansion into forests and pasture degradation are an important factor in climate change. The presence of livestock in the vast majority of the world's ecosystems affects biodiversity. In developed and rapidly developing countries, it is often a major source of water pollution (Gerber, 2005).

The future of the livestock-environment interface will be shaped by how we resolve the balance of two competing demands: for animal food products on the one side and for environmental services on the other. Both demands are driven by the same factors: increasing populations, growing incomes and urbanization. The natural resource base within which they must be accommodated is finite. Therefore, the continuing expansion of the global livestock sector must be accomplished while substantially reducing livestock's environmental impact.

This paper draws from a report submitted to the 20th session of the Food and Agriculture Organisation of the United Nation (FAO) Committee On Agriculture (COAG), held on 25 to 28 April 2007 in Rome. It summarizes recent studies on these issues and suggests possible lines of action for dealing with the environmental challenges posed by the sector.

Environmental impacts

Land and land-use change: humanity's largest land use

Livestock's land use includes (a) grazing land and (b) cropland dedicated to the production of feed. Considering both, livestock represent approximately 70 percent of all agricultural land.

The total land area occupied by *livestock grazing* is 3.4 billion hectares equivalent to 26 percent of the ice-free terrestrial surface of the planet. A large part of these areas are too dry or too cold for crop use, and are only sparsely inhabited. While the grazing area is not increasing at a global scale, in tropical Latin America there is rapid expansion of pastures into valuable ecosystems, with 0.3 to 0.4 percent of forest lost to pastures annually. Ranching is a primary reason for deforestation.

About 20 percent of the world's pastures and rangeland have been degraded to some extent, and this percentage may be as high as 73 percent in dry areas (UNEP, 2004). The Millennium Ecosystem Assessment (MEA, 2005) has estimated that 10 to 20 percent of all grassland is degraded, mainly by livestock. However, some of the dryland grazing ecosystems have proved to be quite resilient and degradation has shown to be partly reversible (Steinfeld *et al.*, 2006).

The total area dedicated to *feed crop production* amounts to 471 million hectares, equivalent to 33 percent of the total arable land. Most of this total is located in Organization for Economic Co-operation and Development (OECD) countries, but some developing countries are rapidly expanding their feed crop production, notably maize and soybean in South America. Again, a considerable part of this expansion is taking place at the expense of tropical forests. It is expected that future growth rates of livestock output will be based on similar growth rates for feed concentrate use (FAO, 2006). Without the necessary precautions, intensive feed production can result in land degradation, including soil erosion and water pollution (Steinfeld *et al.*, 2006).

Gaseous emissions and climate change

Recent estimates of global Green House Gas (GHG) emissions from different sources by the Intergovernmental Panel on Climate Change (IPCC), the United Nations Framework Convention on Climate Change (UNFCCC) and the 'Stern Review', show that land use changes due to deforestation result in 18.3 percent of total GHG emission while agriculture accounts for 13.5 percent (of which agricultural soils is 6 percent, livestock and manure 5.1 percent) and the transportation sector 13.5 percent (of which road transport is 10 percent).

Considering different forms of emissions throughout the livestock commodity-chains, GHG estimates for the livestock sector are substantial (Steinfeld *et al.*, 2006). GHG emissions occur at the level of feed production (e.g. chemical fertilizer production, deforestation for pasture and feed crops, cultivation of feed crops, feed transport and soil organic matter losses in pastures and feed crops), animal production (e.g. enteric fermentation and methane and nitrous oxide emissions from manure) and as a result of transport of animal products. Using such methodology, livestock contribute about 9 percent of total anthropogenic carbon-dioxide emissions, but 37 percent of methane and 65 percent of nitrous oxide emissions. The commodity-chain methodology used in this paper is not used by the IPCC and, therefore, emissions may be attributed in a different manner.

Carbon dioxide is released when previously-forested areas are converted into grazing land or arable land for feed. Therefore, expansion of pasture and cropland at the expense of forests releases significant amounts of carbon dioxide into the atmosphere as does the process of pasture and arable land degradation, often associated with a net loss of organic matter. Carbon dioxide releases resulting from fossil fuel consumption used for the production of feed grains (tractors, fertilizer production, drying, milling and transporting) and feed oil crops need also be attributed to livestock. The same applies to the processing and transport of animal products. Methane is emitted from rumen fermentation and from livestock waste when stored under anaerobic conditions, for example in so-called lagoons. Yet another category is constituted by nitrous oxide emissions from intensive feedcrop production and related chemical fertilizer application.

Regarding polluting gaseous emissions not linked to climate change, livestock waste emits a total of 30 million tonnes of ammonia. Livestock account for 68 percent of total ammonia emissions (Steinfeld *et al.*, 2006).

Technical options are available to mitigate gaseous emissions of the sector. Carbon-dioxide emissions can be limited by reducing deforestation and the sector can contribute to carbon sequestration through a range of practices including: restoring organic carbon in cultivated soils, reversing soil organic carbon losses from degraded pastures and sequestration through agro-forestry. Improved livestock diets as well as better manure management can substantially reduce methane emissions, while careful nutrient management (i.e. fertilization, feeding and waste recycling) can mitigate nitrous oxide emissions and ammonia volatilization. Furthermore, the use of biogas technology is a way to reduce emissions from manure management while increasing farm profit (e.g. savings on energy bill, electricity trading) and providing environmental benefits, such as reduced fossil fuel consumption).

Water

The livestock sector is a key factor in increasing water use and water depletion. The share of livestock sector is about 8 percent of global water use. The major part of this water is used for irrigation of feed crops, representing 7 percent of the global water use. The water used for product processing and drinking and servicing is less than one percent of global water use, but it often is of great importance in dry areas; for example, livestock drinking requirements represent 23 percent of total water use in Botswana.

Important amounts of water are used for irrigating pasture and feed crops, mostly in developed countries. Through the compacting effect of grazing and hoof action on the soil, livestock also have a determining, and often negative, impact on water infiltration and water erosion.

Water quality can be affected by livestock through the release of nitrogen, phosphorus and other nutrients, pathogens and other substances into waterways, mainly from intensive livestock operations. The fact that the livestock sector is industrializing, in a number of concentrated locations, separates the sector from its supporting land base and interrupts the nutrient flows between land and livestock, creating problems of depletion at the source (land vegetation and soil) and problems of pollution at the sink (animal wastes, increasingly disposed of into waterways instead of back on the land).

Livestock have a major role in water pollution though the release of animal manure into freshwater resources. These wastes are not controlled for sediments, pesticides, antibiotics, heavy metals and biological contaminants. Livestock land-use and animal waste-management appear to be the main mechanism through which livestock contribute to the water depletion process.

Multiple and effective options for mitigation exist in the livestock sector that would allow the reverse of current water-depletion trends. Mitigation options usually rely on three main principles: reduced water use (e.g. through more efficient irrigation methods and animal cooling systems), reduced depletion, (e.g. through increased water productivity and mitigated pollution from waste management and feed crop fertilization) and improved replenishment of the water resources through better land management.

Biodiversity

Livestock affect biodiversity in many direct and indirect ways. Livestock and wildlife interact in grazing areas, sometimes positively, but more often negatively. Livestock help to maintain some of the open grassland ecosystems, but animal disease concerns pose new threats to wildlife.

Pasture expansion, often at the expense of forest, has vast negative consequences on some of the most valuable ecosystems in Latin America, while rangeland degradation affects biodiversity on all continents. Crop area expansion and intensification for livestock feed undoubtedly affect biodiversity negatively, sometimes with dramatic consequences (soybean expansion into tropical forests). Water pollution and ammonia emissions, mainly from industrial livestock production, reduce biodiversity, often drastically in the case of aquatic ecosystems. Pollution, as well as over-fishing for fishmeal as animal feed, affects biodiversity in marine ecosystems.

Livestock's important contribution to climate change will clearly have repercussions on biodiversity, while the historic role of livestock as a driver and facilitator of invasions by alien species continues – in particular by way of introduction of exotic pasture seeds and livestock diseases.

Differences between species, products and production systems

There are huge differences in the environmental impact between the different forms of livestock production, and between species.

Cattle provide a multitude of products and services, including beef, milk, hides and draught power. In mixed farming systems, cattle are usually well integrated in nutrient flows and can have a positive environmental impact. In many developing countries, cattle and buffaloes still provide animal draught for field operations, and in some areas, animal traction is on the increase (parts of sub-Saharan Africa), potentially substituting for fossil fuel use. Livestock also consume crop residues some of which would otherwise be burned. However, cattle in extensive livestock production in developing countries are often only of marginal productivity. As a result, the vast majority of feed is spent on the animal's maintenance, leading to resource inefficiencies and often high levels of environmental damage per unit of output, particularly in overgrazed areas.

The dairy sector is more closely connected to the supporting feed resource than is the case for other forms of market-oriented livestock production. Most milk operations tend to be close to areas of feed supply because of their daily demand for fibrous feed, allowing for nutrient recycling. However, excessive use of nitrogen fertilizer on dairy farms is one of the main causes of high nitrate levels in surface water in OECD countries. There is also a risk of soil and water contamination by manure runoff and leaching from large-scale dairy operations.

Beef is produced in a wide range of intensities and scales. At both ends of the intensity spectrum considerable environmental damage occurs. On the extensive side, cattle are involved in degradation of vast grassland areas and are a contributing factor to deforestation (pasture conversion). The resulting carbon emissions, biodiversity losses and negative impacts on water flows and quality constitute major environmental impacts. On the intensive side, feedlot size is often vastly beyond the capacity of surrounding land to absorb nutrients (Steinfeld *et al.*, 2006). In the feedlot stage the conversion of concentrate feed into beef is far less efficient than into poultry or pork, and therefore beef has significantly higher resource requirements per kilogram of meat than pork or poultry. However, taking the total life-cycle into account, including the grazing phase, concentrate feed per kilogram of growth is lower for beef than for non-ruminant systems (CAST, 1999).

The production of sheep and goats is usually extensive, except for small pockets with feed lots in the Near East and West Asia, and North America. The capacity of small ruminants, in particular goats – to grow and reproduce under conditions otherwise unsuitable for any form of agricultural production – makes them useful and very often essential to poor farmers pushed into these environments for lack of alternative livelihoods. Because of their adaptive grazing, sheep and goats can affect land cover and the potential for forest re-growth. Under overstocked conditions, they are particularly damaging to the environment through degradation of vegetative cover and soil.

Extensive pig production, based on the use of household waste and agro-industrial by-products as feed, performs a number of useful environmental functions by turning biomass of no commercial value – which otherwise would be waste – into high-value animal protein. However, extensive systems are incapable of meeting the surging urban demand in many developing countries, not only in terms of volume but also in sanitary and other quality standards. The ensuing shift towards larger-scale grain-based industrial systems has been associated with geographic concentration, leading to nutrient overload of soils and water pollution. Furthermore, most industrial pig production in the tropics and sub-tropics uses waste-flushing systems involving large amounts of water.

Poultry production has been the system most subject to structural change. In OECD countries production is almost entirely industrial, while in developing countries it is already predominantly industrial. Although industrial poultry production is entirely based on feed grains and other high-value feed material, it is the most efficient form of production of food of animal origin (with the exception of some forms of aquaculture), and has the lowest land requirements per unit of output. Poultry manure is of high nutrient content, relatively easy to manage and is widely used as fertilizer and sometimes feed. Other than for feed crop production, the environmental damage, though sometimes locally important, is of a much lower scale than for the other species.

In conclusion, livestock-environment interactions are often diffuse and indirect; and damage occurs at both the high and low end of the intensity spectrum, but is probably highest for beef and lowest for poultry.

What needs to be done?

Major corrective measures need to be taken to address the environmental impact of livestock production that will otherwise worsen dramatically, given the projected expansion of the livestock sector. However, growing economies and populations combined with increasing scarcity of environmental resources and rising environmental problems are translating into a growing demand for environmental services, such as clean air and water, and recreation areas. Increasingly, this demand will broaden from immediate factors of concern, such as reducing the nuisance factors of flies and odours, to the intermediate demands of clean air and water, and then to the broader, longer-term environmental concerns, including climate change, biodiversity, etc. At the local level, markets will undoubtedly develop for the provision of such services; this is already the case for water in many places. At the global level, the emergence of such markets is uncertain although promising models already exist, for example carbon trading.

Encouraging efficiency through adequate market prices

Current prices of land, water and feed resources used for livestock production often do not reflect true scarcities. This leads to an overuse of these resources by the livestock sector and to major inefficiencies in the production process. Any future policy to protect the environment will, therefore, have to introduce adequate market pricing for the main inputs, for example, by introducing full-cost pricing of water and grazing fees.

A host of tested and successful technical options are available to mitigate environmental impacts. These can be used in resource management, in crop and livestock production, and in post-harvest reduction of losses. However, for these to be widely adopted and applied, appropriate price signals which more closely reflect the true scarcities of production factors, and correcting the distortions that currently provide insufficient incentives for efficient resource use are required. The recent development of water markets and more appropriate water pricing in some countries, particularly water-scarce ones, are steps in that direction.

Correcting for environmental externalities

Although the removal of price distortions at input and product level will go a long way to enhancing the technical efficiency of natural resource use in the livestock production process, this may often not be sufficient. Environmental externalities, both negative and positive, need to be explicitly factored into the policy framework, through the application of the 'provider gets – polluter pays' principle.

Correcting for externalities, both positive and negative, will lead livestock producers into management choices that are less costly to the environment. Livestock holders who generate positive externalities need to be compensated, either by the immediate beneficiary (such as with improved water quantity and quality for downstream users) or by the general public (such as with carbon sequestration from reversing pasture degradation).

While regulations remain an important tool in controlling negative externalities, there is a trend towards taxation of environmental damage and incentives for environmental benefits. It may gain momentum in future, tackling local externalities first but increasingly also trans-boundary impacts, through the application of international treaties, underlying regulatory frameworks and market mechanisms. Government policies may be required to provide incentives for institutional innovation in this regard.

Accelerating technological change

A number of technical options could lessen the impacts of intensive livestock production. Concerning feed cropping and intensive pasture management, good agricultural practices can reduce pesticide and fertilizer losses. Conservation agriculture and other forms of resource preserving technologies can restore important soil habitats and reduce degradation. Combining such local improvements with restoration or conservation of an ecological infrastructure at the landscape level may offer a good way of reconciling the conservation of ecosystem functioning and the expansion of agricultural production. Improvements in extensive livestock production systems can also make a contribution to biodiversity conservation, for example by adopting silvopastoral systems and planned grazing management that limits overgrazing of plants and increases biodiversity and quantity of forage, soil cover and soil organic matter thus reducing water loss to evaporation, runoff and sequestring carbon dioxide. Options exist to increase production and achieve a variety of environmental objectives.

Improved and efficient production technologies exist for most production systems. However access to information and technologies and capacity to select and implement the most appropriate ones are restraining factors, which can be reduced through interactive knowledge management, capacity building and informed decision making at policy, investment, rural development and producer levels. Technological improvements need to be oriented towards optimal integrated use of land, water, human, animal and feed resources. In the livestock sector, the quest for optimizing efficiencies will be through feeding, breeding and animal health. Research and management of feed crop production needs to aim at higher yields in more locally adapted eco-friendly production systems and socio-economic research for rural development needs to provide a better understanding of the external factors that enable realization of the improvements in the first two sectors

Reducing negative environmental and social impacts of intensive production

An estimated 80 percent of total livestock sector growth comes from industrial production systems. The environmental problems created by industrial systems mostly derive from their geographical location and concentration. In extreme cases, size may be a problem: sometimes units are so large (a few hundred thousand pigs, for example) that waste disposal will always be an issue, no matter where these units are located.

What is required therefore is to bring waste generated into line with capacity of accessible land to absorb that waste. Industrial livestock must be located as much as possible where cropland within economic reach can be used to dispose of the waste, without creating problems of nutrient loading – rather than geographically concentrating production units in areas favoured by market access, or feed availability. Policy options to overcome the current economic drivers of the peri-urban concentration of production units include zoning, mandatory nutrient management plans, financial incentives and facilitation of contractual agreements between livestock producers and crop farmers. Regulations are also needed to deal with heavy metal and drug residue issues at the feed and waste levels, and to address other public health aspects such as food-borne pathogens.

Whether industrialized or more extensive livestock production systems, they need to strive to obtain lowest possible emissions with full waste treatment adapted to local conditions. This requires close coordination and integration with other development activities like bioenergy, transport, urban and peri-urban development, forestry and others. Associated additional costs need to be absorbed across various economic sectors.

In parallel, there is a need to address the environmental impacts associated with production of feed grain and other concentrate feed. Feed is usually produced in intensive agriculture, and the principles and instruments that have been developed to control environmental issues there need to be widely applied.

Diversifying extensive grazing with the provision of environmental services

Grazing systems need to be intensified in areas where the agro-ecological potential so permits, in particular for dairy production, and where nutrient balances are still negative. In many OECD countries, excess nutrient loading is a major issue in grass-based dairy farming. Reductions in the number of livestock have been imposed, sometimes with very positive results. However, the vast majority of extensive grazing lands are of low productivity. Grazing occupies 26 percent of the ice-free terrestrial surface but the contribution that extensive grazing systems make to total meat production is less than 9 percent.

In a world with around 9 billion people by 2050 and a growing middle class, there will be a growing demand for environmental services; therefore extensive systems will have the opportunity to include the provision of environmental services as an important, and sometimes predominant, purpose. This can be facilitated by payments for environmental services or other incentives to enable livestock producers to enhance resource sustainability.

The central argument here is that the opportunity cost for livestock to use marginal land is changing. Livestock used to occupy vast territories because there was no viable alternative use, whereas other usages (e.g. biodiversity conservation, carbon storage, bio-fuels) are now competing with pasture in some regions. Water-related services will likely be the first to grow significantly in importance in future, with local service provision schemes the first to be widely applied. Biodiversity-related services (e.g. species and landscape conservation) are more complex to manage, because of major methodological issues in the valuation of biodiversity, but they already find a ready uptake where they can be financed through tourism revenues. Carbon sequestration services, through adjustments in grazing management or abandonment of pastures, will also be difficult. However, given the potential of the world's vast grazing lands to sequester large amounts of carbon and to reduce emissions, mechanisms must be developed and deployed to use this potentially cost-effective avenue to address climate change.

Suggesting a shift from some of the current negative grazing practices to environmental-service oriented grazing raises two questions of paramount importance: how to distribute profits from environmental services and how to deal with the poor who currently derive their livelihoods from extensive livestock? Their numbers are considerable. Livestock provide an important source of livelihood in poor countries, for example in Mauritania (where it provides 15 percent of GDP), the Central African Republic (21 percent) and Mongolia (25 percent).

Not all environmental services of sustainable livestock production will be easily paid for through immediate product pricing. Alternative employment generation and social safety nets are some of the needs to ensure sufficient knowledge and labour to maintain marginal but important production areas, which however require effective integration/collaboration with other rural development activities, particularly in countries where poverty and lack of public resources and governance result in unsustainable land use. More encompassing external assistance will be required in countries where global assets such as biodiversity, climate and food security are concerned but where the economic potential for other sectors is limited.

The challenge ahead

As an economic activity, the livestock sector generates about 1.4 percent of the world's GDP (2005); it accounts for 40 percent of agricultural GDP. With a 2.2 percent global growth rate for the last ten years (1995 to 2005) globally and 5.5 percent growth rate in developing countries, the livestock sector is growing faster than agriculture as a whole whose share of overall GDP is declining. However, the livestock sector is much more important than its modest contribution to the overall economy would suggest. Livestock provide livelihood support to an estimated 987 million poor people in rural areas (LID, 1999). Livestock products in moderation are also an important element of a diverse and nutrient-rich diet.

These diverse aspects of livestock's importance inform national decision-making for the sector. The different national policy objectives of food supply, poverty reduction, food safety and environmental sustainability take on different levels of importance depending on factors such as stage of development, per capita income and general policy orientation of a country. Furthermore, the objectives of sector policies and the instruments used to achieve environmental goals should be tailored according to the farming systems and relative stakeholders they target.

A key aspect is that, compared to its economic performance, the environmental impacts of the livestock sector are not being adequately addressed. The problem therefore lies mainly with institutional and political obstacles and the lack of mechanisms to provide environmental feedback, ensure that externalities are accounted for and embed the stewardship of common property resources into the sector.

The first challenge, therefore, is to raise awareness among stakeholders of the scale of the environmental problem. Environmentally-motivated action currently focuses on the functions and protection of specific ecosystems. As we have seen, the mobility of the livestock industry allows its relocation without major problems becoming apparent. However, the pressure on the environment is usually shifted elsewhere, and manifests itself in different forms. For example, intensification may reduce pressure on grazing lands but increase pressure on waterways. Thus, another challenge is to add a sector perspective to the analysis of environmental issues.

The complexity of livestock-environment interactions and their many manifestations make concerted actions more difficult. Investment and production choices are driven by a variety of factors, many of which are external to the livestock/agriculture sector as such. This is also true of many other environmental and development issues and is a major reason why environmental policy-making lags behind other areas. In this sense, the livestock sector is driven by its own set of policy objectives and decision-makers find it difficult to address economic, social, health and environmental objectives simultaneously. Frequently they also lack the tools, information access and platforms to initiate and implement such complex decision making processes.

The livestock sector is industrializing at the level of production units and and food chains while also remaining an important livelihood source for large numbers of small-scale and marginalized producers in many parts of the world. The fact that so many people depend on livestock for their livelihoods and health limits the available options to policy-makers, and involves difficult and politically sensitive decisions on trade-offs. Policy-makers therefore need to address the multiple objectives of livestock development: affordable supply of high value food; food safety; livelihoods and environmental soundness. Perhaps the biggest challenge is therefore to build the institutions and cross-sectoral capacities, both nationally and internationally, to address the complex environmental issues with a sense of urgency, while considering social and public issues.

Reducing livestock's environmental impacts has a net cost that will need to be borne by the sector and the consumers. Addressing the environmental challenge thus raises issues of smallholder competitiveness, reduced purchasing power of the non-affluent consumers and reduced national competitiveness, especially in those countries that do not compensate increasing production costs by increased transfers to producers. Despite these difficulties, the impact of livestock on the local and global environment is so significant that it needs to be addressed with urgency. The challenge for policy makers is to balance the interests of consumers and producers, and to ensure an equitable outcome of this process.

Expecting the livestock sector to deliver on all fronts is ambitious. The policy framework for the livestock sector, as for other areas, is characterized by a large number of trade-offs that need to be balanced at the national and local level. For example, a large commercial expansion of the sector, benefiting from economies of scale and with upgraded food safety standards, creates barriers to smallholder producers. Many simply will not have the financial and technical means to compete and will be forced out of business. Likewise, distortions and externalities can be corrected but the costs of higher input prices and environmental controls will have to be passed on to the consumer, in the form of higher prices for meat, milk and eggs. Balancing these trade-offs and arbitration among stakeholders is a further challenge to policy makers.

Simultaneously with other measures, the reduction of expected demand from the livestock sector would also ease environmental pressure and costs and should find entry into policies in both developed and developing countries while assuring adequate nutritional needs [health] and security of the various population groups.

Given the planet's finite natural resources, and the additional demands on the environment from a growing and wealthier world population, it is imperative for the livestock sector to move rapidly towards far-reaching change. Four lines of action are suggested:

First, the strive for efficiency gains in resource use for livestock production must continue, on the basis of much-required price corrections for inputs, and replacing current suboptimal production with advanced production methods – at every step from feed production, through livestock production and processing, to distribution and marketing. Policy-makers are called upon to steer and facilitate this process.

Second, there is a need to accept that the intensification of livestock production is an inevitable characteristic of the structural change process that is ongoing for most of the sector. The key challenge is to make this process environmentally acceptable by facilitating the right location so as to enable waste recycling on cropland, and applying the right technology, especially in feeding and waste management. Locating industrial livestock units in suitable rural environments and not in congested peri-urban or otherwise favoured settings allows for the availability of land area and recycling of nutrients. When making changes to the structure of livestock market chains in order to mitigate adverse environmental effects, it will at the same time be important to consider social impacts.

Third, extensive land-based production will continue to exist. However, decision-makers will need to adjust grassland-based production so as to include the provision of environmental services as a major purpose, and probably as the most important one in vulnerable areas. Policy makers need to provide a framework for the delivery landscape maintenance, biodiversity protection, clean water and eventually carbon sequestration from extensive grazing systems, in addition to production of conventional livestock commodities.

Last, but certainly not least, for the suggested changes to occur, there is an urgent need to go beyond existing policy frameworks at the local, national and international level. A strong political will and urgency, established together with potential actors and beneficiaries, are required to initiate action and investment in creative ways to avert the environmental risks of continuing 'business as usual.'

References

CAST, 1999. Animal Agriculture and Global Food Supply. Council for Agricultural Science and Technology (CAST) ISBN 1-887383-17-4.

FAO, 2006. World Agriculture: towards 2030/2050, Interim Report. Rome, Italy.

Gerber, P, P. Chilonda, G. Franceschini and H. Menzi, 2005. Geographical determinants and environmental implications of livestock production intensification in Asia. Bioresource Technology 96, 263-276.

LID, 1999. Livestock in poverty focused development. Crewkerne: Livestock In Development.

Millennium Ecosystem Assessment, 2005. Ecosystems and human well-being: synthesis. Washington D.C. Island Press.

Steinfeld, H., P. Gerber, T. Wassenaar, V. Castel, M. Rosales and C. de Haan, 2006. Livestock's long shadow Environmental issues and options. FAO, Rome.

UNEP, 2004. Land degradation in drylands (LADA): GEF grant request. Nairobi, Kenya.

Assessment and reduction of heavy metal input into agro-ecosystems

H. Eckel, U. Roth, H. Döhler and U. Schultheiß

Keywords: heavy metals, heavy metal balances, animal manures, input reduction measures, supplementation

Introduction

It is well known that a considerable share of the inputs of heavy metals into soils is a consequence of agricultural activities, namely fertilisation and the use of organic residual materials like livestock manure, sewage sludge or compost (Wilcke and Döhler, 1995; Nicholson, 2002). Additionally inputs not deriving from agriculture, such as atmospheric deposition, can contribute significantly to the heavy metal load on farmed land.

Still there are a lot of gaps in the knowledge of heavy metal pathways onto farms and hence eventually into the soil. This refers to the relative importance of the various input pathways as compared to the total input, as well as to the options at hand to achieve a metal input reduction.

In view of the activities of the European Commission to develop a soil protection policy on European level (European Commission, 2002), European co-operation in heavy metal related research gains in importance.

The AROMIS Concerted Action was set up by KTBL and 23 research institutions from across Europe, representing EU-25 Member States, Accession Countries and Associated States aiming to provide a cross national assessment of heavy metals in European agriculture.

The main objectives of AROMIS can be summarised as follows:
- Identification of the pathways of heavy metals in agro-ecosystems and evaluation of the relative significance of the various metal input and output pathways.
- Provision of information on current legal regulations relating to heavy metals in agriculture.
- Description of possible technical and legal measures for reducing heavy metal inputs and assessment of the potential effectiveness and practicality of selected measures.
- Identification of future research and technology transfer demands.
- Creation of Europe-wide contacts between research institutions to link heavy metal related research activities in Europe and enhance the exchange of knowledge on ecological, economic, technical, and legal aspects.

Description of work

The project was divided into three major sections:
1. Data collection
   Collection of data and background information on heavy metals in agriculture in the participating countries. This includes the sources of heavy metals, the legal regulations on European and national level and the research activities in this field. To handle this data a Oracle-based database was set up.
2. Heavy metal balances
   An essential part of the project were heavy metal balances on farm level, calculated to demonstrate the inputs, outputs and internal flows of heavy metals and enable the simulation of the effect of input reduction strategies. For this an MS-Excel-based balance tool was developed which allows the calculation of farm level balances. For a number of selected countries, where the data availability was sufficient, balances were calculated taking into account typical or model farms for livestock farming and crop production in the relevant country or region.
3. Input reduction measures
   Based on the balance results in the next step of the project options for the reduction of the heavy metal input were described and evaluated regarding performance as well as economic and ecological consequences.
   In addition to the reduction measures the future research and development needs were identified, derived from the results of the assessment of the input reduction measures and the data gaps which became apparent when setting up the data base.

A list of the heavy metals taken into account in the project is provided in Table 1.

Table 1. List of heavy metals taken into account in the AROMIS project.

Cadmium (Cd)	Lead (Pb)
Chromium (Cr)	Nickel (Ni)
Copper (Cu)	Zinc (Zn)

Results

Heavy metal balances

Input pathways

Atmospheric deposition was found to be an important source of heavy metals (especially Cd, Ni and Pb) to agricultural land. However, different monitoring methods were used by the AROMIS participants to measure deposition and the data were not directly comparable between countries.

Mineral fertilisers, in particular P-fertilisers, were an important source of Cd and sometimes contributed significantly to Cr and Ni inputs. For the farms balanced in the AROMIS study, Cd and Cr inputs with mineral fertilisers (where used) contributed on average 30 and 40% of total inputs, respectively.

Organic residuals such as composts, sewage sludge, or industrial wastes could be an important source of many heavy metals. There was considerable variation in the quality of sludges and composts both within and between the different countries. The application of sewage sludge could lead to significant metal surpluses at the farm level, the size depending on the metal concentration of the material applied. This would be the case even if sludges with metal contents much lower than those permitted by Council Directive 86/278/EEC were used.

At the national scale, organic wastes were found to be a relatively minor source of heavy metals to farmed land because the area to which they were applied was relatively small in most countries. Nevertheless, high metal input rates at the field level indicated that use of these organic wastes should be carefully monitored.

For livestock farms, one of the most important metal input pathways was with livestock feeds. Heavy metals are present in livestock basal diets (e.g. cereals, grass) at background concentrations and may be added to certain feeds as supplementary trace elements (Cu and Zn) to maintain health, normal development and performance, for welfare reasons or as growth promoters. Other metals (e.g. Cd, Pb) are not deliberately added to livestock feeds, but may be present as contaminants in the basal feeds or mineral supplements.

As an example Figure 1 illustrates the share of the various inputs for a pig farm located in Germany (for the values see Table 2).

Only a small fraction of the metals in livestock feeds are retained in animal tissues, with most being excreted and eventually spread to soils with the manure (Schenkel, 2002a). Elevated levels of Cu and Zn were found in pig and poultry manures as a result of feed supplementation and medicinal use described above. Cattle manure metal concentrations tended to be lower than for pigs and poultry. A considerable proportion of the cattle feedstuff metals were derived from homegrown feeds, indicating that they were cycling within the farm and should not be considered as net inputs.

The use of Cu-disinfectants ($CuSO_4$) in cattle farming was found to contribute up to 40% of the total Cu input into these systems (UBA, 2004).

Output pathwuys

The main metal output pathways considered in the farm gate balances were leaching and crop uptake, provided that the crops were exported from the farm and not used as home-grown feeds (see example in Figure 2 and Table 2).

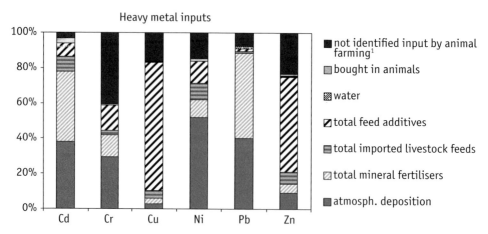

Figure 1. *Share of the various inputs for a pig farm located in Germany (sows, pig rearing and finishing, 80.5 livestock units, 34 ha of land used for feedstuff production).*
[1]*not identified input by animal farming = (heavy metal flow via animal manure + heavy metal flow via animal products) minus (heavy metal flows via homegrown feeds, imported livestock feeds and feed additives, and, if necessary, other inputs, which end up in the manure and the animal products and would thus be counted twice).*

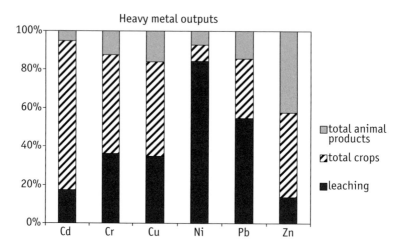

Figure 2. *Share of the various outputs for a pig farm located in Germany (sows, pig rearing and finishing, 80.5 livestock units, 34 ha of land used for feedstuff production).*

Export of metals with animal products such as meat, milk or eggs is limited (except for Zn) even in farms with high intensity animal husbandry. The latter observation is mainly due to the limited transfer of metals from feed to animal tissue or products like milk, eggs etc. Compared to heavy metal levels in crops, levels in animal products are usually one to two orders of magnitude lower.

Table 2. Heavy metal balance for a pig farm located in Germany (sows, pig rearing and finishing, 80.5 livestock units, 34 ha of land used for feedstuff production).

	Cd	Cr	Cu	Ni	Pb	Zn
input [g/ha·yr]						
atmosph. deposition	0.77	12.80	12.00	18.79	25.19	136.58
irrigation	0.00	0.00	0.00	0.00	0.00	0.00
total mineral fertilisers	0.81	5.35	13.48	3.63	30.62	75.72
total organic fertilisers	0.00	0.00	0.00	0.00	0.00	0.00
total imported livestock feeds	0.18	1.07	16.32	3.44	0.54	100.27
total feed additives	0.15	6.15	298.17	4.55	0.92	804.72
water	0.00	0.00	0.28	0.00	0.85	6.82
bought in animals	0.00	0.00	0.09	0.02	0.01	1.41
seeds	0.06	0.37	2.13	0.47	0.11	19.61
not identified input by animal farming[1]	0.06	17.55	65.57	5.34	4.68	344.84
total input	2.03	43.29	408.05	36.24	62.93	1489.96
output [g/ha·yr]						
leaching	0.25	0.94	12.70	12.70	2.41	27.94
total crops	1.13	1.34	18.05	1.25	1.36	91.01
total animal products	0.08	0.33	5.78	1.10	0.63	87.03
animal manure exported from the farm	0.00	0.00	0.00	0.00	0.00	0.00
others 1	0.00	0.00	0.00	0.00	0.00	0.00
others 2	0.00	0.00	0.00	0.00	0.00	0.00
total output	1.46	2.60	36.52	15.05	4.40	205.97
load change [g/ha·yr]	0.57	40.69	371.53	21.19	58.53	1283.99
internal flows [g/ha·yr]						
total home-grown feeds	0.48	0.54	12.48	0.59	0.71	126.65
total other internal flows	0.49	1.89	9.26	1.17	1.49	34.69
animal manure used on the farm itself	1.29	26.88	396.03	14.00	7.72	1324.14

[1]not identified input by animal farming = (heavy metal flow via animal manure + heavy metal flow via animal products) minus (heavy metal flows via homegrown feeds, imported livestock feeds and feed additives, and, if necessary, other inputs, which end up in the manure and the animal products and would thus be counted twice).

Metal load change on farm level

The construction of farm gate heavy metal balances has allowed heavy metal flows on farms across Europe to be compared. Differences in fertiliser management and the quality of feedstuff as well as the feeding regime dominated the ultimate metal load change. Although some negative net metal load changes were found (mostly in arable crop farms without manure or sludge application and use of low-Cd phosphate fertiliser), load changes were mostly positive and strongly depended on the farm type. Table 2 shows the heavy metal balance of the same pig farm as in Figures 1 and 2.

As Table 2 shows there is quite a large surplus especially for Cu and Zn which is a result of the supplementation of feedstuff with these metals. For the same reason the amount of Cu and Zn in the manure is high, resulting in high internal flows. Internal flows are defined as the metals circling within the farm, from livestock manure to the soil and the feed being grown, and back to the animal.

Net negative load changes tended to occur in crop rather than in livestock production farms. In crop production this is mostly due to the exclusive use of mineral fertilisers low in metals, which in combination with elevated leaching losses (e.g. in Northern Europe) can cause important negative load changes. The maximum positive load changes in crop farming were observed in farms using organic fertilisers, especially with sewage sludge, with the exception of Cu for which the highest values occurred in special cropping (grapes and tomato).

In livestock farms negative load changes are also related to high leaching losses in combination with low input systems. This is often due to high internal flows (i.e. the feedstuffs are mainly produced and used on the farm, the manure is also used on the farm itself, e.g. in extensive dairy farming), minimising the need to purchase feedstuffs or fertilisers. On the other hand in farms that import large quantities of feed additives or supplemented compound feeds and use the manure on the farm itself, large positive load changes are common, especially for Zn and Cu – the balance results for these two trace elements reflecting clearly the differences in feeding management between cattle and pig or poultry husbandry.

For Zn and Cu the load changes in livestock production farms often exceeded those of crop production farms, although high load changes can also occur on crop production farms if sewage sludge, manure or fungicides are used. For Cd, Cr, Ni and Pb the differences are less pronounced. Load changes of Cd in plant production farms can exceed those from livestock farms if high-Cd phosphate fertilisers are used.

To assess whether or not the net load changes reported here could lead to unwanted effects, field gate balances should be constructed for each scenario to indicate how soil, crop and groundwater quality will change with time. The data that were used to construct the balances within the AROMIS project confirm that certain loads (either input or output) need to be measured on the site itself or modelled according to site specific conditions instead of using default values from other farms (or even countries); this holds true especially for leaching losses or crop uptake by commercial crops or home-grown feeds.

The differences between the balanced farms (wide range of farm sizes in terms of area and/or stocking rate, use of model/real farms, measured/default metal concentrations, estimated/ modelled/determined mass fluxes etc.) make it difficult to draw universally valid conclusions, even within one farm type. This is the case especially in livestock production, where a meaningful interpretation of the balance results is only possible, if up to now unidentified inputs in livestock farming are carefully analysed.

Nonetheless, the evaluation of the farm balances set up in the AROMIS network is an effective way to illustrate input patterns and tendencies in European agriculture, both in crop and livestock farming. If more detailed data were available, the balances could give a more accurate assessment of the mass and metal fluxes on a farm system. Those balances, which are based on robust data, could be used for model calculations leading to concrete conclusions on soil quality, metal input reduction potentials etc.

Input reduction measures

Minimising soil metal accumulation is an issue of sustainable management and soil protection should be balanced against practical and economic considerations. The options available can be grouped as those that:
- restrict the metal content of inputs to the system;
- limit the application of metal materials such as manures and wastes;
- introduce new technologies.

The control of some of the current sources may only be possible by one of these whilst others may lend themselves to two or more. Change may be brought about by Direct or Indirect Statutory Measures or by voluntary initiatives. Within the European Union there is the need to consider a common legal framework. For any measures the political and practical constraints of enforcement must be considered and recognition given to the problems facing the farming industry. The input reduction measures considered in detail during the AROMIS project included:
- reduced trace element supplementation of livestock feedstuff;
- control of Zn oxide given to weaner pigs;
- control of the use of metal containing footbaths for ruminants;
- control of the heavy metal content of mineral fertilisers (in particular Cd in phosphatic fertilisers);
- adapted fertilisation management (choice and combination of fertilisers).

For many of the sources of metals used in agriculture there are already measures available to effect control (Schenkel, 2002b; UBA, 2004). However there are cost/benefit reasons why these may not be the most suitable ways of protecting soils in the future. Alternative solutions may need to be found or consideration given to a possible waste management solution by removing the metal containing materials from the farm or preventing the contamination in the first place.

The effects of selected input reduction measures were assessed based on data from model as well as existing farms using the AROMIS balance tool. Reduced Cu and Zn supplementation was evaluated for a typical pig production farm (see section 'Animal nutrition for details')

The replacement of $CuSO_4$ for hoof disinfection was calculated for a dairy farm and a range of fertilisation scenarios were evaluated for an arable farm.

Fertilisation: Based on the maximum nutrient supply as defined in the Dutch MINAS system (mineral accounting system) a range of scenarios was calculated showing the influence of fertilisation management on metal inputs. In the scenario exclusively using mineral fertilisers, Cd, Cr and Zn inputs into the farming system are dominated by fertilisation as compared to the input by atmospheric deposition. The same holds true for all six metals in the scenario using sewage sludge (here metal inputs were highest for all metals) and for Cu and Zn in the livestock slurry scenarios, especially the one using pig slurry.

Animal nutrition: A survey of available literature on the options to reduce supplementation showed, that for most animal categories there is some scope for reducing the level of Cu and Zn in the diet even below the new MPL´s (maximum permitted levels) as defined in the Commission Regulation (EC) No 1334/2003 (KTBL, 2002), even though there are uncertainties regarding the options for ruminants.

Model calculations to evaluate the effect of a reduced supplementation of Cu and Zn in pig production on metal input on farm level were carried out, based on data of a typical pig production farm. Three different scenarios were calculated based on the maximum allowances for Cu and Zn according to the Council Directive on additives in feedingstuffs (70/524/EEC – basis for scenario), the EC-regulation 1334/2003 (EC1334, in force from January 2004), and the proposal of the German Standing Committee for Feedstuffs for the revision of EC Directive (FMA) (see Table 3).

In the three scenarios, based on different levels of supplementation, a reduction of 27 to 62% for Cu and 24 to 49% for Zn could be achieved (Tables 4 and 5). It becomes clear that the implementation of regulations restricting the use of Cu and Zn in animal nutrition is an effective measure to limit the amount of metals in the manure.

Since feed supplements are often contaminated with other, undesirable metals (UBA, 2004), reducing the supplementation of Cu and Zn might as well have an effect on those other metals. The extent of this effect could only be estimated.

Animal hygiene: In an example dairy farm, used to evaluate the input of Cu by hoof disinfectants, the inputs with this product were 334 g/ha yr (= 240 g/livestock unit yr) corresponding to 35% of the total Cu input. Use of an alternative to $CuSO_4$ reduced the Cu-surplus for this farm by about 45% (from 908 g/ha yr to 574 g/ha yr). Such measures however, must be accompanied by increased hoof care for the animals to avoid infections and thus an increased use of antibiotics.

Whereas for some input reduction measures the available knowledge is sufficient for realisation in practice, for others the basis for implementation is not given and more research and development activities are required.

Table 3. Former (EC 70/524 EEC), current (EC 1334/2003) and previously proposed (FMA, Petersen 2002) maximum permitted levels of Cu and Zn as additives in animal feed.

Livestock category	Cu [mg/kg (88% DM)]			Zn [mg/kg (88% DM)]		
	70/524 EEC	EC 1334/2003	FMA 2000	70/524 EEC	EC 1334/2003	FMA 2000
Pigs						
Rearer/grower	175	170	30[1]	250	150	100
Finisher	100	25	20	250	150	100
Breeding sow	35	25	20	250	150	100
Poultry	35	25	20	250	150	120
Ruminants						
Dairy cattle	35	35	10[2] 15[3]	250	150	120
Beef cattle	35	35	30	250	150	100
Ewes	15	15	15	250	150	120

[1] up to 2 months.
[2] milk replacer.
[3] other complete feeds.

Table 4. Reduction of Cu and Zn input on farm level.

Reduction scenario	Cu reduction [%]	Zn reduction [%]
1: 70/524/EEC → EC 1334	27	33
2: 70/524/EEC → FMA	62	49
3: EC1334 → FMA	49	24

Table 5. Calculated heavy metal content in pig manure based on maximum permitted levels of supplementation in the feeds.

	Calculated metal content in pig manure [mg/kg dm]	
Scenario	Cu	Zn
70/524/EEC (scenario basis)	251	886
70/524/EEC	182	576
EC1334	91	422

Research demand/demand for future action

The project has highlighted the lack of reliable data to allow a full evaluation of the input and output pathways of heavy metals in agro-ecosystems, metal behaviour in soils and changes in soil metal concentration over time.

The AROMIS consortium advocates the creation of a European-wide network for heavy metal research and monitoring, representing the main agricultural production regions in Europe and covering the range of conditions for agricultural production throughout Europe. Special emphasis should be placed on:
- livestock feeds (homegrown and particularly imported compound feeds and mineral premixes);
- livestock manures;
- organic residues recycled in agriculture.

Additionally systematic trials to investigate the effect of low metal input systems on animal health and welfare and on economic performance should be carried out. Furthermore the transfer of information on the options to reduce the input of heavy metals is considered necessary, e.g. through demonstration projects at farm level on selected farms.

Acknowledgements

The authors wish to thank the European Commission for the financial support of the AROMIS project as well as all AROMIS project partners for their contributions. Special thanks to Dr. Anne Bhogal, Bartosz Golik, Dr. Harald Menzi, Dr. Fiona Nicholson, Dr. Paul Römkens and Dr. Roger Unwin.

The AROMIS Final Report (KTBL 2005. Assessment and reduction of heavy metal input into agro-ecosystems. KTBL-Schrift 432, 232 p.) can be ordered at www.ktbl.de/shop or directly contacting h.eckel@ktbl.de. The AROMIS Heavy Metal Database is accessible at http://daten.ktbl.de/aromis.

References

KTBL, 2002. Fütterungsstragien zur Verminderung von Spurenelementen/Schwermetallen in Wirtschaftsdüngern (Strategies to reduce the concentration of trace elements/heavy metals in livestock manures), KTBL-Schrift 410, KTBL, Darmstadt.

Nicholson, F., 2002. Heavy metal content of animal manures and implications for soil fertility, Report of DEFRA project No SP0516, DEFRA, United Kingdom.

Petersen, U., 2002. Futtermittelrechtliche Vorschriften über Spurenelemente und unerwünschte Stoffe. In: Fütterungsstrategien zur Verminderung von Spurenelementen/Schwermetallen in Wirtschaftsdüngern. KTBL-Schrift 410: 36-41, KTBL, Darmstadt.

Schenkel, H., 2002a. Stoffwechseleffekte und Umweltwirkungen einer gezielten Spurenelementsupplementierung. In: Fütterungsstrategien zur Verminderung von Spurenelementen/Schwermetallen in Wirtschaftsdüngern. KTBL-Schrift 410: 31-35, KTBL, Darmstadt.

Schenkel, H., 2002b. Minderungsmöglichkeiten und Konsequenzen für Produktqualität, Leistung, Gesundheit und Umwelt (in German). In: Fütterungsstrategien zur Verminderung von Spurenelementen/Schwermetallen in Wirtschaftsdüngern. KTBL-Schrift 410: 131-138, KTBL, Darmstadt.

UBA, 2004. Erfassung von Schwermetallströmen in landwirtschaftlichen Tierproduktionsbetrieben (Assessment of heavy metal flows in animal husbandry and development of a strategy to reduce heavy metal inputs into agro-ecosystems by animal manures; final report including English summary). UBA-Texte 06/04, 130 p., Umweltbundesamt, Berlin, Germany.

Wilcke, W. and H. Döhler, 1995. Schwermetalle in der Landwirtschaft (Heavy metals in Agriculture). KTBL Arbeitspapier 217, 98 p.

Copper and zinc accumulation in sandy soils and constructed wetlands receiving pig manure effluent applications

J.M. Novak, A.A. Szogi and D.W. Watts

Keywords: copper, zinc, soil, constructed wetlands

Introduction

The shift in animal production management from small farrow-to-finish operations to confined animal feeding operations (CAFOs) and the concentration of manure production into limited geographic areas have caused concern about the long term ability of soil to assimilate nutrients contained within the manure. Manure applications can rapidly increase soil quality and productivity on most soils. However, soil quality and productivity may be impaired in the long-term unless precautions are taken with respect to the loading and frequency of manure applications. Watersheds with intensive CAFOs frequently contain soils that have excessive amounts of plant macronutrients (N and P; Kellogg *et al.*, 2000). Similarly, over-application of animal manures has also caused plant micro nutrients (Cu, Zn, etc.) accumulation (Senesi *et al.*, 1999; Bolan and Adriano, 2003; Sistani and Novak, 2006). State nutrient management plans may limit manure applications on soils containing excessive nutrients, therefore alternative methods to treat animal wastes may be needed when land is a limiting factor. The purpose of this paper is to discuss how (1) the recent increase in CAFOs and their localization in a few North Carolina Coastal Plain counties have caused concern about sustainability of soil manure applications; (2) long-term application of pig lagoon effluent coupled with a low crop Cu and Zn nutrient requirement causes accumulation in soil; and (3) treatment of pig lagoon effluent using constructed wetlands can result in Cu and Zn concentration reductions prior to land application.

Confined animal feeding operations and manure management

Traditionally, pig production in the USA has been concentrated in the Midwestern states of Iowa, Illinois, and Minnesota. In recent decades, pig production has undergone a major shift towards localization of production facilities to limited geographic areas of the USA. The centralization of pig production into small rearing areas is referred to as CAFOs. The economic advantage of CAFOs is ease of animal management as well as reduced feed, water, and manure hauling transportation expenses. Individual CAFOs can now house thousands of pigs in confined structures. North Carolina is one area of the USA that has experienced consolidation of pig operations through a reduction of 20,000 operations to approximately 2,000 (NCDA, 2007). This reduction in total number of operations has resulted in larger enterprises with more pigs per operation. Prior to 1990, the pig population in North Carolina was between 1.8 and 2.5 million (Figure 1). After 1990, the pig population swelled from approximately 2.5 million head to almost 10 million head (NCDA, 2007). In fact, North Carolina pig population now ranks second among all states behind Iowa which leads with over 16 million head.

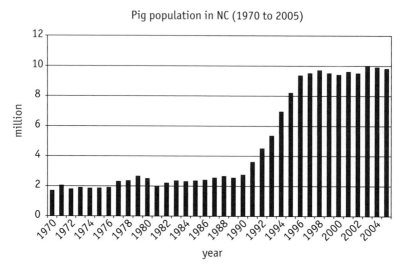

Figure 1. Pig population in North Carolina (1970 to 2005) (NCDA, 2007).

The majority of CAFOs in North Carolina are found in four counties of the Coastal Plain region. Pig CAFOs in these four counties have thrived due to corporate investments, proximity of processing plants, and long history of pig rearing by local farmers (Furuseth, 2001). These four counties account for 59% of North Carolina's entire pig population and are located on only 6% of North Carolina's total land area. Duplin and Sampson counties have the highest pig population (> 2 million head per county) in the state. In fact, these two counties alone account for 44% of North Carolina entire pig population (NCDA, 2007).

Adoption of pig CAFOs has also concentrated pig manure production. Pig manure is flushed from the CAFOs and is stored and treated in anaerobic lagoons to allow for solid/liquid separation and for microbial processing to lower nutrient concentrations (MWPS-18, 2004; Figure 2, left photo). The solids settle to the bottom of the lagoon and at some point must be removed because of lagoon manure storage capacity restrictions. The manure solids are enriched in N, P and trace elements relative to the liquid effluent (Baker, 1996). The sludge is transported and applied to cropland using spreaders equipped to handle slurries (Baker, 1996). The supernatant lagoon effluent is also periodically removed and used as a fertilizer and irrigation source. Using a spray boom, it is surface applied to fields containing row crops or perennial grasses (Figure 2, right photo). Surface application of pig waste allows for recycling of plant macro and micro nutrients as well as irrigation of crops. Nutrients in the manure are taken up by row crops or perennial grasses and removed from the system when the grain or hay is harvested.

Figure 2. Storage of pig manure in a lagoon and spray application of liquid fraction to Bermuda grass field in North Carolina.

Trace element composition of soil amendments and animal manures

Trace elements (i.e. Cu and Zn) can enter the soil by applying a number of amendment sources including commercial fertilizers, liming material, sewage sludge, and animal manures. Commercial fertilizers, liming material, and sewage sludge may contain various concentrations of trace elements as impurities in the parent rock or from differences in urban refuse and treatment processes (Table 1). Applications of these amendments to fields can deliver from low to very high relative amounts of trace metals (Table 1).

Both Cu and Zn are delivered to fields during animal manure applications (Table 2). Livestock requires trace elements in their daily food rations because they are essential for maintaining various physiological processes (Hostetler *et al.*, 2003; Bolan and Adriano, 2003). Because

Table 1. Range of Cu and Zn concentrations in fertilizers, limestones and sewage sludges and their quantity supplied to a field using an application rate of 10 Mg ha^{-1} (dry weight) of each source (USEPA, 1999; Senesi et al., 1999).

Source[1]	Concentration in source (mg kg^{-1})				Delivered to field (g ha^{-1})			
	Cu		Zn		Cu		Zn	
DAP	<1.5	3.2	71	2,193	<15	32	710	21,930
MAP	<1.5	1.6	60	91	<15	16	600	910
TSP	1.6	13	75	696	16	130	750	6,960
Limestone	<0.3	125	<1	450	3	1,250	<10	4,500
Sewage sludge	16	23,700	58	50,000	160	237,000	580	500,000

[1]DAP = diammonium phosphate, MAP = monoammonium phosphate, and TSP = triple super phosphate.

some animal feed stocks contain inadequate supplies, Cu and Zn are added to feeds to ensure an optimal supply and to minimize health disorders. Unfortunately, trace metals are frequently supplemented in amounts that exceed the daily recommended intake amount (Jondreville *et al.*, 2003). Not all of the Cu and Zn consumed by animals are absorbed by the animal's digestive tract. Consequently, the manure is often Cu and Zn enriched (Table 2). Pigs can excrete approximately 80 to 95% of the total daily Cu and Zn contained in dietary supplements (Parkinson and Yells, 1985; Brumm, 1998). As shown in Table 2, manure from different livestock often contains more Zn than Cu. With respect to pigs, this is explained by higher Zn supplements added to feedstocks (up to 3,000 mg kg^{-1}) compared to Cu (250 mg kg^{-1}, Hill *et al.*, 2000). Land disposal of manures to agricultural fields can deliver from low to very high relative concentrations of Cu and Zn (Table 2).

Table 2. Range of Cu and Zn concentrations in various fresh manure sources and their quantity supplied to a field using an application rate of 10 Mg ha^{-1} (dry weight) of each source (NCSU, 2007).

Source	In manure sources (mg kg^{-1})				Delivered to field (g ha^{-1})			
	Cu		Zn		Cu		Zn	
Dairy	3.4	9	14	27	34	90	140	470
Beef	2.3	9	13.5	30	23	90	135	300
Pig	5	44.5	37	155	50	445	370	1,550
Broiler	11.5	---	42	---	115	---	420	---
Turkey	12.5	17.5	115	700	125	175	1,150	7,000

Copper and zinc accumulation in soils

Background total Cu and Zn concentrations in sandy soils of the Coastal Plain region of North Carolina are typically low (14 to 20 mg kg^{-1}). The concentrations are low due to lack of Cu and Zn-bearing minerals in the sandy marine parent material as well as the high degree of mineral weathering (Franklin *et al.*, 2003). On one hand, Cu- and Zn- enriched pig manure applications to sandy Coastal Plain soils have an agronomic benefit by increasing Cu and Zn concentrations which serve as crop micro nutrients (Barker and Zublena, 1995). Many row crops and perennial grasses require trace amounts of Cu and Zn per metric ton of yield. For example, Zublena (1991) reported that Coastal Bermuda grass (*Cynodon dactylon* L.) removes 0.009 and 0.218 kg of Cu and Zn, respectively, per 7.3 Mg ha^{-1} yield of aboveground biomass. On another hand, fields may receive intensive rates of manure applications because land for pig manure application around CAFOs in North Carolina is often limited. In a 10-yr study, Novak *et al.* (2004) estimated that a manure overloaded Coastal Plain spray field in Duplin Co., North Carolina received from 580 to 4,330 m^3 ha^{-1} yr^{-1} of pig lagoon effluent (Table 3).

Table 3. Mean topsoil (0 to 15 cm) plant available and total Cu and Zn concentrations in a North Carolina field after yrs of pig manure effluent applications (standard deviation in parentheses)[1].

Element	Yrs	Range of effluent applications $m^3 ha^{-1} yr^{-1}$	Plant available[2] $mg kg^{-1}$	Total $mg kg^{-1}$
Cu	0	0	0.48 (0.30)a	1.25 (0.58)a
	4	4330 to 860	2.37 (1.39)a	6.20 (2.22)ac
	10	860 to 580	3.37 (1.88)a	8.52 (3.57)bc
Zn	0	0	1.69 (1.39)a	8.10 (0.93)a
	4	4330 to 860	5.89 (3.13)ac	17.34 (5.39)ac
	10	860 to 580	7.42 (1.68)bc	20.09 (3.32)bc

[1]After Novak *et al.* (2004).
[2]Mean Cu and Zn concentrations within a column tested using a 1-way ANOVA; means followed by a different letter are significantly different at the $P < 0.05$ level of rejection.

These high application rates increased both plant available (Mehlich 3 extraction, Mehlich, 1984) and total Cu and Zn concentrations in topsoils.

Fortunately, not all of the trace elements delivered to soils are plant available. Trace elements become plant available through acidification reactions that can dissolve solid phases or through chelation-type reactions with soluble organic compounds secreted by plant roots, microorganisms (Bohn *et al.*, 1979), or from dissolution of soil organic matter (Stevenson, 1994). Chelation of trace elements by soluble organic compounds can increase leaching to ground and surface water sources (Bohn *et al.*, 1979), and some studies have linked consumption of trace element-enriched water with adverse human and mammalian health effects (Senesi *et al.*, 1999). Both Cu and Zn can become sequestered through reactions with minerals and oxide phases thus contributing to higher concentrations in the total pool (Bohn *et al.*, 1979).

The mean concentrations of plant available Cu and Zn measured in topsoil from this spray field were far below the phytotoxic concentration threshold level for most agronomic crops grown in North Carolina soils (Mehlich 3 extractable, > 60 mg kg^{-1} and > 120 mg kg^{-1} for Cu and Zn, respectively; Tucker *et al.*, 2003). This means that Cu and Zn phytotoxic concentration threshold levels most probably have not been exceeded in North Carolina Coastal Plain soils; however, if this spray field is cultivated with a crop that has a lower Cu and Zn sensitivity level compared to values reported by Tucker *et al.* (2003), then lower crop yields may occur. For example, the mean topsoil Zn concentration of 7.42 mg kg^{-1} is approaching the threshold limit (12 to 20 mg kg^{-1}; Keisling *et al.*, 1977) for some Zn-sensitive peanut varieties (*Arachis hypogea*. L). In the future, if the buildup of Zn reaches a critical phytotoxic threshold level, yields of Zn-sensitive crops may be reduced.

Constructed wetlands used to remove nutrients from pig manure effluent

When land availability is limited and nutrient management plans limit manure applications, other treatment systems are needed that will filter nutrients from animal manures. Constructed wetlands, as an integral component of on-farm manure management systems, are much less land intensive than the typical approach of land application (Hunt and Poach, 2001). This technology is perceived as being affordable operationally and has been used around the world for over 30 yrs (Kadlec and Knight, 1996). Typically, manure effluent is treated by flowing through cells colonized with vegetation that removes N by plant and microbial assimilation processes and P by sequestration reactions to sediments (Figure 3).

In a pilot project study, Szogi *et al.* (2003) assessed constructed wetland systems for their ability to filter N; they also examined the wetland's ability to sequester Cu and Zn after 6 yrs of operation (Watts *et al.*, 2001). These scientists measured inflow and outflow effluent volumes and their Cu and Zn concentrations over 2-yrs of the 6 yrs span of effluent treatment (Figure 3). Over the 2-yr monitoring period, approximately 1,126 m^3 of effluent passed through the two wetland cells. They found that the annual mean outflow Cu and Zn concentrations were significantly lower than inflow concentrations (Table 4). Reductions from 82 to 86% between mean annual Cu and Zn inflow vs. outflow concentrations were measured, implying a high degree of sequestration by this constructed wetland system. Wetlands can remove trace elements from waste water via formation and precipitation with oxides and sulfides within the sediment (Tarutis and Unz, 1995) and by plant uptake (Mitsch and Wise, 1998). Plant uptake of trace elements in constructed wetlands accounts for only a small portion removed (Mitsch and Wise, 1998); however, precipitation with oxides is considered the most important assimilation mechanism (Stark *et al.*, 1996).

Figure 3. Constructed wetland system used to treat nutrients in pig waste effluent.

Table 4. Annual Cu and Zn mean concentrations measured in effluent (n = 20 to 24, standard deviation in parentheses)[1].

Element[2]	1997		1998	
	Inflow mg L[-1]	Outflow mg L[-1]	Inflow mg L[-1]	Outflow mg L[-1]
Cu	0.57 (0.28)a	0.08 (0.05)b	0.84 (0.48)a	0.12 (0.09)b
Zn	0.81 (0.43)a	0.15 (0.08)b	1.14 (0.76)a	0.16 (0.12)b

[1]Unpublished data.
[2]Inflow vs. outflow concentrations tested using a *t*-test and means followed by a different letter are significantly different at *P* < 0.001.

The spatial and temporal patterns of Cu and Zn accumulation in sediments of the constructed wetlands were also considered due to possible saturation of Cu and Zn to sediment in the top 0 to 5-cm depth. Saturation of the top few cm of sediment could cause reductions in the wetland's Cu and Zn assimilation capacity. After 6 yrs of annual sampling sediments in three quadrants by depth of both in-series constructed wetlands, they verified that Mehlich 3 (M3) extractable Cu and Zn accumulated with time; but that the Cu and Zn were mainly sequestered by surface sediments (0 to 5 cm deep) in the first one-third of Cell 1 near the inlet (Figure 4, Watts *et al.*, 2001). There was additional sediment surface area available in down system zones for sequestration processes to occur. This shows the promise of additional Cu and Zn sequestration over time.

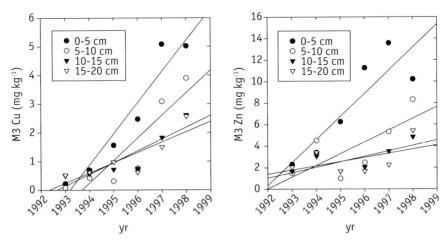

Figure 4. Regression analyses of sediment Cu and Zn concentrations vs. yr of constructed wetland system operation (Watts et al., 2001).

Conclusions

Widespread interest in trace element accumulation in soils has emerged over the past few decades because of applications of amendments and animal manures enriched with trace elements. An example case in the Coastal Plain region of North Carolina was presented to show that shifts in agricultural production from operations with low animal populations into CAFOs have concentrated animal manure production and created a practice of manure disposal on limited land area. Daily rations for livestock in these CAFOs are fortified with trace elements such as Cu and Zn to insure animal health. Because of incomplete trace element assimilation by the digestive system of livestock, their manures will contain trace elements. Physiological differences between livestock will further cause variations in Cu and Zn concentrations in manures.

Manure applications are a benefit to soils because they contain both macro and micro nutrients. Although, Cu and Zn are plant micro nutrients, low plant uptake of these nutrients has resulted in their accumulation in soils. Research has shown that repeated application of pig manure effluent to the same field has caused both Cu and Zn accumulation in sandy soils above background levels. The Cu and Zn levels in these soils, however, are far below established phytotoxic thresholds levels. The mean Zn concentration, while much lower than established phytotoxic levels for most row and pasture crops, might be approaching the sensitivity level for some Zn-sensitive crops. In the future, this may cause a soil quality concern for some fields that have Zn-sensitive crops as part of agricultural rotations.

If nutrient management plans limit land disposal of manures, an alternate method to treat animal manure effluent is by use of an on-site constructed wetland. Nutrient removal from pig manure effluent occurs by uptake through vegetation and microbial processes, as well as by sequestration to sediments within the constructed wetland. Preliminary research has shown that constructed wetlands treating pig manure effluent can be effective in reducing Cu and Zn concentrations. Assessment of spatial and temporal data within the wetland systems also suggests that Cu and Zn sorption occurred in the top few cm of sediments near the effluent inlet. The wetland's Cu and Zn assimilation capacity was not maximized after 6 yrs of operation, implying that wetlands may be a long-term solution to trace element removal in swine manure effluent.

References

Baker, J.C., 1996. Lagoon design and management for livestock waste treatment and storage. North Carolina Cooperative Extension Service, Publication No. EBAE 103-83. (available at: www.bae.ncsu.edu).

Barker, J.C. and J.P. Zublena, 1995. Livestock manure nutrient assessment in North Carolina. In: Proc. 7th Intern. Symp. Agric. Wastes. pp. 98-106. American Society of Agricultural Engineers, St. Joseph, Michigan.

Bohn, H.L., B.L. McNeal and G.A. O'Connor, 1979. Soil Chemistry. John Wiley and Sons, New York.

Bolan, N.S. and D.C. Adriano, 2003. Distribution and bioavailability of trace elements in livestock and poultry manure byproducts. Crit. Rev. Environ. Sci. Technol. 34, 291-338.

Brumm, M.C., 1998. Sources of manure: Swine. In: Animal Waste Utilization, Effective Use of Manure as a Soil Resource. J.L. Hatfield and B.A. Stewart (eds.), pp. 1-49. Ann Arbor Press, Michigan.

Franklin, R.E., L. Duis, B.R. Smith, R. Brown and J.E. Toler, 2003. Elemental concentrations in soils of South Carolina. Soil Sci. 168, 280-291.

Furuseth, O.J., 2001. Hog farming in eastern North Carolina. Southeastern Geographer 41, 53-64.

Hill, M.G., G.L. Cromwell, T.D. Crenshaw, C.R. Dove, R.C. Ewan, D.A. Knabe, A.J. Lewis, G.W. Libal, D.C. Mahan, G.C. Shurson, L.L. Southern and T.L. Veum, 2000. Growth promotion effects and plasma changes from feeding high dietary concentrations of zinc and copper to weaning pigs (regional study). J. Anim. Sci. 78, 1010-1016.

Hostetler, C.E., R.L. Kincaid and M.A. Mirando, 2003. The role of essential trace elements in embryonic and fetal development in livestock. Vet. J. 166, 125-135.

Hunt, P.G. and M.E. Poach, 2001. State of the art for animal wastewater treatment in constructed wetlands. Water Sci. Technol. 44, 19-25.

Jondreville, C., P.S. Revy and J.Y. Dourmad, 2003. Dietary means to better control the environmental impact of copper and zinc by pigs weaning to slaughter. Livestock Prod. Sci. 84, 147-156.

Kadlec, R.H. and R.L. Knight, 1996. Treatment Wetlands. Lewis Publ. Boca Raton, Florida.

Keisling, T.C., D.A. Lauer, M.E. Walker and R.J. Henning, 1977. Visual, tissue, and soil factors associated with Zn toxicity of peanuts. Agron. J. 69, 765-769.

Kellogg, R.L., C.H. Lander, D.C. Moffit and N. Gollehon, 2000. Manure nutrients relative to the capacity of croplands and pastureland to assimilate nutrients: Spatial and temporal trends for the United States. USDA-NRCS-ERS Public. No. NPS00-0579. United Stated Dept. of Agriculture, Washington, DC.

Mehlich, A., 1984. Mehlich 3 soil test extractant: A modification of Mehlich 2 extractant. Commun. Soil Sci. Plant Anal. 15, 1409-1416.

Mitsch, W.J. and K.M. Wise, 1998. Water quality, fate of metals, and predictive model validation of a constructed wetland treating acid mine drainage. Water Res. 32, 1888-1900.

MWPS-18, 2004. Manure characteristics. Section 1. 2nd ed. Midwest Plan Service, Iowa State University, Ames, Iowa.

NCDA, 2007. Statistics-Livestock, Poultry and Dairy. North Carolina Department of Agricultural, Agricultural Statistics Division. (available at: www.ncagr.com/stats/index.htm).

NCSU, 2007. Animal and poultry manure characteristics. North Carolina State University, Raleigh, North Carolina. (available at: www.bae.ncsu.edu)

Novak, J.M., D.W. Watts and K.C. Stone, 2004. Copper and zinc accumulation, profile distribution, and crop removal in Coastal Plain soils receiving long-term, intensive applications of swine manure. Trans. ASAE 47, 1513-1522.

Parkinson, R.J. and R. Yells, 1985. Copper content in soil and herbage following pig slurry application to grassland. J. Agric. Sci. 105, 183-190.

Senesi, G.S., G. Baldassarre, N. Senesi and B. Radina, 1999. Trace element inputs into soils by anthropogenic activities and implications for human health. Chemosphere 39, 343-377.

Sistani, K.R. and J.M. Novak, 2006. Trace metal accumulation, movement, and remediation in soils receiving animal manure. In: Trace Elements in the Environment. M.V. Prasad, K.S. Sajwan, and R. Naidu (eds.), CRC Press, New York.

Stark, L.R., F.M. Williams, W.R. Wenerick, P.J. Wuest and C. Urban, 1996. The effects of substrate type, surface water depth, and flow rate on manganese retention in mesocosm wetlands. J. Environ. Qual. 25, 97-106.

Stevenson, F.J., 1994. Humus Chemistry. 2nd edition. John Wiley and Sons, New York.

Szogi, A.A., P.G. Hunt and F.J. Humenik, 2003. Nitrogen distribution in soils of constructed wetlands treating lagoon wastewater. Soil Sci. Soc. Am. J. 67, 1943-1951.

Tarutis, W.J. and R.F. Una, 1995. Iron and manganese release in coal mine drainage wetland microcosms. Water Sci. Technol. 32, 187-192.

Tucker, M.R., D.H. Stokes and C.E. Stokes, 2003. Heavy metals in North Carolina soils: Occurrence and significance. North Carolina Department of Agricultural and Consumer Services, Raleigh, North Carolina. (available at: www.ncagr.com).

USEPA, 1999. Background report on fertilizer use, contaminants and regulations. EPA 747-R-T8-003. Office of Pollution Prevention and Toxics, Washington, DC.

Watts, D.W., J.M. Novak, A.A. Szogi, K.C. Stone, P.G. Hunt, F.G. Humenik and J.M. Rice, 2001. Copper and zinc accumulation in a constructed wetland treating swine wastewater. p. 144. 2001 Agronomy Abstracts, American Society of Agronomy, Madison, Wisconsin.

Zublena, J.P., 1991. Soil facts: Nutrients removed by Crops in North Carolina. North Carolina Coop. Ext. Serv. Bull. AG-439-16, Raleigh, North Carolina.

Livestock manure management and treatment: implications for heavy metal inputs to agricultural soils

F.A. Nicholson and B.J. Chambers

Keywords: manure storage, manure treatment, heavy metals, soils

Introduction

The group of contaminants known as heavy metals or potentially toxic elements (PTEs) includes zinc (Zn), copper (Cu), nickel (Ni), lead (Pb), cadmium (Cd), chromium (Cr), arsenic (As) and mercury (Hg). Heavy metals enter soils from a range of diffuse and agricultural sources, with leaching losses and plant uptake usually relatively small compared with the total quantities entering soils. As a consequence, they tend to slowly accumulate in topsoils over time. This could have long-term implications for the quality of agricultural soils, including phytotoxicity at high concentrations, the maintenance of soil microbial processes (e.g. nitrogen fixation, biological activity), and the transfer of zootoxic elements to the human diet from increased crop uptake or soil ingestion by grazing livestock (e.g. Cd and Pb), (Nicholson *et al.*, 2003).

It was recently estimated that *c.*30% of total Zn and Cu inputs to agricultural soils in England and Wales in 2004 were from livestock manures, comprising some 1700 t of Zn and 540 t of Cu (Nicholson and Chambers, 2006). The rate of Zn addition to soils at the field level from pig and laying hen manures was estimated at *c.*2.5 kg/ha/yr, equivalent to around 60% of the Zn input rate from sewage sludge (biosolids). Similarly, field Cu input rates from pig manures were *c.*1.5 kg/ha/yr or 50% of the Cu input rate from sewage sludge (Figure 1). These data highlight the important contribution of livestock manure additions (in particular pig and laying hen manures) to the heavy metal burden on some agricultural soils.

Heavy metals are present in livestock diets at background concentrations and may be added to livestock feeds for health and welfare reasons, or as growth promoters. Copper is added to growing pig diets as a cost-effective method of enhancing performance, and is thought to act as an anti-bacterial agent in the gut (Rosen and Roberts, 1996). Zinc is also used in weaner pig diets for the control of post-weaning scours. Most heavy metals consumed in feed are excreted in the faeces or urine, and will thus be present in livestock manure. Manures may also contain heavy metals from drinking water and soil ingestion, added bedding materials (e.g. straw), corrosion or licking/biting of galvanised metal housing components, and from footbaths containing Cu or Zn for sheep and cattle hoof disinfection that are disposed into manure stores. Typical metal concentrations in different livestock manures from England and Wales are shown in Table 1.

This paper describes some commonly used methods of manure storage and treatment, and explores how these may influence heavy metal inputs to agricultural soils from this important source.

Table 1. *Typical concentrations of heavy metals in livestock manures (mg/kg DM) and quantities applied annually (Mt DM) in England and Wales (Nicholson et al., 1999).*

Manure type	Quantity applied	Zn	Cu	Ni	Pb	Cd	Cr	As
Cattle slurry	1.51	170	45	6.0	7.0	0.3	6.0	2.0
Pig slurry	0.20	650	470	14.0	8.0	0.4	7.0	2.0
Cattle farmyard manure	8.01	68	16	2.8	2.4	0.2	2.0	1.2
Pig farmyard manure	1.05	240	168	5.2	3.2	0.2	2.4	0.8
Layer manure	0.29	583	90	10.0	9.0	1.3	5.7	0.3
Broiler litter	1.33	217	32	4.0	3.3	0.6	2.0	0.5

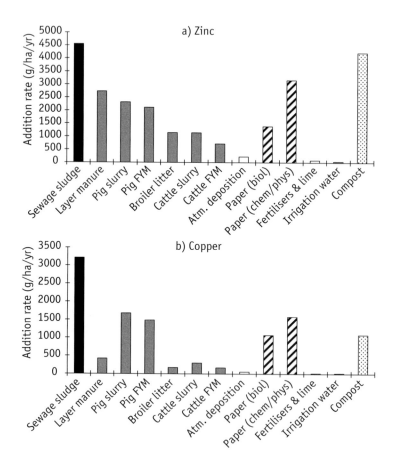

Figure 1. *Zinc (a) and copper (b) addition (input) rates to agricultural soils in England and Wales for 2004 (g/ha/yr).*

Manure storage and treatment methods

There are a number of reasons why livestock farmers are increasingly employing manure storage or treatment technologies. Firstly, European legislation has been introduced to limit manure nutrient (nitrogen and phosphorus) inputs and the timing of applications to land (e.g. EU Nitrates Directive, 91/676/EEC). Secondly, there has been increased pressure from food retailers and buying groups to reduce the potential risks of pathogen transfer from livestock manures to food crops through the use of treatment technologies. And thirdly, the economic climate is driving farmers to improve their operational efficiency. In addition, farmers often experience strong pressure from local residents to control odours during manure storage and land spreading, which can be achieved using some treatment technologies.

The most widely used 'treatment' method for solid manures is storage in field heaps or on concrete until the farmer has the opportunity to spread the manure to land. Composting is a relatively straightforward technology that involves managing the solid manure (usually by regular turning) to achieve elevated temperatures (>55 °C) in the heaps or windrows. This is a cost-effective way to reduce the pathogen content of the manure and will also produce a stabilised product with reduced volume, although there will be losses of nitrogen (N) and carbon dioxide (CO_2) during the composting process. Liquid slurry storage in earth-banked lagoons, above-ground circular tanks or below-ground tanks/structures is still the most common slurry 'treatment' employed by livestock farmers. However, on some large-scale livestock units (e.g. dairy/pig units) more complex forms of mechanical, biological or chemical treatment may be practiced.

Mechanical slurry treatment is largely based on separation technologies such as screening, centrifuging and sedimentation. Screening processes involve passing the slurry through a screen to retain the solid material, with finer screens retaining a higher proportion of the suspended solids, albeit with a reduction in capacity. Settlement techniques may be speeded up by the addition of flocculants or raising the temperature, but overall centrifuging tends to produce the best separation (Burton, 2006). Separation techniques generally produce a solid fraction with a dry matter content in the 20-30% range, so a second thickening or drying step is sometimes required, which may involve composting to produce a stabilised product. For each technology there is a wide range of equipment marketed commercially which varies both in performance and cost. The solid fraction produced by all the mechanical separation techniques contains the insoluble components of the slurry (e.g. organic matter and organically bound N, P and heavy metals) and is easier and cheaper to handle and transport than unseparated slurry. In contrast, the liquid fraction largely contains soluble nutrients and, because of the reduced dry matter content, can be easier to handle at land spreading than unseparated slurry.

The most widespread biological slurry treatment is anaerobic digestion (the breakdown of organic material by micro-organisms in the absence of oxygen) for biogas production (mostly methane – CH_4 and CO_2), often with a centrally located plant treating the slurry produced by a number of farms. Anaerobic digestion can reduce slurry odours and microbial pathogen numbers but will not reduce the nutrient or heavy metal content of the slurry. High initial installation costs have tended to discourage uptake of this technology, unless supported by economic incentives or subsidies. There is extensive Danish experience of both farm scale and community facilities and a similar approach is being implemented across several European countries (e.g. Germany).

Aerobic digestion (the breakdown of organic material by micro-organisms in the presence of oxygen) creates a reduced odour, pathogen-free and useable product; but operational costs are high, due to the energy requirements. Optimising plant operation is critical as ammonia (NH_3) losses can occur where aeration intensity is too high, and CH_4 and nitrous oxide (N_2O) losses at low pH and under other sub-optimal conditions. Ongoing research in France may provide improved technology (Martinez *et al.*, 2003; Daumer *et al.*, 2005). Community facilities for aerobic digestion already form an integral part of the Nitrate Vulnerable Zone (NVZ) Action Programme in parts of Brittany and Belgium.

Chemical treatments are those where a substance (usually an acid or alkaline product) is added to slurry. Acids can be effective at reducing odour and NH_3 emissions, whereas alkaline products raise the slurry pH to 'kill' microbial pathogens.

None of the treatment methods described above remove heavy metals from manure, although metals may be redistributed between the solid and liquid fractions, or their behaviour altered by changes to manure physico-chemical properties (e.g. pH, precipitation reactions). Electroremediation is one of the few treatment methods that will remove heavy metals from manures (Dach and Starmans, 2006). Although electroremediation has been used with some success to treat heavy metal contaminated soils, the technology is in its infancy for use with manures and is currently neither a practical nor a cost-effective solution for farmers. Silicates (e.g. Zeolite) are sometimes used to remove or stabilise soluble nutrients in slurry or separated liquid slurry by ion exchange processes (to correct a perceived nutrient imbalance or to prevent NH_3 emissions), and heavy metals can also be removed using this technique (Krason and Knud-Hansen, 2004). Other methods include the use of natural or constructed wetlands to treat the effluent from fields where manure has already been applied, usually at very high rates. Heavy metals (and nutrients) are taken up by wetland vegetation or adsorbed to soil particles thus preventing them reaching sensitive ground and surface waters. However, this approach is only viable where there is a sufficient area of suitable land available to construct a wetland system.

Implications for heavy metal inputs to soils

Whilst manure storage and treatment technologies will not remove heavy metals from manures *per se*, these practices can have important implications for heavy metal inputs to agricultural soils.

Storage, and especially composting, will decrease the total solids content of solid manures (due to oxidation of the organic matter and losses as CO_2) and may also lead to N losses via ammonia volatilisation or denitrification (as nitrous oxide and di-nitrogen gasses). If the stored/composted manure is then spread to supply a maximum rate of N to the soil (e.g. 250

kg/ha total N, which is the maximum field spreading rate in NVZs in England), then the heavy metals added per unit N applied will exceed those from an equivalent application of 'fresh' manure. For example, laying hen manure (30% dry matter) typically contains 16 kg total N/t fresh weight (fw) and 180 g Zn/t fw (MAFF, 2000; Nicholson *et al.*, 1999). Application to land at a rate of 250 kg total N/ha would add *c.*2.8 kg Zn/ha. However, laying hen manure that has been stored for one year prior to land application would typically be reduced in N content by around 30% (Nicholson *et al.*, 2004). Thus application of the stored manure at the equivalent N rate would add *c.*4.0 kg Zn/ha (Figure 2a). Similar N losses can occur during the storage of liquid slurries, with the heavy metals becoming more concentrated per unit of N remaining compared with 'fresh' slurry. Thus encouraging farmers to store manure, in particular pig and laying hen manures, may lead to higher metal loading rates at the field-level (although farm-level inputs would not be affected). Note that if the manures were applied on a P basis (i.e. to supply a maximum rate of P rather than N) then these differences in metal addition rates would not be as apparent as P losses during manure storage are much smaller.

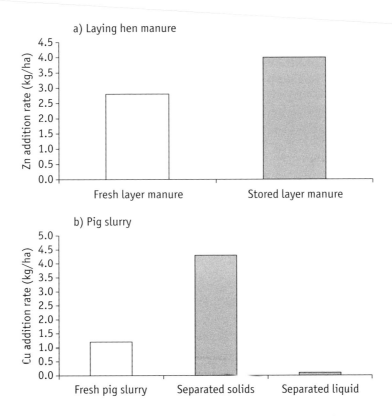

Figure 2. *Effect of solid manure storage and slurry separation on heavy metal addition (input) rates to soils (kg/ha).*

One effect of mechanical slurry separation is to redistribute heavy metals into solid and liquid manure fractions. For example, Burton (2006) reported that >90% of the Zn and Cu in fattening pig slurry was insoluble and would therefore be present in the solid fraction after treatment. Similarly, Chambers *et al.* (1999) showed there was a strong relationship between the dry matter content of slurry and heavy metal concentrations (expressed on a fresh weight basis), Figure 3. Consequently, fields where the separated solid manure fraction was applied would have higher heavy metal loading rates than those receiving unseparated slurry, whilst fields irrigated with the liquid fraction would have lower heavy metal loadings than those receiving unseparated slurry. For example, pig slurry (4% dry matter) typically contains 4 kg N/t fw and 19 g Cu/t fw (MAFF, 2000; Nicholson *et al.*, 1999). Application to land at a rate of 250 kg total N/ha would add *c.*1.2 kg Cu/ha. However, pig slurry that had been separated prior to land application would have a solid fraction with a dry matter content of 20% typically containing around 5 kg N/t fw and 90% of the Cu, and a liquid fraction (3% dry matter) typically containing 3.6 kg N/t fw and 10% of the Cu (K. Smith,

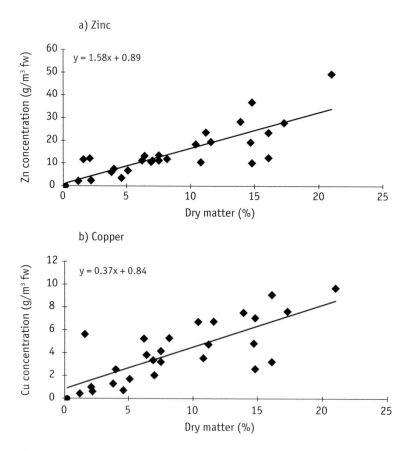

Figure 3. Relationship between cattle slurry dry matter content (%) and a) slurry Zn concentration and b) slurry Cu concentration (g/m³ fw). Based on data summarised by Chambers et al. (1999).

ADAS, pers comm; Burton, 2006). Thus application of the solid fraction at a rate of 250 kg N/ha would add c.4.3 kg Cu/ha, whilst application of the liquid fraction would only add c.0.1 kg Cu/ha (Figure 2b).

The effect of biological treatment technologies (aerobic and anaerobic digestion) on the balance of metal and N concentrations in the digestate is not easy to predict, as it will depend on the process used and the efficiency with which it is operated. In the case of chemical treatments (e.g. acidification, alkaline stabilisation) the manure characteristics and soil environment may be altered following land spreading, with important implications for the form and behaviour of the metals (e.g. uptake by plants, impacts on soil organisms).

Summary and conclusion

Heavy metal inputs can have long-term implications for the quality of agricultural soils including phytotoxicity to crops, the maintenance of soil microbial processes and the transfer of zootoxic elements to the human diet. Around 30% of total Zn and Cu inputs to agricultural soils are estimated to be from livestock manures. Livestock manures may contain heavy metals from a number of sources including dietary supplements, veterinary medicines, soil ingestion and via the disposal of footbaths into manure stores. Whilst most manure storage and treatment technologies will not remove heavy metals from manures *per se*, these practices can have important implications for heavy metal inputs to agricultural soils. Substantial losses of N often occur during the storage of solid manures and liquid slurries, with the heavy metals becoming more concentrated per unit N remaining compared with 'fresh' manure. Thus encouraging farmers to store manures, in particular pig and laying hen manure, may lead to higher metal loading rates at the field-level where manures are applied on an N basis. Slurry treatment, using mechanical separation technologies, will change the partitioning of heavy metals between the solid and liquid fractions. At a given N (or P) application rate, this will increase the metal addition rate from the solid fraction but decrease it from the liquid fraction compared to an unseparated slurry.

The measurement of heavy metal concentrations in untreated/treated manures, as well as those of the receiving soil (particularly those receiving repeated applications of pig/laying hen manures) is therefore important to protect the long-term fertility and quality of agricultural soils, and to protect groundwater or surface water quality which may be adversely affected by heavy metal leaching or run-off.

References

Burton, C., 2006. The contribution of separation technologies to the management of livestock manure. Proceedings of the 12th RAMIRAN International Conference: Technology for Recycling of Manure and Organic Residues in a Whole-Farm Perspective. Vol 1. p43-48

Chambers, B.J., F.A. Nicholson, D.R. Soloman and R.E. Unwin, 1999. Heavy metal loadings from animal manure to agricultural land in England and Wales. Proceedings of the 8th Conference in the FAO ESCORENA Network on Recycling of Agricultural, Municipal and Industrial Residues in Agriculture – RAMIRAN 98, p475-483.

Dach, J. and D.A.J. Starmans, 2006. Electroremediation of heavy metals from liquid manure. Proceedings of the 12th RAMIRAN International Conference: Technology for Recycling of Manure and Organic Residues in a Whole-Farm Perspective. Vol 2. p165-168

Daumer, M.-L., F. Béline and F. Guiziou, 2005. Traitement biologique des lisiers de porcs en boues activées. Guide technique à l'usage des concepteurs, exploitants et organismes de contrôle des stations. Cemagref Editions 2005.

Krason, J. and C.F. Knud-Hansen, 2004. Natural zeolites – remedy for concentrated animal feeding operations and sustainable agriculture. Proceedings of the 11th RAMIRAN International Conference: Sustainable Organic Waste Managements for Environmental Protection and Food Safety. Vol 2. p265-268

MAFF, 2000. Fertiliser Recommendations for Agricultural and Horticultural Crops (RB209). Seventh edition. The Stationery Office, Norwich.

Martinez, J., F. Guiziou, P. Peu and V. Gueutier, 2003. Influence of treatment techniques for pig slurry on methane emissions during subsequent storage. Biosystems Engineering 85 (3), 347-354.

Nicholson, F.A. and B.J. Chambers, 2006. Quantifying heavy metal inputs to agricultural soils in England and Wales. Proceedings of the 12th RAMIRAN International Conference: Technology for Recycling of Manure and Organic Residues in a Whole-Farm Perspective. Vol 2. p41-43

Nicholson, F.A., B.J. Chambers, J.R. Williams and R.J. Unwin, 1999. Heavy metal contents of livestock feeds and animal manures in England and Wales. Bioresource Technology 70, 23-31.

Nicholson, F.A., S.R. Smith, B.J. Alloway, C. Carlton-Smith and B. J. Chambers, 2003. An inventory of heavy metal inputs to agricultural sols in England and Wales. The Science of the Total Environment 311, 205-219.

Nicholson, F.A., B.J. Chambers and A.W. Walker, 2004. Ammonia emissions from broiler litter and laying hen manure management systems. Biosystems Engineering 89 (2), 175-185.

Rosen, G.D. and P.A. Roberts, 1996. Comprehensive Survey of the Response of Growing Pigs to Supplementary Copper in Feed. Field Investigations and Nutrition Service Ltd., London.

Consideration of heavy metals in manure recycling strategies in South East Asia in the nutrient flux model NuFlux

H. Menzi

Keywords: manure management, nutrient balance, trace elements, environmental impacts

Introduction

Development of livestock production in Asia

In developing countries and especially in South-East Asia (SE-Asia), livestock production is growing at unprecedented rates. This process is driven by the increasing demand for animal products that is a consequence of population growth, urbanisation and income growth (De Haan *et al.*, 1998; Steinfeld *et al.*, 2006). According to Delgado *et al.* (1999), the annual growth rate for meat production was 8.4% in China and 5.7% in SE-Asia between 1982 and 1994. Between 1997 and 2020 the average annual growth of consumption is predicted to be 3.0% in China, 3.3% in SE-Asia and 2.8% for the developing world (Delgado *et al.*, 2002).

The livestock sector is responding to this surge in demand for livestock products with drastic transformations (De Haan *et al.*, 1998; Steinfeld *et al.*, 2006; Gerber and Steinfeld, 2008), namely a concentration of livestock close to peri-urban areas, an intensification, industrialization and specialization of the production and an increasing share of monogastrics (pigs and poultry) of the total livestock population (Gerber *et al.*, 2005; Menzi and Gerber, 2005). The new specialised livestock operations are to a large extent on farms without or with little own cropland. Especially the liquid animal waste is, therefore, often not utilised, but discharged to the environment (Menzi and Gerber, 2005). This leads to a considerable pressure on the environment, especially eutrophication and pollution of surface waters, ground water pollution and accumulation of nutrients and heavy metals in soils where livestock wastes are discharged or used in excessive amounts. Furthermore, it is associated with considerable human and animal health risks (e.g. diarrhoea, disease transmission, contribution to antibiotic resistance).

Activities of the LEAD initiative

The rapid change of livestock production in Asia and its consequences has caught the attention of the Livestock, Environment and Development Initiative (LEAD). The LEAD initiative is an inter-institutional project with the secretariat in FAO. The work of the initiative targets at the protection and enhancement of natural resources as affected by livestock production and processing in the context of poverty reduction and public health enhancement, through better policy formulation for appropriate forms of livestock development. Aiming at enhanced policy making, the LEAD initiative develops decision-support tools to assess livestock-environment interactions, for early warning, activity targeting and decision making.

Between 2000 and 2004 the LEAD initiative conducted a project to develop and implement new systems of Area-Wide Integration (AWI) of specialised livestock and crop activities in Thailand, Guangdong Province of China and Vietnam. The project aimed at developing ways to integrate livestock and crop production on a regional scale, to thus enable a sustainable manure management without loosing the economies of scale. In the framework of national projects in the three countries the current situation of livestock production and its environmental consequences was analyzed and options for the improvement of manure management were evaluated (Fang *et al.*, 2000; Rattanarajcharkul *et al.*, 2000; Dan *et al.*, 2005; Hoa *et al.*, 2005). The project primarily aimed at manure recycling rather than treatment. In this framework NuFlux was developed

In the context of the AWI project it was realized that tools are needed that help to quantify nutrient fluxes resulting from livestock production systems. Such tools would be useful to analyse the present situation, evaluate development scenarios, monitor the future development and support farmers in the improved management of manure and fertilisation. To satisfy the variable needs of these applications the model NuFlux was developed, that calculates all relevant fluxes of intensive livestock production and manure management. These fluxes include fresh, dry and organic matter of liquid and solid manure, crop nutrients (nitrogen – N, phosphate – P_2O_5, potash – K_2O, magnesium – Mg) and the heavy metals copper (Cu) and zinc (Zn). P_2O_5 and K_2O were used because they are the form commonly used to express phosphorous (P) and potassium (K) in fertilization. Copper and Zn were included because manure recycling strategies always have to consider that at the present level of Cu and Zn in pig and poultry feed, and consequently in the manure, an accumulation of these elements will take place in the soils on which manure is applied. In the long term this accumulation can harm soil fertility as well as the quality of ground- and surface waters in case of shallow groundwater systems.

Since 2006, the AWI project is succeeded by the Livestock Waste Management in East Asia Project in Thailand, Vietnam and Guangdong Province of China, which aims both at the demonstration of different manure management systems on pilot farms and the preparation of relevant policy frameworks in the project countries.

This publication shall discuss the structure of the model NuFlux, its potential use and the experience made so far with its use as a tool to asses the fluxes of nutrients and trace elements of livestock production and the nutrient and trace element balance of agriculture.

The NuFlux model

'Terms of reference' for the model

An analysis of the needs of the projects and the different project participants provided the following requirements for the nutrient flux and manure management model:
- A tool to provide a robust indicator of the livestock pressure on the environment at various scales (e.g. region, province, farm), on the basis of the sparse available information. It should be adapted to various farming system and the conditions in different countries.

- A tool to support the improvement of fertilisation, especially the proper substitution of part of the mineral fertiliser with manure.
- A planning aid for manure distribution to different crops on farm and area-wide level.
- A user-friendly and easy to update tool which can cope with different levels of detail and reliability of input data (e.g. regional calculations with standard values and farm-specific calculations where relevant management variables for pig, poultry and crop production are available).

General approach of the model

The model is designed to calculate fluxes for pig and poultry in more detail than for ruminants, because monogastrics contribute more than two thirds of the manure in most project regions and because detailed information on ruminant rations are difficult to get.

The model starts the calculations with the most reliable available information: production parameters for livestock (beginning and end weight, amount and composition of feed used), nutrient demand recommendations or nutrient uptake for different crops etc. As shown schematically in Figure 1, the steps for the calculation are then (1) nutrient excretion, (2) amount and composition of fresh manure, (3) amount and composition of manure available for crops, (4) nutrient demand of crops, (5) nutrient balance. A special module provides a tool for planning the distribution of the different types of manure to different crops.

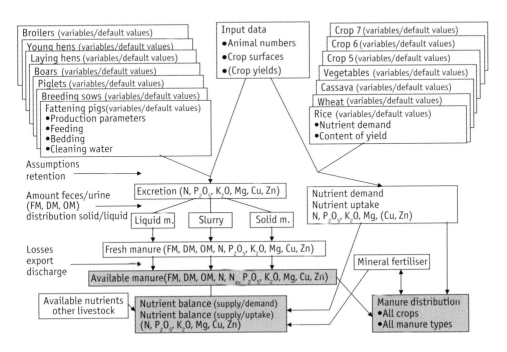

Figure 1. Schematic structure and path of calculation of the model.

For the nutrient balance, both a supply/demand balance and an import/export balance can be calculated. The supply/demand balance is recommended as standard for nutrients, because it reflects the nutrient amounts commonly dealt with (crop requirements or actual amount applied). It is therefore easier to get the necessary input data and it is more robust to lacking or unreliable inputs than the import/export balance, where the composition of the crop product has to be known. For Cu and Zn an import/export balance is calculated in any case. Further details about the nutrient balance approach are discussed by Menzi and Gerber (2006).

The model is equipped with regional default values for all the relevant variables concerning livestock and crop production. These default values which characterise a standard or average management for each project region were put together with local experts. The whole set of default values can be replaced when introducing the model in a new region. If more specific information is available, each default value can be replaced in user-friendly input screens.

Input data livestock production

Detailed calculations of the nutrient fluxes and the manure quantity and quality are done for the following livestock categories: Breeding sows, piglets, fattening pigs, boars, laying hens, young hens, broilers. For each of these categories two production systems or intensities (e.g. high-yielding international breeds and native breeds) can be differentiated. For other livestock categories default values for the available manure nutrients (excretions minus losses) are integrated. Such default values are currently built-in for dairy cattle, other cattle, horses, donkeys/mules, sheep and goats. The list can be changed easily.

The list of relevant management variables that are taken into account for each pig and poultry category includes the following elements:
- Production parameters: e.g. beginning and end weight, duration of rotation, feed conversion ratio.
- Housing systems: systems that produce only liquid manure (slurry, a mixture of urine, faeces and water), systems with liquid and solid manure separately, systems with only solid manure (a mixture of faeces, urine and litter.
- Feed ration: Choice between one feed and phase feeding; amount used per rotation and composition (crude protein, P, K, Mg, Cu, Zn) of each feed.
- Bedding material: share of animals with bedding material and average amount used per animal per day; choice of three bedding materials.
- Cleaning water use (for pigs only): amount of water used per animal per day or amount used per day for the whole heard (with or without drinking water).

For the amount of fresh excreta per herd animal per day (fresh, dry and organic matter) assumptions had to be made on the basis of the experience from Switzerland and the project partners. They will have to be validated and improved with measurements in the regions where the model is used.

Input data crop production

For the nutrient requirements official national or local recommendations were used. These vary considerably between crops and countries. In Thailand, for example, the requirements per hectare range from 14 kg N and 9 kg P_2O_5 for rubber to 300 kg N and 100 kg P_2O_5 for pineapple with an average around 60 kg N and 50 kg P_2O_5. In Vietnam the average is around 90 kg N and 75 kg P_2O_5, and in Jiangshu Province of China 180 kg N and 75 kg P_2O_5. It is quite evident that these considerable differences between countries can only partly be explained by differences in yield. Average yields in Jinagshu Province of China are higher than in Thailand, but the fact that for example the nutrient requirement recommendations for rice are five times higher in Jiangshu whereas those of P_2O_5 are only two times higher can only be explained by a different fertilization strategy. While a reluctant nutrient use (especially N) and a minimization of the load to the environment appears important in Thailand, high yields are a major aim in Jiangshu.

For each project region, all relevant crops are included. A list of at most 49 crops is possible. For each crop the nutrient demand recommendations for N, P_2O_5, K_2O and Mg are included. To allow import/export balances, the nutrient and heavy metal uptake and the yield can also be entered. The nutrient demand is given per hectare or local surface unit (e.g. rai in Thailand).

For P_2O_5, K_2O and Mg in the manure it was assumed that 100% is plant available at medium to long term. To account for losses through volatilization and leaching it was assumed that 70% of the total N in liquid manure (60% in China) and 35% in solid manure are plant available. These values are empirical assumptions agreed upon by the national experts and is based on a good manure management aiming at a minimization of environmentally relevant N losses.

The actual total amount of mineral fertiliser used (N, P_2O_5, K_2O, Mg, Cu, Zn) must be entered for balance calculations.

Other variables and assumptions

Assumptions were necessary for the distribution of the excreted fresh, dry and organic matter and each element to liquid and solid manure in the corresponding housing system as well as for the losses in houses and during storage. To design the model these assumptions were made on the basis of the experience from Switzerland (e.g. Walther *et al.*, 2001, Menzi *et al.*, 1997) and the project partners. They will have to be validated and improved with measurements in the regions where the model is used.

Outputs of the model

The output of the model includes the following parts, which can be jumped to individually:

- Nutrient and heavy metal balance: total balance in kg or tons and in percent of the crop nutrient demand and uptake. For easy interpretation a colour code with 6 classes is used to show the balance in percent of the crop demand or uptake.
- Amount and composition of manure: For each of the three pig and poultry manures the total amount (fresh, dry and organic matter) and the composition (dry matter, N, P_2O_5, K_2O, Mg, Cu, Zn) in kg ton^{-1} is given.
- Summary of intermediate results, manure export and discharge, etc.
- Summary of inputs: animal numbers, crop surface, nutrient demand of crops.
- Summary of all default values.

Languages and translation

To allow an easy translation of the whole model (not program code), all the text is referenced to a translation sheet (about 1000 lines). The whole model can be changed to any language included in this translation sheet with a mouse click. At present English, Thai and Chinese are integrated. Further translations are planned.

Different user modes

The model is equipped with three user modes:
- General user mode: For calculations with default values. Only animal numbers and crop surfaces must be entered. No practical background is necessary. This mode is useful for regional balances (e.g. for scenario calculations at the policy level).
- Advanced user mode: All the default values for livestock and crop production can be changed to more specific data. This approach is recommended for farm-specific calculations and for more detailed scenarios. An agricultural background is recommended.
- Expert user mode (password protected): Meant for the 'national operator', who can change the default settings.

The main screens, the output and the calculation procedure are the same in all modes. This congruency is important to assure that compatible results and conclusions can be achieved.

Programming tools and user manual

The model is a Microsoft Excel file with Visual Basic applications. It can be run on any machine equipped with Microsoft Office 97 or higher. The model can be distributed in a compressed version on one floppy disk or via internet. A detailed user manual is available in English.

Reliability of the model

It was not yet possible to do a sensitivity analysis of the model. Only a general discussion of the accuracy is therefore possible.

The key determinants for the accuracy of the model are the reliability of the available input data. The results on nutrient excretions are probably already within an accuracy around +20% because they are mostly based on information about typical farm management, which is judged as quite reliable and is differentiated for different regions. On the other hand the results on manure quantities and composition are not yet reliable, because practically no information was available on the distribution of the excreta to solid and liquid manure, losses from lagoons, etc. for the project regions or comparable conditions. Unfortunately validation measurements which would be needed to improve the model were not yet possible. First such measurements have started in 2007 in collaboration with the Asian Institute of Technology (AIT) in Thailand.

Past applications of NuFlux

So far NuFlux was mainly used in the framework of the AWI project in Thailand, Vietnam, China and Mexico to calculate regional nutrient balances and manure quantities and composition for some individual project farms.

The results on nutrient excretions were also extensively used in more detailed GIS-applications (nutrient balance maps, development of decision making tools for policy makers, etc.) in Thailand and Vietnam.

Selected results

Excretions

Table 1 gives an overview of the annual excretions per fattening pig place and per sow including piglets up to about 25 kilos in the different countries. Apart of the native breeds in China and the conventional system in Vietnam nutrient excretions per place are quite comparable in the three countries. The differences can be explained by differences in the production intensity (feed conversion ratio and length of rotation) and feed composition. Substantial differences are observed however for the heavy metals. They reflect the large differences in the content of these elements in the feed as reported by the national experts (Table 2). As such the Cu and Zn content of the ration in Thailand is within the range of the actual animal requirements. Slightly higher concentrations are used in Jiangshu Province of China whereas growth promoting doses are used in Vietnam. A field check is, however, needed to see if these values reported by experts really correspond with actual concentrations used by the farmers and if the values collected in the years 2000-2003 are still valid.

Land requirement for manure recycling

Evidently the crop surface needed for the recycling of the manure of one pig not only depends on excretions but also on what crops need and on the application strategy. Important aspects of the application strategy include issues related to whether the manure is dosed according to its N or P_2O_5 content, what proportion of the nutrient requirements are covered with manure and if surpluses are tolerated or not. In the discussions with the national experts it was

Table 1. Annual excretion of total nitrogen (N_{tot}), phosphate (P_2O_5), potash (K_2O), magnesium (Mg), copper (Cu) and zinc (Zn) per fattening pig place and per sow including piglets to 20-25 kg in Thailand, Jiangshu Province of China and Vietnam (Ho Chi Minh City). Values calculated with the NuFlux model based on the default values for livestock management assembled together with national experts. In China and Vietnam production is differentiated between imported breeds/native breeds and modern/conventional systems, respectively.

Country	System	N_{tot}	P_2O_5	K_2O	Mg	Cu	Zn
		kg	kg	kg	kg	g	g
Fattening pig							
Thailand		12.5	5.8	4.8	1.2	5	40
China (Jiangshu)	imported breeds	13.0	10.0	5.0	2.0	20	80
	native breeds	8.0	5.0	6.0	1.2	nv	nv
Vietnam HCMC	modern	10.0	4.9	4.3	1.6	120	80
	conventional	0.3	9.6	4.4	1.6	120	80
Sow including piglets							
Thailand		29.0	16.0	12.0	2.3	7	100
China (Jiangshu)	imported breeds	34.0	22.0	11.0	4.0	40	170
	native breeds	37.0	24.0	12.0	5.0	nv	nv
Vietnam HCMC	modern	25.6	14.2	8.7	3.1	220	140
	conventional	27.4	18.2	9.2	3.3	220	140

nv = no values

generally agreed upon that it is advisable to dose pig or poultry manure according to its P_2O_5 content because a dosing according to the N content would often lead to an over-fertilization with P_2O_5. In all countries it was agreed that 100% of the P_2O_5 requirements could be covered with manure because sufficient recycling capacity is usually the limiting factor.

Assuming the excretions given in Table 1 and the average nutrient requirements discussed above, about 0.12 hectares crop surface would be needed to recycle the manure of one fattening pig place in Thailand. Thus a farm with 10,000 fattening pigs would need 1200 ha. Even if the farm would separate liquid and solid manure and sell the solid manure, it would still need about 640 ha for recycling the liquid manure. If two or even three crops are grown per year the total area required is reduced accordingly.

Heavy metal load

Assuming that the average crop P_2O_5 requirement is fully covered with pig manure, the annual input per hectare would be about 21 g for Cu and 340 g for Zn, both in Thailand and in Jiangshu Province. A quantitative heavy metal balance can not be calculated because no

Table 2. Copper (Cu) and zinc (Zn) content of the pig rations reported by the national experts and used for the calculations of excretions.

		Cu ppm	Zn ppm
Fattening pig			
Thailand	starter phase	5	100
	grower phase	4	60
	finishing phase	3	50
China (Jiangshu)		20	100
Vietnam HCMC	below 60 kg	120-250	120-150
	above 60 kg	200-250	80-120
Sow including piglets			
Thailand	sows	5	50
	piglets	5	100
China (Jiangshu)	sows	15	100
	piglets	40	110
Vietnam HCMC	sows	120-150	80-120
	piglets	250	120

data on the heavy metal uptake of different crops was available yet in the project countries. Nevertheless, assuming that it would be less or equal to the approximately 75 g ha^{-1} Cu and 300 g ha^{-1} Zn assumed by Menzi and Kessler (1998) for arable crops in Switzerland, this input would not lead to an accumulation of Cu and only to a rather insignificant accumulation for Zn, even if atmospheric deposition is also accounted for. However, manure analysis results from Thailand indicate that at least the Cu content of the pig manure might in some cases be higher than estimated on the basis of the information received.

In Vietnam the same calculation reveals an annual input per hectare of nearly 2 kg for Cu and around 1.2 kg for Zn. Such an input leads to a considerable accumulation.

Potential use of nutrient flux models

How far can NuFlux be used?

In the present version of the model only the nutrient and heavy metal balance module is suitable for practical application. For the other modules detailed validation and adaptation is still necessary before an official release and a wider application is justified. Nevertheless, the testing of these modules by pilot users would be very valuable to furnish recommendations for improvements.

In the framework of the Livestock Waste Management in East Asia Project it is planned to integrate NuFlux into a larger decision support tool for the following applications:
- To evaluate the present manure management and environmental impact of a farm or of farms within an area.
- To evaluate different potential manure management options and select the most appropriate one.
- To support good manure management and to supply relevant information about livestock production and its environmental impact for policy makers and spatial planning projects.
- To support capacity building activities.

Workshops conducted recently with stakeholders in Thailand, Vietnam and China indicated that there is a keen interest in such a tool at the policy, research, extension service and education level. However, for farmers simple printed recommendation (leaflets, booklets) will be more appropriate.

When discussing the potential applications of NuFlux (alone or in the context of a wider decision support tool) in more detail we have to differentiate between the different elements or modules:
- The nutrient and heavy metal balance.
- The manure flux tool that calculates amounts and composition of the different types of manures that have to be handled.
- The intermediate results that show different nutrient and heavy metal fluxes.
- The manure distribution tool.

Potential use of nutrient balances

The land livestock balance is a common approach to assess how much crop area is needed to recycle the manure of a given farm or in an area, or if the livestock density on a farm or in an area corresponds to the carrying capacity of the land (if manure is recycled as fertilizer) (Menzi and Gerber, 2006). In the present version of NuFlux the nutrient balance can be used in the following two ways:
- By comparing the total nutrient load from manure and mineral fertilizer with the actual crop requirements or uptake, the total nutrient load on the environment can be evaluated. This evaluation can not differentiate if an eventual surplus will have an impact on the environment in the form of accumulation in the soil, nutrient losses to surface or ground water or as emission into the atmosphere. However, NuFlux could be linked with other models focussing on the environmental impact of a nutrient or heavy metal surplus (e.g. Römkens *et al.*, 2008).
- When the amount discharged is taken into account, the nutrient load in this discharge and the remaining nutrient balance of the manure recycling activities can be assessed.

In principle the nutrient balance approach could be used for the following applications:
- Assessment of the total load to the environment without differentiation between different impacts.

- Assessment of how much crop land (and grassland) is needed for recycling the manure or determination of the possible size of a livestock operation depending on the recycling potential on the farm or in an area.
- Assessment, if the present manure and fertilizer use of a farm corresponds with the crop nutrient requirements or if there is a surplus or deficit.
- Tool to organize the logistics of manure distribution in programs for area-wide collaboration of livestock and crop producers.
- Assessment of how many animals can be allowed in an area in the context of building or farm operation permits or spatial planning.

Potential users of this tool include larger livestock farmers wanting to recycle their manure on their own land or on neighbouring farms (as a planning and management tool), extension services or farmers training institutions (as an awareness raising and educational tool), managers of programs for area-wide integration of livestock and crop production as well as local, provincial and national authorities (for spatial planning).

Potential use of heavy metal balances

Heavy metal balances can primarily be used as an awareness raising tool about the potential environmental threat of heavy metals in the livestock feed. Such balances would have to be accompanied by visualized examples of the detrimental effect that a heavy metal accumulation in the soil can have on soil fertility. The clear aim of such activities should be to sensitize different stakeholders that heavy metal levels in livestock rations should be limited to the trace element requirements of the animals and that heavy metals should not be used as growth promoters or in prophylactic rates because such practice will endanger the sustainable future development of agriculture. A framework of what is acceptable from the environmental point of view must therefore be established. Other model calculations describing in more detail what happens to the metals in the soil (uptake, leaching, etc.) are needed for this.

Potential use of manure fluxes

A model providing information on the amount and composition of different manures resulting from intensive livestock production could be used by large livestock farmers and managers of programs for area-wide integration of livestock and crop production for the following applications:
- Planning the dimensions of manure storage and treatment facilities.
- Evaluating the nutrient value of different types of manure.
- Evaluating the amount of manure needed to cover the nutrient demand of different crops.
- Planning aid for programs of area-wide integration of livestock and crop production.
- Input for cost calculations.

Based on the information on amount and composition of manure further modules for planning manure management could be developed (e.g. distribution of manure to different crops, fertilizer plans etc.).

Outlook and conclusions

Models that estimate mater fluxes (nutrients, organic matter etc) can be a powerful tool in the context of livestock and manure management both for awareness raining and as management tools. They allow an easy simulation and visualization of fluxes and their potential impact. In the case of balances such model calculations are often more reliable than measurements. However, for a more detailed flux analysis that is needed to estimate the amount and composition of manure, extensive validation activities are required before reliable tools for on-farm application can be released.

In crop fertilization the focus has traditionally been mainly on nutrients (macro-elements) and organic matter. Nevertheless, especially in the situation of rapid growth of intensive livestock production with minimal environmental restrictions that we are faced with today in Asia and Latin America trace-elements (heavy metals) are increasingly gaining importance as a threat to the environment and thus to sustainable development. The trace elements therefore merit an increased awareness not only from the animal physiology side but also in the context of ecological considerations. Copper and Zn are certainly the first elements to consider closely in such considerations. However, any other trace elements intentionally introduced in animal rations such as cobalt (Co), chromium (Cr) and arsenic (As) should also be included in such considerations. NuFlux can provide information on the accumulation of trace elements in agriculture. However, to assess the environmental relevance of this accumulation and to provide guidance on the limits of trace element use in livestock production NuFlux should be linked with other models looking at processes in the soil.

The model was developed for countries in Asia and Latin America. Nevertheless, with the necessary adaptations, it might also be a valuable tool in Western countries. The results can even be expected to be more reliable because of better input data and because better assumptions backed by research will be available.

References

Dan, T.T., T.A. Hoa, L.Q. Hung, B.M. Tri, H.T.K. Hoa, L.T. Hien, N.N. Tri, P. Gerber and H. Menzi, 2005. Animal waste management in Vietnam problems and solutions. In: Bernal, M.P., R. Moral, R. Clemente, and C. Paredes (Eds.), Proc. 11th conference of the 'Recycling Agricultural, Municipal and Industrial Residues in Agriculture Network' (RAMIRAN), Murcia, Spain, October 6-9 2004, volume 2, 337-340.

De Haan, C., H. Steinfeld and H. Blackburn, 1998. Livestock and the environment, finding a balance. European Commission Directorate-General for Development, Food and Agricultural Organization of the United Nations.

Delgado, C.L., M. Rosegrant, H. Steinfeld, S. Ehui and C. Courbois, 1999. Livestock to 2020: The next food revolution. Paper 28. Washington, International Food Policy Research Institute.

Delgado, C.L., M.W. Rosegrant and S. Meijer, 2002. Livestock to 2020: The revolution continues. Paper presented at the World Brahman Congress, Rockhampton, Australia, April 16, 2002.

Fang, Y., J.S. Yang, S.S. Kkaer, P. Gerber, B.S. Ke and H. Menzi, 2000. Area-wide integration of specialised livestock and crop production in Jiangsu Province, China. In: Proc. 10th conference of the 'Recycling Agricultural, Municipal and Industrial Residues in Agriculture Network' (RAMIRAN), Gargnano, Italy, September 6-9 2000, 287-293.

Gerber, P.J. and H. Steinfeld, 2008. Worldwide growth of animal production and environmental consequences. In: P. Schlegel, S. Durosoy and A.W. Jongbloed (Eds.), Trace elements in animal production systems, Wageningen Academic Publishers, Wageningen, the Netherlands, pp.21-32.

Gerber, P., P. Cilonda, G. Franceschini and H. Menzi, 2005. Geographical determinants and environmental implications of livestock production intensification in Asia. Biores. Tech. 96, 263-276.

Hoa, T.A., T.T. Dan, T.D. Luan, P. Gerber and H. Menzi, 2005. Economic analysis of various animal waste treatments in selected Provinces in Vietnam. In: Bernal, M.P., Moral, R., Clemente, R. and C. Paredes (Eds.), Proc. 11th conference of the 'Recycling Agricultural, Municipal and Industrial Residues in Agriculture Network' (RAMIRAN), Murcia, Spain, October 6-9 2004, volume 2, 229-302.

Menzi, H. and P. Gerber P., 2005. The land livestock balance approach and it's implications for intensive live-stock production in South-East Asia. In Rowlinson, P., Wachirapakorn, C., Pakdee, P. and M. Wanapat (Eds.), Proc. International Conference 'Integrating Livestock-Crop Systems to Meet The Challenges of Globalisation', Khon Kaen, Thailand, November 14-18 2005. British Society of Animal Science, 131-144.

Menzi H. and P. Gerber, 2006. Nutrient balances for improving the use-efficiency of non-renewable resources: experiences from Switzerland and Southeast Asia. In Frossard, E., Blum, W.E.H. and B.P. Warkentin (Eds.), 'Function of soils for human societies and the environment'. Special publication of the Geological Society (London) 266, 171-181.

Menzi H., R. Frick and R. Kaufmann, 1997. Ammoniakemissionen in der Schweiz: Ausmass und technische Beurteilung des Reduktionspotentials. Schriftenreihe der Forschungsanstalt für Agrarökologie und Landbau (FAL), 26, 107 pp.

Menzi H. and J. Kessler, 1998. Heavy metal content of manures in Switzerland. In Martinez J. an d Maudet M.N. (Eds.), Proc. 8th conference of the 'Recycling Agricultural, Municipal and Industrial Residues in Agriculture Network' (RAMIRAN), Rennes, France, May 26-29 1998, volume 1, 495-506.

Rattanarajcharkul, R., W. Rucha, S. Sommer and H. Menzi, 2000. Area-wide integration of specialised livestock and crop production in Eastern Thailand. In: Proc. 10th conference of the 'Recycling Agricultural, Municipal and Industrial Residues in Agriculture Network' (RAMIRAN), Gargnano, Italy, September 6-9 2000, 295-300.

Römkens, P.F.A.M., Moolenaar S.W., Groenenberg J.E., Bonten L.T.C., de Vries W., 2008. Copper and Zinc in feed (additives): an essential burden? In: P. Schlegel, S. Durosoy and A.W. Jongbloed (Eds.), Trace elements in animal production systems, Wageningen Academic Publishers, Wageningen, the Netherlands, pp.115-136.

Steinfeld, H., Gerber, P., Wassenaar, T., Castel, V., Rosales, M. and C. de Haan, 2006. Livestock's long shadow, environmental issues and options. Food and Agricultural Organisation of the United Nations (FAO), Rome, 390 pp.

Walther, U., H. Menzi, J.P. Ryser, R. Flisch, B. Jeangros, W. Kessler, A. Maillard, A.F. Siegenthaler and P.A. Vuilloud, 1994. Grundlagen für die Düngung im Acker- und Futterbau. Agrarforschung 1/7, 1 40.

Impact of aquaculture on aquatic environment: trace minerals discharge

S.P. Lall and J.E. Milley

Keywords: aquaculture, trace minerals, cadmium, biofoulants

Introduction

Aquaculture involves cultivation of aquatic organisms (fish, molluscs, crustaceans, unicellular algae, macroalgae, and higher plants), using extensive or intensive methods to increase their production to a level above that obtained naturally in a particular aquatic environment (freshwater, brackish and seawater). Commercial aquaculture is an ancient practice, although large scale farming is relatively recent. The earliest known treatise on aquaculture is the 'Classic of Fish Culture' in 500 BC by Fan Lei, a Chinese politician turned fish-culturist who developed pond production of carp. Oyster culture is known to have been practiced in Japan and Greece approximately 2000 years ago. Seaweed culture is much more recent, the earliest known text being published in 1952 in Japan.

Aquaculture experienced rapid expansion in the 1950's at an annual compounded rate of 3% of total production and reached 54.8 million tonnes in 2003 (FAO, 2004). The total global fisheries landing was 146.3 million tonnes with aquaculture contributing to 37.4% of that total. Asia was the largest aquaculture producer, followed by Europe, North America, South America and Africa. The major aquaculture-producing countries include China, India, Japan, Indonesia, Thailand, USA, Philippines, Korea, Taiwan, Greece, Canada, Norway and Chile. Inland aquaculture, mostly carp culture, currently accounts for two-thirds of the total production (excluding seaweed production), but marine farming is growing rapidly. Approximately 195 species are cultured, including finfish (92), crustaceans (23), molluscs (36) and aquatic plants (44), in freshwater, brackish water and seawater under a wide range of environmental conditions and culture systems (ponds, tanks, raceways and cages). Farmed fish and shellfish contribute to more than 16% of the human diet. The major species currently cultured include carps, scallops, clams, oysters, mussels, marine shrimp, salmonids (Atlantic salmon, Coho salmon, Chinook salmon, rainbow trout and arctic char) and tilapias. Cultured herbivorous fishes include various Chinese and Indian carps, some tilapia species, mullets and milkfish.

Aquaculture systems are classified by the density of animal biomass in a given area of water, which determines the density of management input required. Extensive systems rely on natural productivity of aquatic organisms for food supply. Intensive systems require a high degree of management including manufactured feeds for the species farmed. Semi-intensive systems fall between the two extremes and depend on the supply of feed and natural productivity of ponds. In areas where natural fresh, brackish water are abundant, fish are lightly stocked and they feed on natural preys (zooplankton and fish, etc.), plants, dead and decaying animal material, microbes and other benthic organisms. The growth of photosynthesizing plankton such as algae and diatoms and zooplankton can be stimulated with inorganic or organic fertilization to increase natural productivity. A number of fish species such as tilapia and carp are amenable to production in semi-intensive systems where feeding of formulated feeds compliments the level and quality of natural foods. Unlike in extensive systems, higher stocking density of fish can be supported and feeding commercial feeds partially fulfils their nutrient and energy requirements. Intensive systems depend on high levels of fish production in tanks, raceways, ponds and cages and because the large number of animals per unit area requires a high amount of total feed input to support the biomass, a high turnover of water is needed to remove large volumes of animal wastes. This does not allow for significant natural productivity of aquatic organisms to provide sufficient nutrients or energy for their growth and development and therefore a complete feed must be provided to the cultured species in the form of compounded, dry, moist or semi-moist feed.

Salmonid and marine fish are cultured in cages made of wood, steel or plastic, with the use of plastics becoming more prevalent. The development of the salmon farming industry has resulted in increasingly larger sea cages with more offshore locations and their volumes may vary from a few hundred m^3 for older cage types to over 10,000 m^3 for more modern cages. A fairly diverse group of crustaceans, including lobster, marine shrimp, freshwater prawn, and crayfish are now cultured in ponds. In recent years, farming of several shrimp species of the penaeid famly, such as *Metapenaeus ensis, Penaeus japonicus, P. monodon, P. indicus, P. mergiensis, P. aztecus, P. setiferus, P. schmitti, P. chinesis, P. penicillatus, P. stylirostris* and *P. vannamei* have shown the most significant growth globally. Molluscs such as oysters, mussels, clams and cockles are the most important species farmed in marine and brackish waters. They are also cultivated by a range of technologies from simple capture of juveniles for on-growing in ponds or lagoons, to elaborate procedures involving hatcheries, nurseries and complex grow-out and harvesting systems. The filter feeding habits of these organisms range from browsing on marine algae and epiphytes to aggressive predation.

Marine algae cultured range from microscopic unicells (phytoplankton, endophytes) to the tallest vegetation known, the giant kelps. They are either harvested or cultured in coastal areas for human consumption or animal feed use. Commercially valuable species include some seaweeds e.g., *Ascophyllum nodosum, Laminaria digitata*. Microalgae are also cultured for feeding herbivorous invertebrates in aquaculture nursery operations (e.g. *Isochrysis galbana*).

Of the different fish and crustacean species being cultured, complete information on the quantitative nutrient requirements and feed utilization is limited to less than a dozen freshwater species and some recommendations are available for selected nutrients for a few marine species. Limited research efforts have been directed to investigate the influence of dietary factors and specific nutrient on male reproduction, egg and larval development and survival of offspring but these factors remain the major bottleneck for hatchery production of marine fish species. The aquaculture industry relies heavily on ingredients of marine origin (fishmeal and fish oil) and currently efforts focus on the substitution, at least in part, of such marine resources with terrestrial agriculture products such as plant protein and oil. A significant amount of research effort has been directed to develop nutritional strategies to reduce the nutrient loads into the environment.

Aquaculture and the environment

Major environmental impacts of aquaculture have been associated with high-input high-output intensive systems for carnivorous marine fish and salmonids cultured in raceways and cages. This includes the discharge of suspended solids (faeces and uneaten food), antibiotics, other drugs, disinfectants, algaecides and other chemical compounds leached in water. A major proportion of faecal and feed waste contributes to increased concentrations of suspended and particulate matter in sediments near the immediate vicinity fish net-pens. Minerals and organic enrichment of recipient waters causes the build-up of anoxic sediments, changes in benthic communities (alteration of seabed fauna and flora communities) and eutrophication of water. Major environmental issues surrounding the development of marine finfish aquaculture and its environmental impacts have been reviewed by environmental groups, researchers and government agencies and they are the subject of many publications (Pillay, 1992; Black, 2002; Hargrave, 2005).

Several factors influence the horizontal transport and dispersion of these accumulated materials at the bottom of net-pens. These include water current, tidal flow, residual circulation, presence of turbulence and wind-related physical and environmental factors. Eutrophication of coastal and estuarine waters from aquaculture, as well as other human activities, results in dissolved inorganic and organic nutrients and increased biological demand from oxygen containing material. Dissolved inorganic nutrients released from marine fish culture and regenerated from sediments enriched with organic matter under net-pens stimulate phytoplankton production and increase oxygen demand. Such nutrient enrichments also have the potential to stimulate the development of macroalgal beds and to influence benthic fauna through changes in the rates and nature of deposition of particulate organic matter (Bourget *et al.*, 1994). Eutrophication can alter the ratio between essential nutrients (carbon: nitrogen: phosphorus) for phytoplankton growth and also cause a shift in species distribution. Although the occurrence of harmful algal blooms (HAB) have been linked to fish farms in lakes, other types of plankton blooms in the marine environment have not shown such clear link to the formation of HAB.

There is extensive literature documenting changes in the benthic fauna community structure associated with high levels of nutrients and organic matter additions. An increase in the flux of products from the breakdown of nutrients such as ammonium and dissolved reactive phosphate from the sediment into overlying water (Enell and Löf, 1983) causes changes in the population structure of the benthic microfauna. Only fauna such as nematodes and polychaetes, tolerant of low oxygen conditions and reduced sulfides, are able to survive under the conditions of high organic sedimentation (Duplisea and Hargrave, 1996). Moderate increases in organic matter supply may stimulate macrofauna production and increase species diversity; however increasing higher rates of organic input, diversity and biomass causes a decrease. Most environmental impact studies conducted under sea cages have shown that the local extent of benthic community structure and biomass is limited to less than 50 m. Water depth and current velocity are considered critical factors in determining the patterns of sedimentation around cage sites, and therefore impact of benthic fauna differs at different farm sites.

Organic wastes such as food and faeces cause the sediments to have high organic content. Biogeochemical cycling of elements in these eutrophic fish farm sediments is very rapid (Holmer and Kristensen, 1992). Changes in the sediment are characterized by low redox potential, high content of organic material and accumulation of organic nitrogenous and phosphorus compounds. Particulate organic material settles to the bottom and causes the formation of anoxic sediments. Some of these changes include depletion of oxygen in pore waters and reduction of sulfate to sulfides, a dominant anaerobic process in the sediment of the cage farming areas of coastal waters. The most severe impacts are mainly restricted to the immediate vicinity of the aquaculture operation. In many cases the enhanced activity of sulfate reducing and methanogenic bacteria within the sediment results in out gassing of carbon dioxide and hydrogen sulfide. At some marine cage farms this has caused loss of appetite, gill damage and increased mortalities of fish. In addition, stress induced by exposure to hydrogen sulfide and poor water quality may be responsible for the increase in and persistence of disease observed at some cage fish farms. Away from the sea cages, as organic material flux and oxygen demand increases, animal communities return to background conditions typified by increased species diversity and functionality. In contrast to marine waters, enrichment of the sediments beneath freshwater cage farms is less likely to affect the long-term production potential of a site. Also fish grown in land-based ponds, tanks and raceways (both marine and freshwater) would not be expected to be at risk from different effects of organic enrichment of the sediment ecosystems.

For farmed salmon and trout, mass balance models have been developed for nitrogen and phosphorus, indicating 50% of the nitrogen and 28% phosphorus supplied with the food is wasted in the dissolved form. A significant amount of bound P in fish meals is not digested and settles in sediments. Maximum nutrient release can be estimated from the hydrographic conditions, such as water volume, tidal water exchange and currents, in the immediate vicinity of the farm. In recent years, better site management and production levels, improvements in feeding efficiency, development of highly digestible nutrient-dense diets and feed quality have significantly reduced aquaculture waste and consequently the environmental impact.

The environmental impacts of shrimp cultivation are well known and numerous studies have been conducted to investigate this problem (Dierberg and Kiattisimkul, 1996; Flaherty and Vandergeest, 1998). Environmental impact depends on geography, type of cultivation practice, conditions of natural habitats and natural systems capacity, wastewater generation and treatment, type of feed, therapeutants and feed additives, and geological and hydrological conditions. The early stages of developing large-scale shrimp culture resulted in physical degradation of coastal habitats, for example, through conversion of mangrove forests and destruction of wetlands, salinization of agricultural and drinking water supplies, and land subsidence due to groundwater abstraction.

In the past two decades, however, significant effort has been directed to minimize the adverse effects of solid and soluble wastes in effluent water through changes in feed formulation, better nutrition and feeding strategies, improved aquaculture practices and stringent environmental regulations. Imposing strict controls on the number of cages being used per unit surface area and on the stocking density according to the local conditions and maintaining water quality for sustainable operations have had positive impacts in reducing the impact of aquaculture on environment. There is also development of integrated aquaculture systems using various species in different trophic levels such as seaweed, shellfish, omnivorous and carnivorous fishes for efficient utilization of various forms of nutrients and high productivity (Neori *et al.*, 2004). In many countries, progress has been made in the development of offshore fish farming systems to expand the areas available for aquaculture and to minimize the environmental impact of aquaculture in coastal areas.

Generally, the following nutritional strategies are implemented to minimize environmental impact of aquafeeds on the environment: (a) estimation of correct nutrient balance, particularly for nitrogen and phosphorus, (b) minimizing excessive amounts of nutrients in feed from feed ingredients or additional supplements, (c) selection of feed ingredients with high nutrient bioavailability and low fibre, chitin and other indigestible material content, (d) development of feeds with better feed efficiency, (e) minimization of feed waste by the use of better feeding systems and feeding guides.

Mineral uptake by fish

All aquatic organisms have the ability to absorb and retain minerals from their environment; however, their incorporation in the body is highly selective. Large areas of body surface and the gills of fish are exposed to the external environment of varying salinity ranging from 0 to 35‰ in freshwater and seawater. In freshwater, the active uptake of salts through the gills, along with a low body surface permeability, enables them to maintain blood ion levels that are more concentrated than those of the external water medium. Marine fish lose water through any permeable surface and thereby increase their level of salt. They replace water loss osmotically by drinking seawater. The gut actively absorbs monovalent ions and water into the blood, accumulating most of the divalent ions in the intestine. Excess monovalent ions derived from swallowed seawater and the passive uptake across the body surface are excreted, mainly through the gills.

Mineral uptake and accumulation of trace elements depends on the food source, environment, species, stage of development, and physiological status of the animal. In marine food chains, the mechanism of trace metal transfer at various trophic levels has been reported (Bernhard and Andreae, 1984). Most trace metals show the highest concentration increase at the first trophic level (seawater-phytoplankton). In zooplankton feeding on phytoplankton, mainly Cd, Cu, and Zn concentrations increase. However, plankton-feeding fishes show higher Cu and Zn levels than fish preying on invertebrates and fish feeding on invertebrates and fish. Generally, the concentration of Cd, Cu, and Zn increases along the food chain at the lower levels, reaching a maximum with crustaceans and then decreasing in fish. Limited information is available on the transport of other elements in the marine food chain. Fertilization of fish ponds is common practice in pond culture of warm water fish such as catfish, tilapia and carp. It has been demonstrated that fertilization of ponds with inorganic fertilizers and manures augment the production of natural food organisms and these organisms supply essential trace elements for growth and reproduction.

Unlike some soft tissues of fish such as liver, muscle and kidney, skeletal tissues, otolith and lenses retain metals for a longer duration and their concentration has been used to detect environmental pollution (Dove and Kingsford, 1998). The distribution of many essential and toxic elements in several fish, crustaceans and molluscs has been reported, however, the complete elemental composition is available for only a few fish species and restricted mainly to edible parts (Lall, 1995). The mineral composition of rainbow trout varies with fish size, stage of life cycle, and reproductive status. The concentration of Ca, Cu, Fe, K, Mg, Mn, Na, P, Sr, and Zn was higher in juvenile fish than in adults (Shearer, 1984). A decrease in somatic Mn, Fe, and Zn was found during gonadal maturation in female but not male fish.

Metals enter the hydrosphere from either natural processes or pollution through activities such as mining operations, burning of fossil fuels, agriculture, and urbanization. The solubility of trace metals in natural waters is principally controlled by pH, type and concentration of ligands and chelating agents, and oxidation state of the mineral components and the redox environment of the system. Soluble forms are usually ions (simple or complex) or un-ionized organometallic chelates or complexes. The ability of fish to regulate abnormal concentrations varies with the species. Certain fish and crustaceans are able to excrete high proportions of excessive metal intake and thus regulate the concentration in the body at relatively normal levels (reviewed by Bury *et al.*, 2003). Generally the major excretion routes are faeces and urine. Newly hatched fry and fingerlings are poorer regulators of these metals than nonessential minerals such as Hg, Cd, and Pb. Experimental studies on the sub-lethal effects of several metals conducted under laboratory conditions show morphological, physiological (growth, swimming performance, respiration, and reproduction), and behavioural changes in several aquatic organisms. The toxicity mechanisms of metal ions include blocking of essential biological functional groups of enzymes, displacing the essential metal ion in the biomolecule (enzyme or protein), and modifying the active conformation of the biomolecule (Handy, 1996).

Impact of dietary trace elements on the environment

Fish feeds require mineral supplementation due to poor availability of trace elements and phosphorus from feed ingredients. Feeds of carnivorous fish may contain a high proportion of fish meal and other marine by-products. These feed ingredients contain a high concentration of minerals particularly calcium, phosphorus and iron. Bioavailability studies have shown that supplementation of fish meal-based diets with zinc and manganese is necessary for optimum growth, bone mineralization and to prevent cataract formation in fish (reviewed by Lall, 2002). Inorganic and organic compounds of zinc, iron, copper, manganese and selenium are commonly used as supplements in feeds. The discharge of minerals from uneaten food and excretion of urine and faeces by fish has a direct influence on the enrichment of the aquatic environment. Minerals excreted in soluble and particulate forms affect the water quality directly, whereas the particulate form can settle to the bottom of the tank or accumulate in the sediment under fish cages. Soluble P is readily available as a nutrient for the growth of algae leading to algal blooms and fish kills in lakes and coastal areas. Phosphorus settles in sediments and is gradually released as the soluble form during anaerobic or other related biological processes. It appears that the form of minerals in feeds consumed by fish will affect both the amount of soluble and particulate mineral excreted. The breakdown of organically-bound minerals from feed and faeces varies considerably due to differences in the chemical characteristics of feed ingredients and type of mineral supplements used, the environmental conditions (e.g. temperature, oxygen, pH and salinity, water current) and the type of micro-organisms in natural waters.

Elevated levels of Zn and Cu have been found in sediments under the sea cages of salmon farms (Smith *et al.*, 2005; Dean *et al.*, 2007). Experimental studies have demonstrated that the concentration of these trace elements in sediments decrease to the background levels after chemical remediation (Brooks and Mahnken, 2003; Brooks *et al.*, 2003). These studies indicate that metals complexed with free sulfides, particularly copper and zinc are released to the water column during the chemical remediation process. Studies conducted in Canada, Finland and Scotland have shown that Zn sediment levels were elevated in the surface 20 cm, below which the sediment concentration was not affected (Dean *et al.*, 2007; Smith *et al.*, 2005; Uotila, 1991). In the Canadian study, Cu and Zn sediment concentrations were elevated in the upper 10-20 cm of cores collected near fish cages, however, it was relatively low and close to the background levels for these metals in deeper sediments (Smith *et al.*, 2005). In the same study, the highest concentration observed adjacent to farm cages, decreased with increasing distance >200m from the cages.

Copper, zinc and cadmium may be incorporated into the sediment from feed and faeces accumulated, however, these trace elements can also be associated with naturally occurring organic debris. The decomposition of organic material during early stages of diagenesis in the oxidized surface layers influences the pore water concentrations of Cu, Zn and Cd, and results in elevated pore water levels. Organic particles associated metals are remobilized into the dissolved phase, and soluble Cu, Zn and Cd form complexes with organic ligands in pore waters (Petersen *et al.*, 1995; Ponce *et al.*, 2000).

The bioavailability of sediment metals is determined by the abundance of organic compounds such as sulfides and organic material (Casas and Crecelius, 1994; Miller *et al.*, 2000). The bioavailability of some metals may be low around organically enriched sediments of highly reducing farm sites (Brooks and Mahnken, 2003; Brooks *et al.*, 2003). It appears that Zn and Cu enrichment of sediment under marine farm sites affects recolonization by benthic organisms (Morrisey *et al.*, 2000). The mineralization of organic wastes and the reoxidation of sediments during the fallowing of farm sites and recovery of sediment decreases the binding sites and trace elements become more bioavailable and potentially toxic to benthic organisms.

Generally fish feeds are not supplemented with Cd, however, fishery by-products and other ingredients may contribute to a minor amount of Cd in the finished feed. Cadmium is relatively less soluble in reducing pore waters than oxic overlying waters (Smith *et al.*, 2005; Sundby *et al.*, 2004). Results from these studies indicate that the difference in solubility may cause a flux of soluble oxidized species from the overlying water into the anoxic sediment and the formation of insoluble species, resulting in authigenic enrichment of Cd in sediments. Cadmium can be scavenged from the water column and ultimately deposited with organic particles (Thomson *et al.*, 2001). During the burial of Cd in organic matter, remineralisation and subsequent release, it can enhance oxic pore water levels (Gobeil *et al.*, 1997). However, the Cd concentration falls rapidly in anoxic pore waters, and sediment Cd enrichment can occur (Morford and Emerson, 1999; Thomson *et al.*, 2001). Soluble Cd may be immobilized and precipitated in the form of insoluble sulfide species in anoxic conditions (Gobeil *et al.*, 1997; Thomson *et al.*, 2001); however, Sundby *et al.* (2004) suggested that Cd can exist in pore waters of sulphidic sediments as organic and inorganic complexes. It was the stability of such complexes, and not the abundance of sulfides that controls the precipitations of authigenic Cd (Sundby *et al.* (2004). A recent study conducted by Smith *et al.* (2005) showed that in sediment samples collected from adjacent to a marine fish farm, an increase in Cd concentration was the result of not only authigenic enrichment due to anoxic sediment conditions, but also due to an enhancement of Cd precipitation associated with the high level of organic material from the fish farm wastes.

Biofoulants

Several inorganic antifouling compounds from farm vessels, cage unit structure, moorings and nets used in cages, may also accumulate in sediment and enrich organic materials from feeds, feces and other materials present under the cages. Biofoulants have been widely used to control pathogens and algae in fish culture. Many of the biofoulants have copper, in the form of copper sulfate ($CuSO_4$), copper oxide (Cu_2O), or copper thiocyanate (CuSCN) as the active ingredient although some also have zinc (SEPA, 2000) Copper is an essential trace metal for cellular metabolism but can be extremely toxic for aquatic animals at higher concentrations. For crustacean culture, the recommended level is usually lower (0.1-0.5ppm) than for most finfish (0.5-1.0 ppm) (Boyd, 1990). Biofoulants work by creating a toxic layer as the component biocides leach out. Cuprous oxide is used almost exclusively in fish farming industry today and they contain as much as 40% copper by weight. Regulation and legislation of toxic antifouling products for use in the aquaculture industry is either unavailable or lacking in many countries (Braithwaite and McEvoy, 2004).

Constant use of biofoulants can lead to the accumulation of copper and/or zinc in the tissue of the organism and also the surrounding environment. Research related to the use of copper sulfate as a biofoulant in aquaculture activities has focussed on the bioaccumulation of copper in the species of interest in a controlled setting from the prospective of fish or human health. Reddy *et al.* (2006) studied the bioaccumulation of copper in freshwater prawn, *Macrobrachium rosenbergii* and Carvalho and Fernandes (2006) in the neotropical fish *Prochilodus scrofa*. Few studies exist on the impact or load of copper and other trace minerals entering the environment as a result of aquaculture activities.

In addition to copper based antifoulants, organotins are among the most widely used organometallic compounds. Dialkyltin (DBT) has been used as a stabilizer in polyvinyl chloride polymers (PVC) and trialkyltin (TBT) derivatives as anti-fouling agents in boat paint. Both these compounds can enter the aqueous environment from aquaculture activities, DBT by leaching from aquaculture equipment (PVC pipes/tubes, plastic covers from floating boxes) and TBT from leaching from paints. Liu *et al.* (2006) found butyltin residues in seawater and cobia, *Rachycentron canadum* from aquaculture sites in Taiwan.

Conclusions

The successful development of global aquaculture has caused some negative impacts on the aquatic environment in particular cage farming of fish in lakes and coastal areas. The major environmental impacts include the discharge of suspended solids (feces and uneaten food), drugs, disinfectants, algaecides and other chemical compounds leached into the water. Although several nutritional strategies are in place to reduce organic and inorganic enrichment of recipient waters and to minimize the build-up of anoxic sediments that change benthic communities and cause eutrophication of water, fish feeds contain trace elements at a higher concentration than required due to limited information on their requirement and bioavailability from feed ingredients. Elevated levels of zinc, iron, copper, cadmium and manganese have been found in sediments under sea cages and in solid wastes generated by land-based fish farms. An increase in copper levels from antifouling agents has also been detected in sediments. In order to minimize the adverse effects of trace elements on benthic organisms, particularly in their reproduction, recruitment success and survival, there is a need to develop proper models based on mineral bioavailability, feed consumption and excretion, water current and flow and other environmental factors. The necessity to increase research efforts to better define the dietary requirements of trace elements of major concern particularly for zinc, iron, copper and manganese as well as measure their bioavailability and retention in the body of fish is obvious. The best option for now is to minimize zinc, iron, copper, cadmium and manganese concentrations in feeds and to consider use of highly available forms of inorganic and organic trace elements in supplements.

References

Bernhard, M. and M.O. Andreae, 1984. Transport of trace metals in marine food chains. Life Sci. Res. Rep. 28, 143-167.

Black, K., 2002. Environmental impact of aquaculture. Sheffield Academic Press Ltd., Sheffield, United Kingdom. 214 pp.

Bourget, E., L. Lapointe, J.A. Himmelman and A. Cardinal, 1994. Influence of physical gradients on the structure of northern rocky subtidal community. Ecoscience 1, 285-299.

Boyd, C.E., 1990. Water quality in ponds for aquaculture. Alabama Agricultural Experimental Station, Auburn University, Alabama. 482 pp.

Braithwaite, R.A. and L.A. McEvoy, 2004. Marine biofouling on fish farms and its remediation. Adv. Mar. Biol. 47, 215-252.

Brooks, K.M. and C.V.W. Mahnken, 2003. Interactions of Atlantic salmon in the Pacific Northwest environment III. Accumulation of zinc and copper. Fish. Res. 62, 295-305.

Brooks, K.M., A.R. Stierns, C.V.W. Mahnken and D.B. Blackburn, 2003. Chemical and biological remediation of the benthos near Atlantic salmon farms. Aquaculture 219, 355-377.

Bury, N.R., P.A. Walker and C.N. Glover, 2003. Nutritive metal uptake in teleost fish. J. Exp. Biol. 206, 11-23.

Carvalho, C.S. and M.N. Fernandes, 2006. Effect of temperature on copper toxicity and hematological responses in the neotropical fish *Prochiodus scrofa* at low and high pH. Aquaculture 251, 109-117.

Casas, A.M. and E.A. Crecelius, 1994. Relationship between acid volatile sulfide and the toxicity of zinc, lead and copper in marine sediments. Environ. Toxicol. Chem. 13, 529-536.

Dean, R.J., T.M. Shimmield and K.D. Black, 2007. Copper, zinc and cadmium in marine cage fish farm sediments· An extensive survey. Environ. Pollut. 145, 84-95.

Dierberg, F. and W. Klattisimkul, 1996. Issues, impacts and implications of shrimp aquaculture in Thailand. Environ. Manag. 20, 640-646.

Dove, S.G. and M.J. Kingsford, 1998. Use of otoliths and eye lenses for measuring trace-metal incorporation in fishes: A biogeographic study. Mar. Biol. 130, 377-387.

Duplisea, D. E. and B.T. Hargrave, 1996. Response of meiobenthic size-structure, biomass and respiration to sediment organic enrichment. Hydrobiologia 339, 161-170.

Enell, M. and J. Löf, 1983. Environmental impact of aquaculture-sedimentation and nutrient loadings from fish cage culture farming. Vatten 39, 346-375.

FAO, 2004. Review of the state of world fisheries and aquaculture 2004. FAO Fisheries Department, 153 pp.

Flaherty, M. and P. Vandergeest, 1998. Low salt shrimp aquaculture in Thailand: goodbye coastline: hello khon Kaen! Environ. Manag. 22, 817-830.

Gobeil, C., R.W. Macdonald and B. Sundby, 1997. Diagenetic separation of cadmium and manganese in suboxic continental margin sediments. Geochimica et Cosmochimica Acta 61, 4647-4654.

Handy, R.D., 1996. Dietary exposure to trace metals in fish. In: E.W. Taylor (Ed.), Toxicology of aquatic pollution: physiological, cellular and molecular approaches. Cambridge University Press, United Kingdom, pp. 29-60.

Hargrave, B., 2005. Environmental effects of marine finfish aquaculture. The handbook of environmental chemistry, Volume 5 Water Pollution Part M (O. Hutzinger, Ed.). Springer-Verlag, Berlin Heidelburgh, 467 pp.

Holmer, M. and E. Kristensen, 1992. Impact of marine fish cage farming on metabolism and sulphate reduction of underlying sediments. Mar. Ecol. Prog. Ser. 80, 191-201.

Lall, S.P., 1995. Macro and trace elements in fish and shellfish. In: A. Ruiter (Ed.) Fish and Fishery Products. CAB International, Wallington, Oxon, United Kingdom, pp. 187-213.

Lall, S.P., 2002. Minerals. In: J.E. Halver and R.W. Hardy (Eds.), Fish Nutrition, Academic Press, San Diego, CA, pp. 259-308.

Liu, S.-M., M.-P. Hsia and C.-M. Huang, 2006. Accumulation of butyltin compounds in cobia *Rachycentron canadum* raised in offshore aquaculture sites. Sci. Total Environ. 355, 167-175.

Miller, B.S., D.J. Pirie and C.J. Redshaw, 2000. An assessment of the contamination and toxicity of marine sediments in the Holy Loch, Scotland. Mar. Poll. Bull. 40, 22-35.

Morrisey, D.J., M.M. Gibbs, S.E. Pickmere and R.G. Cole, 2000. Predicting impacts and recovery of marine-farm sites in Stewart Island, New Zealand, from the Findlay-Watling model. Aquaculture 185, 257-271.

Morford, J.L. and S. Emerson, 1999. The geochemistry of redox sensitive trace metals in sediments. Geochimica et Cosmochimica Acta 63, 1735-1750.

Neori, A., T. Chopin, M. Troell, A.H. Buschmann, G.P. Kraemer, C. Halling, M. Shpigel and C. Yarish, 2004. Integrated aquaculture: rationale, evolution and state of the art emphasizing seaweed biofiltration in modern mariculture. Aquaculture 231, 361-391.

Petersen, W., K. Wallman, L. Pinglin, F. Schroeder and H.D. Knauth, 1995. Exchange of trace elements at the sediment-water interface during early diagenesis processes. Marine and Freshwater Research 46, 19-26.

Pillay, T.V.R., 1992. Aquaculture and the environment. John Wiley & Sons, New York, NY, 185 pp.

Ponce, R., J.M. Forja and A. Gomez-Parra, 2000. Influence of anthropogenic activity on the vertical distribution of Zn, Cd, Pb and Cu in interstitial water and coastal marine sediments (Cadiz Bay, SW Spain). Ciencias Marinas 26, 479-502.

Reddy, R., B.R. Pillai and S. Adhikari, 2006. Bioaccumulation of copper in post-larvae and juveniles of freshwater prawn *Macrobrachium rosenbergii* (de Man) exposed to sub-lethal levels of copper sulfate. Aquaculture 252, 356-360.

SEPA, 2000. Policy 28: Initial dilution and mixing zones for discharges from coastal and estuarine outfalls. Scottish Environment Protection Agency.

Shearer, K.D., 1984. Changes in elemental composition of hatchery-reared rainbow trout, *Sulmo gairdneri*, associated with growth and reproduction. Can. J. Fish. Aquat. Sci. 41, 1592-1600.

Smith, J.N., P.A. Yeats and T.G. Milligan, 2005. Sediment geochronologies for fish farm contaminants in Lime Kiln Bay, Bay of Fundy. In: Hargrave, B.T. (Ed.), Environmental effects of marine finfish aquaculture. The Handbook of Environmental Chemistry, Vol 5, Springer, Berlin, pp. 221-238.

Sundby, B., P. Martinez and C. Gobeil, 2004. Comparative geochemistry of cadmium, rhenium, uranium, and molybdenum in continental marine sediments. Geochimica et Cosmochimica Acta 68, 2485-2493.

Thomson, J., S. Nixon, I.W. Croudace, T.F. Pedersen, L. Brown, G.T. Cook and A.B. MacKenzie, 2001. Redox-sensitive element uptake in north-east Atlantic Ocean sediments (Benthic Boundary Layer Experiment sites). Earth and Planetary Science Letters 184, 535-547.

Uotila, J., 1991. Metal contents and spread of fish farming sludge in Southwestern Finland. In: Maekinen, T. (Ed.), Marine aquaculture and environment. Nord, 22, pp. 121-126.

Farm-scale nutrient and trace element dynamics in dairy farming

I. Öborn, H. Bengtsson, G.M Gustafson, J. Holmqvist, A.-K. Modin-Edman, S.I. Nilsson, E. Salomon, H. Sverdrup and S. Jonsson

Keywords: element balance, farm systems analysis, organic farming, trace metals

Introduction

Good nutrient management within agricultural production systems is important from an economic, natural resource and environmental perspective. Dairy farming systems are often characterised by large on-farm nutrient flows relating to home-produced feed crops and to any animal manure produced mainly being applied on-farm. Element balance calculations provide a useful tool to establish the degree of imbalance between inputs and outputs at farm- and field-scale (e.g. Smaling *et al.*, 1999; Öborn *et al.*, 2003). The magnitude of flows and balances can be used as an indicator of the potential environmental impact of dairy production systems (Thomassen and de Boer, 2005), and evaluated in relation to existing soil element pools, cropping systems and soil and landscape properties.

This paper summarises some major findings from the project 'Fluxes and balances of nutrients and trace elements in different farming systems' (Öborn *et al.*, 2005), with emphasis on trace element dynamics in dairy farming. The project formed part of 'FOOD 21 – Sustainable food production', an interdisciplinary research programme encompassing the whole agro-food chain (Andersson *et al.*, 2005). The overall objective was to analyse a dairy farming system, assess its sustainability with reference to nutrient and trace element dynamics, and build the capacity for management design with the goal of increasing the sustainability of the farming system. Specific objectives included to (1) quantify pools, fluxes, and balances of nutrients and trace elements in the soil-crop-animal system at farm-scale, (2) assess the degree of element imbalance, (3) clarify the importance of different farm features, such as degree of self-sufficiency and types of feedstuff used, for the occurrence and amounts of specific elements in animal products and manure, and (4) quantify the element losses through leaching and surface run-off at field level. Additional aims were to compare organic management with conventional and to address possible adaptive approaches to management. Annual element balances were established at farm-gate, field and barn level and the FARMFLOW model was developed to enable farm- and field-scale dynamic simulations for longer time periods.

Materials and methods

Description of the study farm

A three-year case study was carried out at the Öjebyn farm in northern Sweden (65°21'N, 21°24'E) where organic (Org) and conventional (Conv) dairy farming had been practised in parallel since 1988 (Jonsson, 2004). Both systems had 40-50 dairy cows and 58 (Org) and 47 (Conv) ha agricultural land, of which 10 and 7 ha respectively were used for grazing (Öborn *et* al., 2005). The difference in allocated land area was based on the fact that lower yields were predicted in the organic system at the start of this full-scale experiment. There were two cow houses, and manure and urine from each farming system were stored separately (Gustafson *et al.*, 2003). The six-year crop rotation included a forage crop (*Avena sativa* L. and *Pisum* sp.) undersown with ley, three years of grass/clover ley (*Phleum pratense* L., *Festuca pratensis* L., *Trifolium pratense* L., *Trifolium repens* L.(only Org)), spring barley (*Hordeum vulgare*, L.) and potatoes (*Solanum tuberosum*, L.) (Bengtsson *et al.*, 2003). The stocking density was 0.9 (Org) and 1.1 (Conv) animal unit ha^{-1}. The average production of energy-corrected milk (3.14 MJ L^{-1}) was 8600 (Org) and 9300 (Conv) kg milk cow^{-1} yr^{-1} (Gustafson *et al.*, 2007). The composition of the feed ration (based on dry matter) in the Org herd was 70% roughage, 13% barley and 17% concentrate, while in the Conv herd the corresponding proportions were 55, 21 and 24%.

Quantification of element flows

All flows to and from the farm and within the farm were recorded on a daily basis during the entire study period. This included the amount and origin of the different feed components and bedding materials used. The amount of silage, barley and straw harvested was weighed before transportation from the field, while loads of manure and urine were weighed before being spread on the field. Samples were taken regularly for dry matter determination and chemical analyses. The procedures for recording, quantification, sampling and analysis are described in detail in Bengtsson *et al.* (2003) and Gustafson *et al.* (2003). The atmospheric deposition was estimated from regional environmental monitoring data, and surface run-off and leaching were quantified by combining soil water chemical data with hydrological modelling of the water flows (Bengtsson *et al.*, 2006).

Mass balance assessments

Element mass balances were established on an annual basis at different scales, i.e. farm-gate, field and barn level, as illustrated in Figure 1.

The farm-gate balances included products purchased in and products sold from the farm (Equation 1), whereas the field balances focused on crop production, including the major biophysical flows to and from the field (soil surface) (Bengtsson *et al.*, 2003; Bengtsson, 2005) (Figure 2). The barn balances were sub-system balances focusing on the animal production component of the farm, including the element flows in the chain from feed to animals and ending up in animal products and animal wastes (manure/urine) (Gustafson *et al.*, 2003; 2007) (Figure 2).

Farm-gate balance = Purchased products – Sold products   (1)

which translates into

[Purchased feed + Mineral fertilisers + Lime + Pesticides + Seeds + Heifers + Water + Bedding material] – [Milk + Cash crops + Cows for slaughter + Calves].

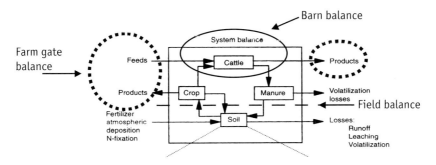

Figure 1. Conceptual model of element flows on a dairy farm using three different approaches for element balances (modified from Oenema and Heinen, 1999).

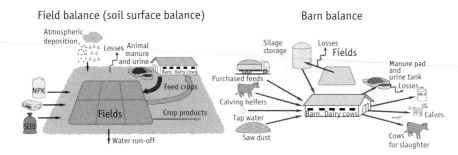

Figure 2. Illustration of input and output flows included in the field balance calculations (left; Bengtsson, 2005) and the barn balance calculations (right; Gustafson et al., 2003).

Results and discussion

Farm-gate balances

During one year, farm-gate balances were established for 14 elements, seven macronutrients (Ca, K, Mg, N, Na, P and S) and seven trace elements (Cd, Cu, Fe, Mn, Mo, Se and Zn) (Table 1). All trace elements except Cd are micronutrients for dairy livestock and are thus added as minerals in the feed. The inflow of macronutrients was relatively small, 75-211 kg N ha^{-1} yr^{-1}, 11-22 kg P ha^{-1} yr^{-1} and 26-76 kg K ha^{-1} yr^{-1}, and was higher in the Conv than in the Org system. The farm-gate balance for N showed a surplus of 25 (Org) and 143 (Conv) kg ha^{-1} yr^{-1}, biological N-fixation not included (Öborn, 2004). Phosphorous was the element closest to being in balance and it might be important to monitor the farm-gate balance for P over several years to ensure that there is no depletion of the system. The same applies for K in the Org system.

The inputs of trace elements, especially Fe, Mn, and Zn, largely exceeded the outputs, resulting in a risk of accumulation in the soil (Table 1). The excess was twice as high in

Table 1. Farm-gate inputs, outputs and balances (input-output) of trace elements (Cd, Cu, Fe, Mn, Mo, Se and Zn) (g ha⁻¹ yr⁻¹), and element use efficiency (EUE, %) in conventional (C) and organic (O) dairy farming systems at the Öjebyn Farm, Sweden, during a 12-month period.

Flow/balance	Item	Cd C	Cd O	Cu C	Cu O	Fe C	Fe O	Mn C	Mn O	Mo C	Mo O	Se C	Se O	Zn C	Zn O
Input															
Crop production[1]	Mineral fertiliser	0.01	-	6.1	-	0	-	33	-	0.2	-	0.00	-	4	-
	Lime[2]	-	0.13	-	0	-	0	-	0	-	0	-	0	-	17
Animal production	Heifers	0.00	0.00	0.8	0.6	13	9	0	0	0.6	0.5	0.39	0.29	8	6
	Bedding	0.03	0.02	0.3	0.3	36	28	46	36	0.0	0.0	0.00	0.00	5	4
Feed	Water	0.00	0.00	0.1	0.1	0	0	0	0	0.0	0.0	0.00	0.00	2	2
	Barley	0.06	0.01	8.7	3.0	52	14	26	9	0.5	0.3	0.02	0.00	51	20
	Concentrate	0.21	0.13	17.6	10.9	339	183	281	154	0.8	0.5	2.22	1.11	288	149
	Molasses	0.17	0.05	2.6	0.8	250	75	33	10	0.0	0.0	0.07	0.03	18	6
Total In		0.48	0.21	36	16	690	309	420	209	2.1	1.3	2.7	1.4	376	187
Output															
Crop production	Cash crops (potato)	0.02	0.01	2.4	1.6	0	0	2	2	0.0	0.0	0.00	0.00	5	6
Animal production	Milk	0.00	0.00	0.6	0.4	3	3	1	0	0.4	0.4	0.13	0.09	48	33
	Cows for slaughter	0.00	0.00	0.8	0.6	13	10	0	0	0.7	0.5	0.42	0.30	8	6
	Calves	0.00	0.00	0.2	0.1	3	2	0	0	0.2	0.1	0.08	0.06	2	2
Total Out		0.02	0.01	4	3	19	14	3	2	1.3	1.0	0.6	0.4	63	47
Balance In-Out		0.46	0.20	32	13	671	295	417	207	0.8	0.3	2.1	1.0	313	140
EUE (%) (Out/In)×100		4	5	11	17	3	4	1	1	61	75	23	32	17	25

- = not applied.
[1]Seeds and pesticides (only conventional) not determined.
[2]Not included in the 1-year balance. Lime was added once per 6-year rotation to both the Org and Conv system based on soil analyses, i.e. long-term similar amounts of Cd and Zn were added through lime to the Org and Conv systems.

Conv compared with Org due to different feeding strategies, i.e. a higher proportion of purchased feed (Conv) and on-farm produced roughage (Org) (Gustafson *et al.*, 2007). Since Fe and Mn are major soil constituents this was probably a minor problem, while the long-term accumulation of Zn, and potentially also Cu, could be more of a concern. Selenium added to the concentrate was the major inflow, while import and export of living animals also contributed to the Se flows. The main input sources of Cd were concentrate, molasses and lime (occasionally applied to the fields), where it occurs as a contaminant. Molybdenum was the trace element with the highest removal (outflow) compared with inflow, 61% (Conv) and 75% (Org) being utilised.

The farm-gate balances expressed per unit area of land (Table 1) led to lower values for the Org system, where more land was allocated for the same size of dairy herd as in the Conv system. Farm-gate balances were also established based on the amount of milk being produced in the system, and then the differences between the systems were smaller, although Conv still showed a greater excess than Org (Öborn *et al.*, 2005). However, if the excess is to be used as indicator of potential risk for losses and/or soil accumulation/depletion, area-based values might be more useful than production-based values.

Barn balances

Silage K concentrations were significantly lower in the Org dairy system than in the Conv system. No other significant differences in element (Ca, Cu, Mg, Mn, N, P, S and Zn) concentrations could be detected in feed or animal products. However, differences in feeding strategies and in quantities of input and output flows were reflected by a larger proportion of external element flows in Conv (20-50%) than in Org (10-40%) due to a higher proportion of purchased feed in Conv and lower milk production and higher manure production in Org. Thus there was a higher degree of self-sufficiency through on-farm cycling of nutrients and trace elements to the feed in the Org system (Gustafson *et al.*, 2007).

The recovery of Ca, Mg, P and S was very good, i.e. the elements in the feed could be traced back in animal products and manure/urine (Gustafson *et al.*, 2007; Figure 3). Calculating the element content of manure and urine as [inputs – milk] underestimated the amount of Cu, Mn and Zn and overestimated the amount of N and K compared with the empirical data. This means that there were some sources of Cu, Mn and Zn in the feed-animal manure system that was not taken into account and that there were N and K sinks within the system. A combination of ammonia emissions and losses from silage effluent and manure/urine handling were probable sinks for N and K. There were 30-50% higher outputs of Cu, Mn and Zn in manure/urine compared with the feed inputs. Probable sources could have been metal structures in the cow houses and metal components in the feed and manure handling equipment, such as feeding fences, drinking bowls, feed conveyors, water pipes, metal dung scrapers etc., which at the Öjebyn farm had been in use for a long time (and partly started to become worn out). Galvanized surfaces can have been a major Zn source (but also other metal structures e.g. brass) while Mn probably originated from e.g. steel and iron, and brass and copper pipes could have been possible Cu sources. Other possible non-feed sources of Cu and Zn were foot bath and udder cream. Trace element sources other than feed components

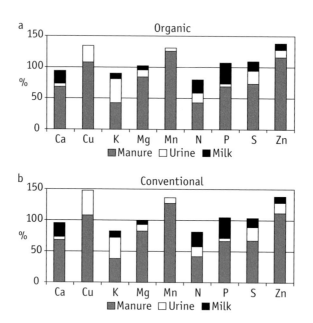

Figure 3. Recovery of feed inputs in the outputs (manure, urine, milk). For Ca, Mg, P, S: In=Out; for K, N: In>Out; and for Cu, Mn, Zn: In<Out (Gustafson et al., 2007).

were also observed by Eckel *et al.* (2005). De Belie *et al.* (2000) reviewed the durability and corrosion of metallic building structures and components in the agricultural environment and found that corrosion rates could be high due to e.g. high concentrations of aggressive gases, acids and salts.

Field balances

The field balances, including major flows to and from the field (soil surface), showed that the on-farm cycling of most elements was much larger than the flows through the farm-gate (Figure 4). For example, the K input was about 100 (Org) and 150 (Conv) kg ha^{-1} yr^{-1} (Figure 4), of which 25 (Org) and 75 (Conv) kg ha^{-1} yr^{-1} came from external sources. For K, the inputs were on average smaller than the outputs in both systems, which meant that there was a negative K balance ranging between -18 and -22 (Org) and -27 and +58 kg ha^{-1} yr^{-1} (Conv). Around 12-14 kg K ha^{-1} yr^{-1} was released from soil minerals through chemical weathering, as estimated by simulation modelling (Öborn *et al.*, 2005), but it is uncertain whether this can compensate for the net out-take in the long-term. The P field balance, calculated as an average for the crop rotation, showed a small excess, i.e. it was slightly positive in all years, 1-6 (Org) and 3-9 (Conv) kg P ha^{-1} yr^{-1}. These were close to the values obtained in the farm-gate balance calculations, indicating that there were no major P sources or sinks (including losses) on the farm.

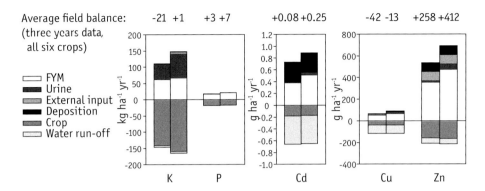

Figure 4. Field balances calculated for Cd, Cu, K, P and Zn as an average for the crops in the rotation (mean value for 3 years of data). Input flows included cattle manure (FYM), urine, external inputs (mineral fertiliser, lime, seeds) and atmospheric deposition, while output flows included removed harvested crops and water run-off (surface run-off and leaching) (Bengtsson, 2005; Öborn et al., 2005).

For Cd, Cu and Zn, atmospheric deposition and leaching were relatively more important input and output flows than for the major nutrients (P, K), for which the agronomic input flows dominated (Figure 4). The Zn field balances showed excess applications in all years (positive balances), more so in the Conv system (390-460 g ha^{-1} yr^{-1}) than in the Org (230-320 g ha^{-1} yr^{-1}). Some of this excess was lost through leaching and surface run-off but the majority, on average 412 (Conv) and 258 g ha^{-1} yr^{-1} (Org), was accumulated in the soil. Bengtsson *et al.* (2006) quantified the trace element surface run-off and leaching at three soil types at the Öjebyn farm during a 5 year period, and estimated that surface run-off were 0.22±0.20 (Cd), 50± 33 (Cu) and 24 ±20 (Zn) g ha^{-1} yr^{-1} whereas the leaching losses were 0.13±0.09 (Cd), 43±18 (Cu) and 16±9 (Zn) g ha^{-1} yr^{-1}.

Farm systems analyses

The FARMFLOW model was initially developed for P (Modin-Edman *et al.*, 2007) and later also parameterised and applied for simulations of Cd and Zn flows (Modin-Edman, 2007) (Figure 5). Using a dynamic farm systems model enabled estimations of long-term effects of nutrient and trace element imbalances at farm and field level, e.g. risks of long-term soil accumulation or depletion. Scenarios have been run for different stocking rates and with uniform/non-uniform distribution of manure between fields in order to evaluate soil P accumulation rates and the risk of losses (Modin-Edman, 2007).

How representative is the Öjebyn farm?

In 2005, the most common (36%) herd size of dairy farms in Europe (EU 15) was between 20-49 dairy cows (SCB, 2007), which is within the 40-50 cow range in the Conv and Org systems at the Öjebyn farm. In Sweden, the average dairy herd size in 2005 was 48 cows (SCB, 2007). The arable crops (barley and potatoes) included in the Öjebyn rotations are also

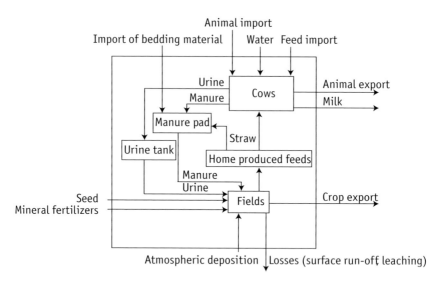

Figure 5. A stock and flow diagram illustrating the FARMFLOW model structure with farm imports, exports and internal cycling of elements within a dairy farming system (Modin-Edman, 2007).

frequently grown in Europe. The very northerly location of Öjebyn (65°N) is very suitable for forage production, as the long day-length in summer partly compensates for the short growing season. Yields of ley and forage crops at the site are well within the Swedish average level, whereas the amount of harvested barley grain is lower (~65% of national average) (Bengtsson *et al.*, 2003; SCB, 2007). In 2006, 7% of Swedish agricultural land was devoted to organic farming and 6% of dairy cows were organically certified or qualified for certification (SCB, 2007). The majority of the organic area was cropped with grass/clover leys (39%) and cereals (25%). The official Swedish target is for organic production to increase to 20% of agricultural land by 2010 (SCB, 2007).

The chemical composition of animal manure can be one way to compare different farm management and feeding systems. Comparisons of trace element concentrations with data from dairy farms in England and Wales (Nicholson *et al.*, 1999) showed that the variation between farms was greater than that between sampling occasions at Öjebyn (Table 2). The average Zn concentrations at Öjebyn were very similar to those measured in England and Wales, and were lower than the Swedish average value. Manganese and Cu concentrations at Öjebyn were within, or just below, the range of the other studies, but lower than their averages.

Table 2. Trace elements in solid cattle manure (FYM) from dairy farms in Sweden (Steineck et al., 1999) and England and Wales (Nicholson et al., 1999), and from the Öjebyn farm. Mean ± standard deviation or mean and range (min-max) is given, mg kg^{-1} of dry matter, n= number of farms (national surveys) or samples (Öjebyn).

Location	System	n	Cd	Cu	Mn	Se	Zn
Sweden	Conventional	17	0.16±0.07a[1]	31±8a	230±32a	0.6±0.3a	174±32a
	Organic	23	0.13±0.06a	29±14a	208±75b	0.5±0.3a	148±66b
England and Wales		6	0.38	38	nd	nd	153
			(<0.10-0.53)	(26-56)			(99-238)
Öjebyn	Conventional	9	0.25[2]±0.03a	21±2a	204±24a	nd	152±14a
	Organic	11	0.21[2]±0.03a	20±2a	185±23a	nd	150±21a

nd = not determined.
[1]Different letters (a, b) indicate statistically significant differences between conventional and organic farming practices (P<0.05).
[2]n=5 (Conv) and 7 (Org) (Cd only analysed year 1).

Conclusions

Lessons learned from the Öjebyn study

The Öjebyn farm is representative of European dairy production and thus the research findings are applicable to other dairy farms in temperate climates. Lessons learned from the Öjebyn study were that:
- Internal flows dominated in the dairy production systems investigated and have to be considered in nutrient management.
- Large on-farm element sinks (N, K) and sources (Cu, Mn, Zn) were indicated and diffuse trace element sources in particular need to be further investigated since they contributed 30-50% of the metals found in manure.
- Diffuse flows through atmospheric deposition (In) and leaching losses (Out) were more important for trace elements than macronutrients, and contributed significantly to the field element balances (Cd, Cu, Zn).
- Future challenges include an increasing demand for organically produced dairy products and hence organically produced feed. Good knowledge and skill in farm-scale nutrient management will be crucial for the sustainability of the production system.
- Element balances are useful as diagnostic tools for detecting farm-scale nutrient and trace element imbalances. Farm-gate, barn and field balances provide complementary information.

Some concluding remarks regarding the element balance approach

Simple element balances can be used to evaluate final goals, whereas dynamic models are necessary in order to predict the pathway leading to those goals. In this project we applied simple element balance calculations to different parts of the farming system, i.e. farm-gate, barn and field scale, and created a model (FARMFLOW) by which dynamic mass balances of elements in the dairy farming system could be analysed. Farm-gate balances are usually based on easily accessible data, but they do not include the internal flows. Barn and field balances, on the other hand, have the potential to include all flows to and from the barn or field but do not distinguish between external and internal sources (e.g. of elements in manure). By combining farm-gate, barn and field balances, we were able to identify internal sinks and sources, which would not have been possible using only one of the element balance approaches. We were also able to take into account how surpluses or deficits of silage were handled in the farming systems, which is important to decrease the variation between years in balance calculations.

Acknowledgements

Thanks to the staff at the Öjebyn farm, Department of Agriculture Research for Northern Sweden, for access to data and good collaboration. This work was carried out within the Food 21 – Sustainable Food Production Programme funded by the Strategic Foundation for Environmental Research (MISTRA) and SLU.

References

Andersson, R., B. Algers, L. Bergström, K. Lundström, T. Nybrandt and P.-O. Sjödén, 2005. Food 21: A research program looking for measures and tools to increase food chain sustainability. Ambio 34, 275-282.

Bengtsson, H., 2005. Nutrient and Trace Element Flows and Balances at the Öjebyn Dairy Farm – Aspects of Temporal and Spatial Variation and Management Practices. Faculty of Natural Resources and Agricultural Sciences, Swedish University of Agricultural Sciences (SLU). Acta Universitatis Agriculturae Sueciae, Doctoral Thesis No. 2005:2.

Bengtsson, H., I. Öborn, S. Jonsson, I. Nilsson and A. Andersson, 2003. Field balances of some mineral nutrients and trace elements in organic and conventional dairy farming – a case study at Öjebyn, Sweden. Eur. J. Agron. 20, 101-116.

Bengtsson, H., G. Alvenäs, S.I. Nilsson, B. Hultman and I. Öborn, 2006. Cadmium, copper and zinc outputs via leaching and surface run-off at the Öjebyn farm in Northern Sweden – temporal and spatial variation. Agric. Ecosyst. Environ. 113, 120-138.

De Belie, N., B. Sonck, C.R. Braam, J.J. Lenehan, B. Svennerstedt and M. Richardson, 2000. Durability of building materials and components in the agricultural environment, Part II: Metal structures. J. agric. Engng Res. 75, 333-347.

Eckel, H., U. Roth, H. Döhler, F. Nicholson and R. Unwin (Eds.), 2005. Assessment and reduction of heavy metal input into agro-ecosystems. KTBL-Schrift 432, Darmstadt.

Gustafson, G.M., E. Salomon, S. Jonsson and S. Steineck, 2003. Fluxes of K, P and Zn in a conventional and an organic dairy farming system through feed, animals, manure and urine – a case study at Öjebyn, Sweden. Eur. J. Agron. 20, 89-99.

Gustafson, G.M., E. Salomon and S. Jonsson, 2007. Barn balance calculations of Ca, Cu, K, Mg, Mn, N, P, S and Zn in a conventional and organic dairy farm in Sweden. Agric. Ecosyst. Environ. 119, 160-170.

Jonsson, S. 2004. Öjebynprojektet – Ekologisk produktion av livsmedel. Slutrapport (The Öjebyn Project – Organic Food Production. Final Report). Department of Agricultural Research for Northern Sweden, Swedish University of Agricultural Sciences (SLU), Öjebyn, Sweden. 44pp (In Swedish with English summary).

Modin-Edman, A.-K. 2007. Simulations of Phosphorous, Zinc and Cadmium Mass Flow in Dairy Farming Systems. Doctoral thesis. Department of Chemical Engineering, Lund University. Reports in Ecology and Environmental Engineering 2007:1.

Modin-Edman, A.-K., I. Öborn and H. Sverdrup, 2007. FARMFLOW-A dynamic model for phosphorus mass flow, simulating conventional and organic management of a Swedish dairy farm. Agric. Syst. 94, 431-444.

Nicholson, F.A., B.J. Chambers, J.R. Williams and R.J. Unwin. 1999. Heavy metal contents of livestock feeds and animal manures in England and Wales. Biores. Tech. 70, 23-31.

Oenema, O. and M. Heinen, 1999. Uncertainties in nutrient budgets due to biases and errors. In: Smaling, E.M.A., O. Oenema and L.O. Fresco (Eds.), Nutrient Disequilibria in Agroecosystems: Concepts and Case Studies. CAB International, Wallingford, United Kingdom, 75-97.

Öborn, I. 2004. Do we risk nutrient imbalances in organic dairy production? In: Steiness, E (Ed), Geomedical aspects of organic farming. The Norwegian Academy of Science and Letters, Oslo, 71-83.

Öborn, I., A.C. Edwards, E. Witter, O. Oenema, K. Ivarsson, P.J.A. Withers, S.I. Nilsson and A. Richert Stinzing, 2003. Element balances as a tool for sustainable nutrient management: A critical appraisal of their merits and limitations within an agronomic and environmental context. Eur. J. Agron. 20, 211-225.

Öborn, I., A.-K. Modin-Edman, H. Bengtsson, G.M. Gustafson, E. Salomon, S.I. Nilsson, J. Holmqvist, S. Jonsson and H. Sverdrup, 2005. A systems approach to assess farm-scale nutrient and trace element dynamics: A case study at the Öjebyn dairy farm. Ambio 34, 301-310.

SCB, 2007. Agricultural Statistics 2007. Statistics Sweden. (www.scb.se/templates/Product____37545.asp accessed August 10, 2007).

Smaling, E.M.A., O. Oenema and L.O. Fresco (Eds.), 1999. Nutrient Disequilibria in Agroecosystems: Concepts and Case Studies. CAB International, Wallingford, United Kingdom. 322 pp.

Steineck, S., G. Gustafson, A. Andersson, M. Tersmeden and J. Bergström, 1999. Plant Nutrients and Trace Elements in Livestock Wastes in Sweden. Swedish Environmental Protection Agency Report 5111, 1-27.

Thomassen, M.A. and I.J.M. de Boer, 2005. Evaluation of indicators to assess the environmental impact of dairy production systems. Agric. Ecosyst. Environ. 111, 185-199.

Trace elements in poultry litter: prevalence and risks

J.B.J. van Ryssen

Keywords: poultry excreta, ruminants, copper, arsenic, pollution

Introduction

Poultry litter is the general term used for the mixture of poultry excreta, bedding material, spilled food, feathers, etc., i.e. the product that has to be disposed of when poultry production units are cleaned. Disposing of the product has become an increasingly difficult problem worldwide because of the expansion and concentration of poultry production units in specific areas (Ribaudo *et al.*, 2003). Some components in poultry litter have the potential to pollute the environment. Unfortunately all methods of disposing of the product seem to pose, directly or indirectly, a threat to the environment and/or to human health and well-being, or are perceived to exist. These individual components constituting poultry litter vary in their contribution to the concentration of trace elements in the litter. Factors contributing to this variation will be highlighted and evaluated in terms of the disposal of litter, with emphasis on the feeding of poultry litter to farm animals.

Factors affecting the trace element composition of poultry litter

Many tables have been published on the elemental composition of the different types of poultry litter (e.g. Van Ryssen *et al.*, 1977; 1993; Kelley *et al.*, 1998; Kpomblekou-A *et al.*, 2002). The sources of trace elements in poultry litter include not only the bird droppings, bedding material, spilled feed, but also drug residues and chemicals that have been used to treat the litter (Kpomblekou-A *et al.*, 2002). Soil can contribute to this if litter became contaminated with soil during collection (Brosh *et al.*, 1998). The contribution to the trace element composition of litter from each of these factors varies, and is not always easy to regulate or control.

Dietary route

Classes of birds and dietary requirements

Mineral requirements of the different classes of birds with different productions purposes differ (Table 1), and consequently the composition of their excreta will differ (Table 2).

Since the trend in poultry production worldwide is away from the inclusion of antibiotics as growth promoters in diets, micronutrients, including the trace elements that have antioxidant properties, are now included in diets at concentrations well above published requirements, to assist in strengthening the immune systems of the bird against infections, and neutralizing free radicals produced in high producing birds (Rebel *et al.*, 2004). This is contrary to any attempts to minimize excretion of minerals, though it is also argued that such high dietary inclusions should be considered as part of the requirements of the bird,

Table 1. Dietary requirements of relevant mineral elements for laying hens and broilers (NRC, 1994).

Elements	Laying hens	Broilers
Calcium (%)	2.7 - 4.1	0.8 - 1.0
Copper (mg/kg)	?	8
Iron (mg/kg)	38 - 56	80
Manganese (mg/kg)	17 - 25	60
Selenium (mg/kg)	0.05 - 0.08	0.15
Zinc (mg/kg)	29 - 44	40

being required to optimize the immune systems of the body. This approach is practised, for instance, in prescribing higher selenium allowances than the standard recommendations to high producing dairy cows to protect them against mastitis (Jukola et al., 1996).

Elements in raw feed ingredients and supplementation

The theoretical objective in formulating a diet is to meet these requirements as closely as possible. The raw feed ingredients of a diet could contain many if not all the nutrients required by the birds. For macro-nutrients, including calcium and phosphorus, nutritionists would usually use computer programmes to match the nutrient profile of a bird's requirements with the nutrients in the raw material and make up any shortcomings through supplementation, though for the minerals, allowing fairly wide margins of safety. However, in the case of trace elements it is not feasible for feed manufacturers to quantify the trace element content of the raw feed ingredients they are using. Most if not all feed manufacturers include mineral and vitamin premixes in their poultry diets irrespective of what elements are present in the raw materials. This 'shotgun' type of approach is practised to ensure that there are no deficiencies. Ruiz et al. (2000) stated: 'Despite the fact that most practical diets contain adequate iron and copper to satisfy the nutritional requirements of broilers, both minerals are still generally supplemented in order to avoid possible deficiencies'. Consequently, many of the elements are supplied in excess of the bird's requirements, as is the situation in New Zealand (Ravindran, 2006), as an example. Within limits these excessive levels do not pose any risks to the health and well-being of the bird. However, most of these surpluses would be voided through the excreta.

Pharmacological use of trace element

Usually, when specific elements are included in diets for reasons beyond the basic nutritional requirements of the bird, high concentrations would be present in the litter. Specific pharmacological uses of drugs containing high levels of trace elements are practised in some countries. These products include coccidiostats and antimicrobial substances.

Copper

According to the NRC (1994) the copper requirements for the different classes of poultry are up to 8 mg/kg of the diet (Table 1). However, for many years cupric sulfate pentahydrate has been included in broiler (as well as pig) diets to act as a growth promoter at inclusions rates of between 100 and 300 mg/kg feed (Fisher *et al.*, 1972; Skřivan *et al.*, 2000). This practice is still permitted in some countries such as the USA. High levels of dietary copper from different sources, including those in organic compounds, are believed to act as antimicrobial substances and fungicides in the digestive tract of the birds, similar to the action of subtherapeutic antibiotics (Ewing *et al.*, 1998; Arias and Koutsos, 2006; Lim *et al.*, 2006). An additional advantage of the added copper compounds is that it inhibits microbial proliferation in litter. Therefore, high levels of copper in manure reduce odours through the suppressing of bacterial activity that breaks down manure with the release of gases that cause odours.

A more recent discovery is that copper supplementation at these so-called therapeutic levels was found to reduce the cholesterol levels in eggs and poultry meat (Pesti and Bakalli, 1998; Skřivan *et al.*, 2000; 2005; Balevi and Coskun, 2004). Other elements and substances can apparently also contribute. Lien *et al.* (2004) concluded that 125 mg Cu/kg and 800 to 1600 µg Cr/kg were the most effective levels of copper and chromium in diets to reduce egg yolk cholesterol levels. Konjufca *et al.* (1997) compared copper with garlic as a means of reducing cholesterol in muscle and found that breast muscle cholesterol was reduced by 24% using a copper compound and 15% by feeding garlic. They also found that copper supplementation was more efficient in reducing meat cholesterol levels than substituting fat with oil. Consequently, the inclusion of such high copper levels in poultry diets has positive health implications for humans. This raises the question of a bird's copper requirements. If production and efficiency are the criteria, < 8 mg Cu/kg would be required in the diet. If minimum cholesterol levels in meat and eggs are the criteria, then the requirement would be ⩾150 mg Cu/kg diet (Pesti and Bakalli, 1998).

However, a consequence of these high levels of dietary copper is high concentrations of copper in poultry excreta, posing an environmental pollution risk and toxicity risk to animals consuming poultry litter. Broiler litter collected in the state of Alabama, USA, had copper concentrations ranging from 25 to 1003 mg/kg (Ruffin and McCaskey, 1998). Westing *et al.* (1985) recorded a mean copper concentration in broiler litter of 593 mg/kg and Banton *et al.* (1987) reported levels of up to 920 mg Cu/kg in broiler litter. The copper concentrations quoted in Table 2 are relatively low, implying that copper sulfate is not used in broiler diets in South Africa. Also, a recent EU directive restricts the maximum level of copper allowed in diets for poultry to 35 mg/kg (Skřivan *et al.*, 2005).

At these high copper intakes Pesti and Bakalli (1998) recorded only slight increases in copper concentrations in the different egg components. Arias and Koutsos (2006) reported that the copper concentration in the liver of birds starts to increase at copper intakes above 200 mg/kg diet, while Van Ryssen *et al.* (1977) recorded a curvilinear increase in Cu concentration, becoming more pronounced at above 150 mg Cu/kg in the diet.

Arsenic

In some countries products containing arsenic are included as feed additives in poultry diets. In the USA about 70% of all broiler diets contain the product, roxarsone (25% As), included as a coccidiostat (Jones, 2007). In Europe the use of arsenical feed additives has been prohibited since 1998. Where permitted, inclusion rates in diets range from 22.7 to 45.4 roxarsone/kg. The concentration of arsenic in litter varies between 14 and 76 mg/kg litter. Most of the arsenic present in poultry litter is in an organic form that is considered non-toxic. However, it is claimed that organic arsenic in litter can change to the more toxic inorganic forms when subjected to certain conditions in the soil (Garbarino *et al.*, 2003). Intensive media and internet campaigns are being launched worldwide against this product and the risks of an accumulation of arsenic in poultry litter and the environment. An article title: 'Arsenic, A roadblock to potential animal waste management solutions' (Nachman *et al.*, 2005) tells it all. However, in a recent publication Jones (2007) reviewed evidence for and against the dangers of arsenicals in poultry diets, disputing many of the claims made against the inclusion of roxarsone in poultry diets. Amongst others, he pointed out that one publication agitating against the use of arsenicals made a calculation error of 7000%.

Arsenic levels in poultry litter reported in the literature are: 35 ± 19 mg/kg (Kunkle *et al.*, 1981), 76 mg/kg (Westing *et al.* (1985), 26 - 30 mg/kg (Kelley *et al.*, 1996), 31.5 mg/kg (Anderson and Chamblee, 2001) and 20.6 mg/kg (Kpomblekou-A *et al.*, 2002).

Inadvertent inclusions

Feed grade mineral supplements usually contain impurities including toxic elements (Ammerman *et al.*, 1977). In most countries some form of feed quality and safety regulations and assessments exists to ensure that raw materials comply with specifications and are not contaminated with heavy metals, pathogens and products such as dioxin. However, it is almost impossible to check every batch of materials going through a feedmill for contaminants. In 2004 a zinc sulfate source of Chinese origin containing between 32 to 64 g cadmium/kg zinc sulfate, was sold to South African feed companies. The same contaminated product was apparently sold also in Norway and France (Coetzee, 2006). The recurrence of such an incidence in Norway was reported quite recently (Feedinfo News Service, 16/05/2007). After the melamine scandal much stricter measures and inspections are insisted upon by importing countries in that suppliers and exporting countries should ensure that their products meet certain quality standards (Feedinfo News Service, 01/06/07).

Production systems

Production systems differ: Some birds are kept in batteries, some on deep litter systems and some are free ranging. In Table 2 the differences in the micro-element content of samples in South Africa are presented. The 'backyard' birds were free ranging in a communal type of set-up. When droppings were collected care was taken not to have any soil contaminating the samples. The broilers were kept on deep litter (one cycle) and the layers were in batteries.

Table 2. Element concentration (mg/kg DM) in excreta from backyard chickens vs. broilers (deep litter, one cycle) and layers (batteries) in South Africa (Van Ryssen et al., 1993).

	Backyard (free-range)	Broilers (litter)	Layers (batteries)
Aluminium	9885	834	1683
Arsenic	41.6	4.9	2.5
Cadmium	1.5	0.32	0.50
Chromium	51.6	11.2	9.2
Cobalt	7.1	1.1	1.4
Copper	36.2	43.6	45.9
Iron	16762	1335	2271
Lead	46.4	0.55	1.17
Mercury	1.60	0.48	1.71
Molybdenum	9.1	1.5	10.4
Vanadium	50.8	12.1	17.9
Zinc	351	254	372

The ash content of these samples was (g/kg DM): 417 in backyard manure, 151 in broiler litter and 353 in layer excreta (Van Ryssen *et al.*, 1993). The striking observation was that the manure from free-ranging birds contained exceptionally high concentrations of most elements, especially the toxic heavy metals, compared to the other classes of birds. It was concluded that they probably consumed large quantities of soil and some of their own excreta. From an animal nutritional point of view, the most significant difference between layer manure and the other classes of litter is the calcium content (not presented in Table).

Deep litter systems

The two main factors affecting elemental composition of poultry litter regarding bedding material are the type of material used and the period of stay on the litter material:

Bedding material could consist of wood shavings or sawdust, peanut hulls, pine chips, rice hulls and others (Kelley *et al.*, 1998; Kpomblekou-A *et al.*, 2002.). 'Soft wood' such as pine contains < 1.0% ash, and the ash is predominantly calcium as calcium carbonate (Campbell, 1990). Consequently, wood shavings as bedding material contribute very little to the mineral content of the litter (Kelley *et al.*, 1996), and would rather dilute the mineral concentration of litter.

In some countries such as the USA, most poultry houses are cleaned after a number of batches of birds has been reared on the same litter material (Kelley *et al.*, 1995). Period of stay on bedding material will affect the proportion of bedding material in the litter. Kunkle *et al.* (1981) confirmed that the concentration of nutrients in broiler litter increased 10 to 15% with each additional flock of birds kept on the litter. In five-flock broiler litter the dietary mineral levels are concentrated 2.75 to 3.25 times (Kunkle *et al.*, 1981). Kelley *ct al.* (1996) reported significantly higher concentrations of Al, Ba, B, Cd, Cr, Fe, Mg, Mn, Ni, K, P, Si and Zn in multi-flock litter than in single-flock litter, though the concentrations of lead and mercury were below detection limits.

However, apart from the accumulation of elements, there is also a build-up of pathogens as well as the infestation of the litter with insect, especially in re-used litter (Kelley *et al.*, 1995). In the production systems where a build-up of pathogens occurs, the inclusion of antibiotics into diets becomes beneficial and essential. Using antimicrobial products containing trace elements such as arsenic, or substances with antimicrobial actions such as copper sulfate, constitutes some of the main problems of excessive mineral elements in poultry litter.

Birds, especially those on deep litter systems recycle a proportion of elements through coprophagy. From tracer studies with radioactive cobalt it has been calculated that chickens kept on bedding ingest 5-25% of the faeces produced by the group. In a parallel group kept on wire netting 3-17% of faeces was still ingested (Bonnafous, 1973, cited by Hörnicke and Björnhag, 1979). Hetland *et al.* (2003) reported that layers on deep litter systems consume a significant amount of litter material. They stated that having access to litter may improve the bird's welfare since it functions as both a dust bath and a feed component.

Additions to the litter

Products containing mineral elements are added to bedding material in poultry houses for a number of reasons. Boric acid is added to litter to control insects such as cockroaches and darkling beetles from proliferating in litter (Sander *et al.*, 1991). Aluminium sulfate (alum) is added to litter to inhibit ammonia volatilization in litter. Alum, ferrous sulfate and calcium hydroxide are effective in precipitating phosphorus when added to litter by reducing phosphorus solubility (Kpomblekou-A *et al.*, 2002). Although most of these products are added to poultry litter intended as fertilizer, it could end up in the animal feed chain.

Disposal and risks

Land application

The application of poultry litter as fertilizer on cultivated lands is the most common method of disposing of litter. The widely recognised problem with manure application to the soil is the overloading of the soil with phosphorus and nitrogen, which can leach out of the soil and contaminate waterways (Cooperband and Good, 2002). Consequently, in many states of the USA regulations require a poultry producer to have an approved plan for poultry litter disposal before erecting a poultry house (Bagley and Evans, 1998). Furthermore, Dozier III *et al.* (2003) pointed out that when litter is applied to soil according to a nitrogen content basis, zinc and copper applied through the litter can exceed crop requirements by 660% and 560%, respectively. Therefore, repeated use of poultry litter as a fertilizer could increase the zinc and copper concentrations in soil. A build-up of zinc in the soil results in phytotoxicity, causing reduced crop yields (Skřivan *et al.*, 2005).

On the positive side, poultry litter used as fertilizer will increase trace elements in the soil, and that will increases the mineral content of forages and, consequently could have a beneficial effect on the nutrition of grazers. Litter could also change the soil pH which affects availability of minerals to plants.

Poultry litter as animal feed

The feeding of poultry litter to livestock is a very controversial topic. Many countries have prohibited the practice. On the other hand, in many countries where it is permissible to be used as an animal feed, it is a saver of animal lives especially during periods of feed shortages such as during droughts. Problems associated with the feeding of poultry litter to livestock have been investigated thoroughly, especially in the USA (Westing *et al.*, 1986; Fontenot 1991). Bagley and Evans (1998) stated that there appears to be no more health risks to animals from consuming broiler litter than from any other source of cattle feed, provided the litter is processed properly. The rumen does an excellent job of breaking down and converting broiler litter into nutrients. As a relatively low quality food source, it supplies mainly non-protein nitrogen and energy. Guidelines have been compiled on the processing and using of poultry litter as a ruminant feed (Fontenot, 1991; Rankins *et al.*, 1993; Ruffin and McCaskey, 1998; Van Ryssen, 2001). However, public perception regarding food safety can be influential and this practice has now also been prohibited in the USA.

An abundance of most minerals is present in poultry excreta (Westing *et al.*, 1985) and it is an excellent source of minerals to animals (Ruffin and McCaskey, 1998). Comparing the concentrations in the poultry litter with levels at which the elements are toxic (Table 3), it is clear that only one or two elements are present at the toxic ranges. Considering that poultry litter would constitute usually only a proportion of a ruminant's total diet, in most situations at the most 30%, most elements would be supplied in excess of requirements but not close to toxic ranges. The concentrations of the typical toxic 'heavy metals', cadmium, mercury and lead, were low in poultry litter. Furthermore, since many chronic toxicity symptoms are induced deficiencies of other minerals, deficiencies are unlikely to occur when poultry litter is included in a diet. Since poultry litter is a rich source of mineral elements for ruminants there is usually no need to supplement any elements in the diets of ruminants when relatively high levels of litter are fed, except for salt (NaCl) when it is used to control intake.

Studies have confirmed that the elements in litter are readily absorbed from the digestive tract of ruminants. Suttle and Price (1976) reported that for Cu, Van Ryssen and Mavimbela (1998) for Se and Ben-Ghedalia *et al.* (1996) for Zn, Mn, Cu, Co and Se. However, there are many factors determining and affecting the bioavailability of mineral elements, such as chemical form, concentration of minerals interacting with one another, etc. In the case of poultry litter it can be assumed that the bioavailability of the elements would be relatively low because the concentrations of elements antagonistic to the absorption of other elements are all high.

Table 3. A comparison between the requirements and upper safe dietary limits of trace elements for ruminants (Puls, 1994) and concentration (mg/kg DM) of elements in poultry litter, published in different reports.

	Requirements sheep*	Publication				Maximum tolerance**
		a	b	c	d	
Aluminium		2200	834	992	1350	< 1200
Arsenic		21	4.9	76	30	< 7 - 100
Cadmium		0.3	0.32	6.1	0.23	< 50
Chromium		3.7	11.2	2.6	2.1	< 40
Cobalt	0.1 - 0.2	0.4	1.08	2.3	0.87	< 30 (10)
Copper	7 - 11	450	44	593	380	< 20*** (25)
Iron	30 - 50	2073	1335	1023	1310	< 4000 (500)
Lead			0.55	1.9		< 5 - 1000
Mercury			0.48			< 4
Manganese	20 - 40	388	317	371	400	< 2000 (1000)
Molybdenum	0.5	0.9	1.5	9.2		< 10 (10)
Selenium	0.1 - 0.2	5.5	0.62	1.8		< 5 (2)
Vanadium			12.1	5.7		< 330
Zinc	20 - 33	399	254	496	340	< 5000 (750)

*NRC (1985).
**Puls (1994) In brackets NRC for sheep (1985).
***Sheep.
a Kpomblekou-A *et al.* (2002).
b Van Ryssen *et al.* (1993) – broiler litter.
c Westing *et al.* (1985).
d Kelley *et al.* (1998) Multiple flocks. Litter exposed to 2-3 flocks.

Practically the only risk factor to ruminants from a trace mineral point of view is the high copper content in the litter of birds in countries where copper compounds are still included as growth stimulants in poultry diets. At high concentrations of copper in poultry litter, Suttle *et al.* (1978) and Rankins *et al.* (1993) reported linear increases in the liver copper concentrations in sheep and steers, respectively, with increasing levels of litter in their diets. On the other hand, at low levels of copper in litter (47 mg Cu/kg litter), Van Ryssen and Jagoe (1981) found no increase in copper retention in the liver of sheep as levels of litter in the diet increased. They concluded that the concentration of elements antagonistic to copper absorption such as iron, zinc, sulfur and molybdenum were sufficiently high in the diet to control copper accumulation in the liver. Sheep are very susceptible to copper toxicity, with British sheep breeds being the worst. It is quite an accepted fact that when poultry litter containing high concentrations of copper is fed to sheep, copper toxicity is likely to occur. Even though cattle are more resistant to copper toxicity than sheep

(Fontenot, 1991), cases of copper toxicity among cattle have been reported regularly in the USA (e.g. Banton *et al.*, 1987). Westing *et al.* (1985) concluded that despite slight increases in the concentrations of some elements in edible tissues, cattle fed broiler litter would not present a hazard to the health of humans consuming the edible tissues. In fact, products from ruminants consuming poultry litter would be rich sources of trace elements such as copper, iron, zinc and selenium.

Little if any coverage has apparently being given in recent literature on arsenic in poultry litter fed to ruminants. The upper safe dietary concentration of arsenic in ruminants varies widely depending on the valence of arsenic and chemical form of the product containing arsenic. Ammerman *et al.* (1977) quoted studies that suggested that the most toxic arsenicals are well tolerated at dietary levels of 10 to 20 mg/kg and the least toxic up to 1000 mg As/kg. Arsenic trioxide is toxic to cattle at an oral dose of 15 - 45 g and sodium arsenite at 1 - 4 g (Ammerman *et al.*, 1977). They pointed out that arsenic is loosely bound to tissue protein and is not cumulative with time. Consequently chronic toxicity due to arsenic is seldom reported. According to Hamilton (2005) arsenic does not accumulate in the body to an appreciable extent nor does it biomagnify in the food chain. A withdrawal period of 5 days is prescribed for broilers on a diet containing roxarsone (Jones, 2007).

The arsenic in litter originating from roxarsone was found to be in the organic form that is practically inert, but can be converted to more toxic forms under anaerobic conditions (Garbarino *et al.*, 2003). Arsenic levels of between 20 and 76, as presented in Table 3 are not within a toxic range for ruminants. However, it is not clear if and to what extent roxarsone would be converted to more toxic inorganic compounds in the rumen.

The concentration of aluminium is high in most poultry litter samples. However, aluminium is not a toxic element to animals. Very high concentrations (2000 mg Al/kg) have slight negative effects on food intake and the absorption of minerals such as phosphorus (Allen *et al.*, 1990), but livestock can tolerate high levels of aluminium in their diets.

Most species of animals have a high tolerance towards high levels of dietary iron. Iron interacts with some of the other elements such as copper, zinc and manganese that are also present at high levels in poultry litter (Skřivan *et al.*, 2005). However, at high levels in the body, iron can act as a pro-oxidant (Pitzen, 1994).

Relatively high levels of zinc are included in poultry diets with the resultant high concentrations of zinc in litter. The antibiotic, zinc bacitracin, has also been used in poultry diets. These high concentrations do not pose a problem to livestock because zinc is not a toxic element to animals. Relatively high concentrations would be desirable in litter fed to animals because of the high concentrations of all the other elements in litter that can suppress zinc absorption. High levels of zinc in litter are a problem when the litter is used as a fertilizer. In plant products high in phytic acid, the bioavailability of zinc is low. According to Matsui and Yano (1998) methods of destroying the phytic acid in plants would improve the bioavailability of zinc in the products, thus ensuring that lower levels of supplemental zinc can be included in diets.

Conclusions

It must be accepted that problems related to toxic mineral elements in poultry litter can not be approached in isolation from other aspects of poultry production, including economic considerations, environmental constraints, practical management considerations and public perceptions. The poultry producer should make every attempt to keep his side clear from dubious practices, and base production practices on sound scientific principles, even though public pressure would probably be a strong driving force in achieving certain goals. This is evident from the statement by Ball (2006) that retailers in the United Kingdom have achieved total dominance over sales of poultry products and as a consequence have demanded and obtained, with the acquiescence of the industry, control over the production process.

To minimize the concentration of trace elements in poultry litter originating from the diet, the trace element content of the raw materials would have to be considered when formulating a diet. This would require the chemical analysis of each batch of feed entering a feed factory. At this stage that is probably not feasible and feed manufacturers have to rely on the supplier of the raw material to provide some guarantees at least for those elements that pose health or environmental pollution risks, though this encompasses broader issues such as international trade and quality control. However, some uncertainties and probably conflicting views remain regarding requirements, e.g. should requirements be based on animal needs under stress, where high levels of trace elements provide antioxidant protection to the body; and should high intakes of an element such as copper be considered acceptable because it facilitates the production of healthy products from the human nutrition point of view?

Using trace element sources with the highest possible bioavailability would not only increase the proportion absorbed, but also reduce the concentration of the element required to be included in the diet, thus reducing the amount excreted. Organic forms of elements qualify for this, and are used increasingly in livestock diets. Feeds can also be treated to improve the bioavailability of minerals, e.g. including phytase in diets to break down phytic acid in feedstuffs.

The feeding of poultry litter to livestock is prohibited in many countries, also in those where the disposal of poultry litter and environmental pollution due to litter are serious problems. Litter as an animal food is therefore not an issue in solving the disposal problem. However, if a poultry producer cannot dispose of the litter from his premises because of problem elements in the litter, it would force feed manufacturers to avoid the inclusion of such substances in their products.

References

Allen, V.G., J.P. Fontenot and S.H. Rahnema, 1990. Influence of aluminum citrate and citric acid on mineral metabolism in wether sheep. J. Anim. Sci. 68, 2496-2505.

Ammerman, C.B., S.M. Miller, K.R. Fick and S.L. Hansard II, 1977. Contaminating elements in mineral supplements and their potential toxicity: A review. J. Anim. Sci. 44, 485-508.

Anderson, B.K. and T.N. Chamblee, 2001. The effect of dietary 3-nitro-4-hydroxyphenylarsonic acid (roxarsone) on the total arsenic level in broiler excreta and broiler litter. J. Appl. Poult. Res. 10, 323-328.

Arias, V.J. and E.A. Koutsos, 2006. Effects of copper source and level on intestinal physiology and growth of broiler chickens. Poult. Sci. 85, 999-1007.

Bagley, C.P. and R.R. Evans, 1998. Broiler litter as a feed or fertilizer in livestock operations. Mississippi State Extension Services. http://msstate.edu/dept/poultry/pub1998.htm.

Balevi, T. and B. Coskun, 2004. Effects of dietary copper on production and egg cholesterol content in laying hens. Br. Poult. Sci. 45, 530-534.

Ball, J.A., 2006. Chicken production in Europe – Where do we go from here? Proc. Aust. Poult. Sci. Symp., Sydney, February 2006.

Banton, M.I., S.S. Nicholson, P.L.H. Jowett, M.B. Brantley and C.L. Boudreaux, 1987. Copper toxicosis in cattle fed chicken litter. JAVMA. 191, 827-828.

Ben-Ghedalia, D., J. Miron and E. Yosef, 1996. Apparent digestibility of minerals by lactating cows from a total mixed ration supplemented with poultry litter. J. Dairy Sci. 79, 454-458.

Bonnafous, R., 1973. Thèse Doctorat Sciences, Toulouse.

Brosh, A., Y. Aharoni, D. Levy and Z. Holzer, 1998. Effects of source and content of ash in poultry litter used in diets for beef cattle. J. Agric. Sci., Camb. 131, 87-95.

Campbell, A.G., 1990. Recycling and disposing of wood ash. TAPPI Journal, 73 (9), 141-146.

Coetzee, C., 2006. Heavy metal guidelines for the feed industry. AFMA Symposium, October 2006, SCIR Conference Centre, Pretoria, South Africa.

Cooperband, L.R. and L.W. Good, 2002. Biogenic phosphate minerals in manure: implications for phosphorus loss to surface waters. Environ. Sci. Technol. 36, 5075-5082.

Dozier III, W.A., A.J. Davis, M.E. Freeman and T.L. Ward, 2003. Early growth and environmental implications of dietary zinc and copper concentrations and sources of broiler chicks. Br. Poult. Sci. 44, 726-731.

Ewing, H.P., G.M. Pesti, R.I. Bakalli and J.F.M. Menten, 1998. Studies on the feeding of cupric sulphate pentahydrate, cupric citrate, and copper oxychloride to broiler chickens. Poult. Sc. 77, 445-448.

Fisher, C., D. Wise and D.G. Filmer, 1972. The effect of copper on the growth of broilers and the interaction of copper with zinc and iron. 4th Wrld Poult. Congr., Madrid 2, pp. 759-764.

Fontenot, J.P., 1991. Recycling animal wastes by feeding to enhance environmental quality. The Professional Animal Scientist 7, 1-8.

Garbarino, J.R., A.J. Bednar, D.W. Rutherford, R.S. Beyer and R.L. Wershaw, 2003. Environmental fate of roxarsone in poultry litter. 1. Degradation of roxarsone during composting. Environ. Sci. Technol. 37, 1509-1514.

Hamilton, J.W., 2005. The facts on arsenic. http://dartmoureedu/~toxmetal/TXQAas.shtml.

Hetland, H., B. Svihus, S. Lervik and R. Moe, 2003. Effect of feed structure on performance and welfare in laying hens housed in conventional and furnished cages. Acta Agric. Scand., Sect. A, Anim. Sci. 53, 92-100.

Hörnicke, H. and G. Björnhag, 1979. Coprophagy and related strategies for digesta utilization. Ch 34 In: Ruckebusch, Y. and Thivend, P. (Eds.), Digestive Physiology and Metabolism in Ruminants. MTP Press Ltd. Int. Med. Publ. pp. 707-730.

Jones, T.T., 2007. A broad view of arsenic. Poult. Sci. 86, 2-14.

Jukola, E., J. Hakkarainen, H. Saloniemi and S. Sankari, 1996. Blood selenium, vitamin E, vitamin A, and β-carotene concentrations and udder health, fertility treatments, and fertility. J. Dairy Sci. 79, 838-845.

Kelley, T.R., O.C. Pancorbo, W.C. Merka, S.A. Thompson, M.L. Cabrera and H.M. Barnhart, 1995. Bacterial pathogens and indicators in poultry litter during re-utilization. J. Appl. Poult. Res. 4, 366-373.

Kelley, T.R., O.C. Pancorbo, W.C. Merka, S.A. Thompson, M.L. Cabrera and H.M. Barnhart, 1996. Elemental concentrations of stored whole and fractioned broiler litter. J. Appl. Poult. Res. 5, 276-281.

Kelley, T.R., O.C. Pancorbo, W.C. Merka, S.A. Thompson, M.L. Cabrera and H.M. Barnhart, 1998. Accumulation of elements in fractionated broiler litter during re-utilization. J. Appl. Poull. Res. 7, 27-34.

Konjufca, V.H., G.M. Pesti and R.I. Bakalli, 1997. Modulation of cholesterol levels in broiler meat by dietary garlic and copper. Poult. Sci. 76, 1264-1271.

Kpomblekou-A, K., R.O. Ankumah and H.A. Ajwa, 2002. Trace and nontrace element content of broiler litter. Commun. Soil Sci. Plant Anal. 33, 1799-1811.

Kunkle, W.E., L.E. Carr, T.A. Carter and E.H. Bossard, 1981. Effect of flock and floor type on the levels of nutrients and heavy metals in broiler litter. Poult. Sci. 60, 1160-1164.

Lien, T.F., K.L. Chen, C.P. Wu and J.J. Lu, 2004, Effect of supplemental copper and chromium on the serum and egg traits of laying hens. Br. Poult. Sci. 45, 535-539.

Lim, H.S., I.K. Paik, T.L. Sohn and W.Y. Kim, 2006. Effects of supplementary copper chelates in the form of methionine, chitosan and yeast on the performance of broilers. Asian-Aust. J. Anim. Sci. 19, 1322-1327.

Matsui, T. and H. Yano, 1998. Formulation of low pollution feeds for animal production. Proc. 8th Wrld Conf. Anim. Prod. Seoul, Korea, pp. 110-118.

Nachman, K.E., J.P. Graham, L.B. Price and E.K. Silbergeld, 2005. Arsenic, A roadblock to potential animal waste management solutions. Envir. Health Perspect. 113, 1123-1124.

NRC, 1985. National Research Council: Nutrient Requirements of Sheep. 6th rev. ed. National Academy in Science, Washington, DC.

NRC, 1994. National Research Council: Nutrient Requirements of Poultry. 9th rev. ed. National Academy in Science, Washington, DC.

Pesti, G.M. and R.I. Bakalli, 1998. Studies on the effect of feeding cupric sulphate pentahydrate to laying hens on egg cholesterol content. Poult. Sci. 77, 1540-1545.

Pitzen, D., 1994. The trouble with iron: Resolved by pro-oxidant : anti-oxidant balancing of dairy rations. Feed Int. pp. 22-23.

Puls, R., 1994. Mineral Levels in Animal Health. Diagnostic Data. 2nd ed. Sherpa Int, Canada.

Rankins, D.L., J.T. Eason, T.A. McCaskey, A.H. Stephenson and J.G. Floyd, 1993. Nutritional and toxicological evaluation of three deep-staking methods for the processing of broiler litter as a feedstuff for beef cattle. Anim. Prod. 56, 321-326.

Ravindran, V., 2006. Broiler Nutrition in New Zealand – Challenges and Strategies. 2006 Arkansas Annual Nutrition Conference. Rogers, Arkansas, USA, 12-14 September 2006. pp. 1-7.

Rebel, J.M.J., J.T.P. van Dam, B. Zekarias, F.R.M. Balk, J. Post, A. Flores Miñambres and A.A.H.M. ter Huurne, 2004. Vitamin and trace mineral content in feed of breeders and their progeny: effects of growth, feed conversion and severity of malabsorption syndrome of broilers. Br. Poult. Sci. 45, 201-209.

Ribaudo, M.O., N.R. Gollehon and J. Agapoff, 2003. Land application of manure by animal feeding operations: Is more land needed? J. Soil Water Conserv. 58, 30-39.

Ruiz, J.A., A.M. Pérez-Vendrell and E. Esteve-Garcia, 2000. Effect of dietary iron and copper on performance and oxidative stability in broiler leg meat. Br. Poult. Sci. 41, 163-167.

Ruffin, B.G. and D.L. McCaskey, 1998. Feeding broiler litter to beef cattle. Circular ANR-557: http://gallus.tamu. edu/waste.bfcattle.html.

Sander, J.E., L. Dufour, R.D. Wyatt, P.B. Bush and R.K. Page, 1991. Acute toxicity of boric acid and boron tissue residues after chronic exposure in broiler chickens. Avian Des. 35, 745-749.

Skřivan, M., V. Skřivanová and M. Marounek, 2005. Effect of dietary zinc, iron and copper in layer feed on distribution of these elements in eggs, liver excreta, soil and herbage. Poult. Sci. 84, 1570-1575.

Skřivan, M., V. Skřivanová, M. Marounek, E. Tumova and J. Wolf, 2000. Influence of dietary fat source and copper supplementation on broiler performance, fatty acid profile of meat and depot fat, and on cholesterol content in meat. Br. Poult. Sci. 41, 608-614.

Suttle, N.F. and J. Price, 1976. The potential toxicity of copper-rich excreta to sheep. Anim. Prod. 23, 233-241.

Suttle, N.F., C.S. Munro and A.C. Field, 1978. The accumulation of copper in the liver of lambs on diets containing dried poultry waste. Anim. Prod. 26, 39-45.

Van Ryssen, J.B.J., 2001. Poultry litter as a feed ingredient: The South African situation. SA-ANIM SCI vol. 2. pp. 1-7. www.sasas.co.za/popular/Popular.html.

Van Ryssen, J.B.J. and H.M. Jagoe, 1981. Retention of trace elements in the liver of sheep fed poultry manure as a ration component. S. Afr. J. Anim. Sci. 11, 273-278.

Van Ryssen, J.B.J. and D.T. Mavimbela, 1998. Broiler litter as a source of selenium for sheep. Anim. Feed Sci. Technol. 78, 263-272.

Van Ryssen, J.B.J., P. Channon and W.J. Stielau, 1977. Minerals and nitrogen in poultry manure. S. Afr. J. Anim. Sci. 7, 195-197.

Van Ryssen, J.B.J., S. van Malsen and A.A. Verbeek, 1993. Mineral composition of poultry manure in South Africa with reference to the Farm Feed Act. S. Afr. J. Anim. Sci. 23, 54-57.

Westing, T.W., J.P. Fontenot, W.H. McClure, R.F. Kelly and K.E. Webb, 1985. Characterization of mineral element profiles in animal waste and tissues from cattle fed animal waste. 1. Heifers fed broiler litter. J. Anim. Sci. 61, 670-681.

Copper and zinc in feed (additives): an essential burden?

P.F.A.M. Römkens, S.W. Moolenaar, J.E. Groenenberg, L.T.C. Bonten and W. de Vries

Keywords: copper, zinc, feed additives, soil quality, surface water, cattle

Introduction

Zinc and copper are a 'natural' part of animal feeding materials like grass and maize (roughage) and of the raw materials that are the basis for feed concentrates. Also, extra copper and zinc is added in quite large amounts to feed concentrates and through mineral supplements. Copper and zinc are essential elements for plants and animals. Essential functions of copper include enhanced growth and feed efficiency (National Research Council, 2005). On the other hand copper is used in foot/hoof baths to control foot rot and other lameness-related problems as well as an anthelmintic to control gastrointestinal parasites in ruminants (Bang *et al.*, 1990). Zinc is added to the diet to increase growth and lactation performance (NRC, 2005). A large part (> 90%) of copper and zinc that is fed to animals leaves the animal through manure and/or urine. In countries with intensive animal husbandry this leads to a large emission of these metals to soils. Especially the application of manure on arable land and grassland as a source of nitrogen and phosphorus results in a large net accumulation of copper and zinc. This has raised the discussion on what essential levels of both metals in feed additives should be in view of environmental effects on soils and ecosystems. At present, there is a substantial gap between what is believed to be essential versus what is allowed according to EU regulations. For copper, nutritional requirements based on animal health range from 4 mg kg^{-1} diet (pigs) to 21 mg kg^{-1} dm for dairy cattle. EU regulations allow copper levels of 35 mg kg^{-1} for ruminants, pigs and poultry except for pigs up to 12 weeks of age where maximally 170 mg kg^{-1} is allowed (EU, 2003). For zinc the recommended levels for cattle (up to 50 mg kg^{-1}) are far below allowed levels (150 mg kg^{-1}; EU, 2003), but feeds of pigs and poultry contain between 90 and 150 mg kg^{-1}.

It is, therefore, not surprising that the actual amount fed is mostly in excess of actual nutritional needs. This in turn results in a large emission of copper and zinc to agricultural soils. A recent overview of heavy metal balances across EU member states (Eckel *et al.*, 2005) shows that net accumulation rates for copper vary from -138 to +908 g ha^{-1} yr^{-1} (median value: 109 g ha^{-1} yr^{-1}) and from -131 to +3523 g ha^{-1} yr^{-1} (median: 389 g ha^{-1} yr^{-1}) for zinc. A negative net accumulation only occurred at sites without application of animal manure (or sewage sludge).

For the Netherlands, average net accumulation rates are in the same order of magnitude albeit that due to high inputs of manure, net accumulation rates in 2000 are higher than the median values reported at the EU level (Zn: 550 g ha^{-1} yr^{-1}; Cu: 200 g ha^{-1} yr^{-1}; Delahaye *et al.*, 2003). These values are based on a nationwide inventory of inputs and outputs and do not consider specific farms. Also these inventories did not account for additional inputs of copper and zinc due to copper in foot/hoof baths and copper and zinc in mineral supplements. A recent inventory at the farm level, nevertheless, shows that national averages are in line with average accumulation rates measured at the farm level (Boer and Hin, 2003). Accumulation rates across 17 individual dairy farms ranged from 17 to 574 g ha^{-1} yr^{-1} for copper (average: 203 g ha^{-1} yr^{-1}) and from -29 to 1187 g ha^{-1} yr^{-1} (average: 507 g ha^{-1} yr^{-1}) for zinc.

The net positive input to soils invariably will lead to an increase in the copper and zinc level in the soil. This in turn leads to increased leaching rates from soil to ground- and surface waters. Especially in countries such as the Netherlands, the average groundwater level is high and effects of increased levels on the quality of ground- and surface waters are expected to occur within decades. At present, the contribution of leaching of metals like Cd, Zn and Cu to the load of metals in surface waters in the Netherlands is substantial and ranges from 20% (Cd) to 40% (Zn) of the total emission to water (Bonten and Brus, 2006). Especially in areas with a high groundwater table such as the peat and clay areas in the western parts of the Netherlands, this already leads to elevated levels of especially Cu, Zn and Ni at levels above the MTR level (Bonten and Brus, 2006).

The continuation of the accumulation of metals in soils from manure, therefore, will lead to even higher levels of metals in surface waters. This can pose new restrictions on the use and disposal of manure or sludge (not relevant in the Netherlands since it is not allowed to be used) since the Water Framework Directive will impose standards for 'good quality' of surface waters (EU Directive 2000/60/EC). At present it is still unclear to what extent the use and disposal of manure leads to unacceptable levels of copper and zinc in soil, ground- and surface waters. Changes in the heavy metal content in soils due to long-term application of manure and or sludge are difficult to measure. A study by Keller and Desaules (2003) shows that within a timeframe of 5 years no significant changes in the zinc and cadmium content were detected despite a large positive farm balance for both metals. Based on the average zinc surplus of 500 g ha^{-1} yr^{-1} (Dutch situation) this would lead to a net change in the soil metal content of 0.36 mg kg^{-1} yr^{-1} (assuming 100% accumulation in the top 10 cm and a bulk density of 1.4 kg/dm^3). A net increase of 1.8 mg kg^{-1} after 5 years, therefore, is hard to detect when average zinc levels are in the order of magnitude of 20 to 50 mg kg^{-1}.

On the long term, however, ongoing accumulation will lead to significantly increased levels in soil, and, even more so, in ground- and surface water. This can be 'predicted' by carrying out dynamic balance calculations (Moolenaar, 1999; Moolenaar and Lexmond, 1999; Moolenaar *et al.*, 1997; De Vries *et al.*, 2004). These developments urge the need to reduce inputs to the soil. Reducing the levels of both elements in feed additives may prove a very effective way to reach lower emission levels.

To assess the changes in soil and water quality in time, a model concept has been developed (De Vries *et al.*, 2002, 2004; Groenenberg *et al.*, 2006) that allows for the calculation of changes in the copper and zinc content in the soil. The model accounts for both inputs (manure, atmospheric deposition, etc.) as well as outputs (crop uptake, leaching). Field data have been collected in several studies (Boer and Hin, 2003; den Boer *et al.*, 2007) and will be used to evaluate the effectiveness of measures to reduce inputs.

This analysis will be used to identify areas where current and future inputs from agriculture will result in unacceptable high levels of copper and zinc in soils. However, this approach does not consider the required input based on the requirements of animals. One of the major 'problems' in the Netherlands is the large amount of animal feed that is imported to meet the nutritional requirements. The intensity of Dutch agriculture is, to a large extent not related to the available surface area. Especially intensive large scale pig (NL: 12 million pigs) and poultry production (NL: 90 million) is possible due to import of feed raw materials through the Rotterdam harbour. This has caused a large imbalance between the amount of manure produced by animals and the amount that can be used as soil amendment (N and P fertilizer). Obviously the majority of metals added to the feed will end up in the manure due to the low absorption rate by the animals.

In this study we will focus on cattle farms, and present data on how balances from the perspective of animal needs (in terms of nutrient requirements) match (or not) those of the farm. Ultimately a farmer wants to grow healthy cattle where food requirements are met by the supply. The question to be addressed, therefore, is to what extent mineral supplements and other food additives are needed to meet the requirements of the animals.

In this paper the following issues will be addressed:
1. An evaluation of the current inputs of copper and zinc on a national and farm level and calculation of farm balances for specific farms in the Netherlands (as well as a limited comparison with other countries).
2. An assessment of the current quality of the soil in the Netherlands (based on actual soil quality standards) regarding copper and zinc as well as predicted changes within 100 years due to current agricultural practices.
3. An assessment of the impact of metal leaching from the soil on the quality of surface waters.
4. An assessment of the degree to which the Cu and Zn supply in cattle breeding meets (or exceeds) the nutritional needs of animals and how this nutritional need can be managed by roughage and/or additives.
5. An assessment of how farm measures (both regarding quality of feed and management of waste) will decrease the load of copper and zinc to soils.

Material and methods

Data collection and farm balances

Data from inputs and outputs were collected at individual dairy farms (Boer and Hin, 2003; Kool *et al.*, 2006). Inputs considered include feed, feed additives, fertilizers, manure, atmospheric deposition and litter. Outputs include metals exported in animal products such as meat, eggs, milk, etc., manure (sometimes manure is exported from the farm), feed (locally grown feed sold to other farmers) and leaching. For certain fluxes (e.g. atmospheric deposition, metals in animal products) fixed values were used. Either because measurements could not be performed or because of an assumed limited variation (atmospheric deposition). Data at the national level were adopted from the Dutch Central Bureau of Statistics (CBS:

www.cbs.nl). Detailed information on the number of farms, animals, amount of manure and heavy metal emission to and from soil was obtained at www.statline.nl.

Models used to predict changes in soil and leaching to ground- and surface waters

To predict changes in soil quality due to accumulation of metals in soils, all inputs and outputs to and from soil have to be quantified. Changes in the metal content with time were calculated for the topsoil as well as deeper soil layers according to Equation 1:

$$\Delta Me\text{-soil} = 0.1 \times [\{atm.dep. + \Sigma(manure + compost + fert.)\} - \{crop\ upt. + leach.\}] / \{z \times 100 \times \rho\} \quad (1)$$

All in- and outputs are expressed in g ha^{-1} yr^{-1}, z is the depth of the layer in dm and ρ equals the bulk density of the soil in kg l^{-1}. For grassland it is assumed that the addition and uptake by crops occurs in the 0 - 10 cm layer, whereas in arable land a depth of 25 cm is used (due to ploughing). Below this depth changes in the metal content are assumed to occur only due to leaching.

Details of the model approaches (for plant uptake and leaching), schematization and inputs of copper and zinc used here are listed in De Vries *et al.* (2004) for zinc and Groenenberg *et al.* (2006) for copper. In short, the model calculates changes in the soil and leaching to ground- and surface water using several model routines. Important ones include a leaching model and the plant uptake model. The relation between soil and soil solution concentration was described by a generic Freundlich equation that accounts for differences between soils:

$$Log[Me_{solution}] = Const_{soil-solution} + a \times log[Org.Mat] + b \times log[clay] + c \times pH + d \times log[Me_{soil}] \quad (2)$$

Coefficients for Equation 2 were derived from a large database containing approx. 1400 data points from soils in the Netherlands (Table 1). Equation 2 allows for the calculation of the dissolved metal concentration in any given soil, depending on soil properties.

Plant uptake for zinc is modelled also using a calibrated model based on data from several field studies. Also here, a Freundlich type equation is used to model zinc uptake considering differences in pH, clay- and organic matter content of the soil:

Table 1. Values for the coefficients in the relationships relating dissolved total concentrations and reactive soil concentrations of zinc and copper, according to Equation 2 after Römkens et al. (2004). Units are mol kg $^{-1}$ and mmol l^{-1} for me-soil and me-solution, respectively, and % for organic matter and clay content.

Metal	Constant (Const)	Org. Mat. (a)	Clay (b)	pH CaCl$_2$ (c)	Me-soil (d)	R^2	se(Y)
Zn	4.69	-0.35	-0.48	-0.54	1.08	0.77	0.51
Cu	1.10	-0.28	-0.27	-0.18	0.87	0.42	0.49

$$Log[Zn_{crop}] = Const_{soil-plant} + a \times pH + b \times log[Org.Mat.] + c \times log[Clay] + d \times logZinc_{soil}] \tag{3}$$

Coefficients for different crops (grass, maize, wheat, sugar beet and potato; De Vries *et al.*, 2004) are shown in Table 2 and were derived by regression using field data obtained at regular arable farms. For copper, no clear relation between soil composition and crop copper content could be derived. Therefore, median values of copper in arable crops were used (field data) in the model calculations. Copper levels used ranged from 4 to 12 mg kg^{-1} dry matter for various crops.

To calculate changes on a regional scale, the Netherlands is divided into more than 6400 individual geographic units, each unit has its own hydrological and soil chemical characteristics (Kroes *et al.*, 2001). Data on leaching fluxes needed to calculate changes in the soil metal content as well as leaching to ground- and surface waters are available for all units. Also soil properties needed in Equation 1 and 2 are available for the entire soil profile (down to 12 m below the surface level (Kroon *et al.*, 2001)). Data of the annual load of copper and zinc at the field level (g ha^{-1} yr^{-1}) were derived from available data on N-inputs that were converted to a corresponding load of both metals depending on the type (and amount) of manure used (De Vries *et al.*, 2004; Groenenberg *et al.*, 2006). Data on crop uptake (kg ha^{-1}) are based on average yield data available for relevant crops.

Table 2. Values for the coefficients in the relationship relating total concentration of Zn in different plants and in soil according to Equation 3. Units are mg kg^{-1} for Zn in soil and plant (dry matter) and % for organic matter and clay.

Crop	Const.	pH-KCl (a)	Org. mat. (b)	Clay (c)	Zn-soil (d)	R^2	se-y$_{est}$
Grass	2.29	-0.13	-0.06	-0.35	0.33	0.20	0.14
Maize	0.90	-0.10	0.28	-0.62	0.90	0.64	0.09
Wheat-grain[1]	1.20	-0.06	0	0	0.37	0.73	0.06
Potatoes	1.11	-0.08	0.12	-0.38	0.45	0.50	0.07
Sugar beet	2.61	-0.38	-0.46	-0.6	1.17	0.63	0.15

[1]Relationships for wheat were also used for other cereals.

Results

National and regional balances for copper and zinc

The total input of copper and zinc to soils in the Netherlands has decreased steadily since 1980 although the decrease in this 25 year period was more pronounced for copper (62% reduction) than for zinc (37% reduction). The reduced input has several causes:

- limitations in the amount of manure to be used as a source of N and P, regulated through the Nitrate Directive;
- changes in the allowed amount of copper and zinc in animal feeds since 1977, and
- a significant decrease in the absolute number of animals held at farms. In the period 1994 - 2005 the total number of cows and pigs in the Netherlands decreased by almost 20% and 15%, respectively (CBS, www.cbs.nl).

It is, therefore, not surprising that the observed decrease in emission to soils is largely due to a reduction of inputs through manure (Figure 1). For copper the reduced emission of manure amounts to 615 tons year[-1] (555 tons year[-1] for zinc) which is equivalent to 73% of the total reduction (63% for zinc). This is roughly equivalent to a load change (decrease) of 300 g ha[-1] for both metals, assuming an equal distribution of manure among the 2 million ha of land used for agriculture (predominantly arable crops and pasture) in the Netherlands.

For copper it should be noted, however, that an important source is not considered in these data. This concerns the amount of copper used in foot baths. Although it is not allowed, farmers discharge the waste in the manure storage facility. Few estimates of the total amount thus added to soil via manure are available but these suggest that it can be a significant contribution. Data from an experimental farm in the Netherlands (de Marke) show that contribution of copper in food baths during 3 consecutive years (2001-2004) equals 39%, 135%, and 35% respectively of the known other sources during these years. If applicable to the Netherlands as a whole, the 35% level (conservative estimate) is equal to an additional copper load of 175 tons year[-1]. Furthermore, the amount of copper and zinc provided to livestock either as a free mineral mix to the feed or via drinking water is not take into account in the statistics.

In Figure 2 the relative contribution of all known sources for copper (excluding foot baths and freely added mineral mixes to the feed or added to drinking water) are shown for 1980

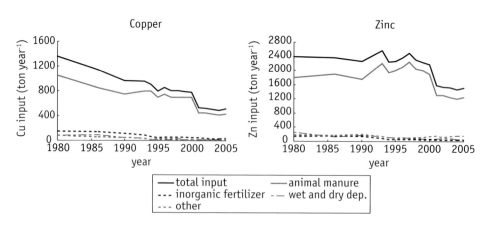

Figure 1. Changes in input of copper (left) and zinc (right) since 1980 to agricultural soils (arable land and grassland, CBS, 2007, www.cbs.nl).

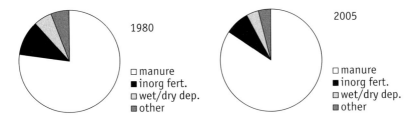

Figure 2. Contribution of manure to the total load of copper to agricultural soils (CBS, www.cbs.nl).

and 2005. Obviously, inputs from manure still dominate the input to soils, in fact the relative contribution has increased. This means that measures to reduce inputs via feed will be effective in reducing the total load as well.

The net accumulation rates for copper and zinc on a national scale are difficult to obtain based on the CBS data. For both metals only crop uptake has been quantified. Output via animal products as well as leaching and run-off are not quantified. In Table 3 an overview of the balances based on the in- and output data is shown.

Although not all outputs have been considered, differences with more specific balances at a national level are small. A study by Delahaye *et al.* (2003) showed that the net load of copper and zinc to the utilized agricultural area equalled 400 and 1099 tons year^{-1}, respectively.

Regional balances for copper across the Netherlands for the year 2005 were derived recently in a modelling study by Groenenberg *et al.* (2006). The results (see Figure 3) are in line with the order of magnitude of the national balance and the experimental data from Boer and Hin (2003). The net copper accumulation rate varies between 31 to 237 g ha^{-1} yr^{-1} depending on soil and farm type. Removal rates by crops and leaching to ground- and surface waters do account for a significant removal of copper from the soil although the input from manure (ranging from 193 to 345 g ha^{-1} yr^{-1}) is still substantially higher. In this study, however, copper in foot baths has not been accounted for. For zinc a similar study was conducted with comparable results (De Vries *et al.*, 2004).

Table 3. Balances for copper and zinc based on national data (in tons year^{-1}).

	Input (total)		Uptake crops		Balance	
	1980	2005	1980	2005	1980	2005
Copper	1360	515	140	95	1220	420
Zinc	2400	1515	700	550	1700	965

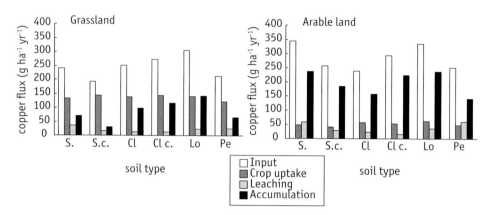

Figure 3. Summary of copper balances as a function of soil type. S: sand, Cl: clay, Lo: loess, Pe: peat. Addition 'c' indicates calcareous soil (Groenenberg et al., 2006).

Evaluation of present quality of soils and surface waters

Despite the considerable load of both metals, inputs of copper and zinc have not resulted (yet) in a significant increase of the soil metal content in the Netherlands. Aside from the fact that changes are small and difficult to measure within 5 or 10 years, monitoring networks are not of sufficient density to detect regional changes in the metal content of soils. In Figure 4 the actual copper and zinc levels in soils in the Netherlands are shown.

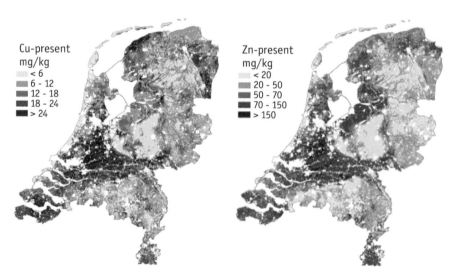

Figure 4. Average present levels of copper (left) and zinc (right) in Dutch soils in mg kg^{-1}.

As stated before, little impact of intensive animal husbandry is observed in terms of measurable accumulation of copper and zinc in soils due to manure application. In general, levels of both copper and zinc are highest in the central-western clay and peat areas. In these soils, metals are retained mostly due to the high clay and organic matter content and higher pH level (in clay soils) compared to sandy soils that prevail in the eastern part of the country. In fact the major source of copper and to a lesser extent zinc in peat soils in the western part of the Netherlands is household waste that has been applied during the Middle Ages. This was done to raise the surface level and get rid of the city waste. In contrast to the limited effect of agriculture on the quality of soil (at present), the impact of soil quality on ground- and surface water quality is substantial (Figure 5).

In Figure 5, predicted dissolved copper and zinc concentrations in surface waters due to leaching from soil are expressed as a function of the ecological threshold level (MTR level, 3.8 and 40 µg l^{-1} for copper and zinc, respectively). These figures illustrate that in low-lying areas with high groundwater tables, impact of leaching from soil on surface water quality is high. Net accumulation of both copper and zinc, therefore, will further aggravate the already high load of both metals to the surface water system.

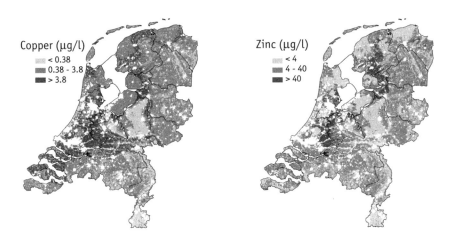

Figure 5. Predicted contribution of leaching from soil to the dissolved concentration in surface waters for copper (left) and zinc (right).

Impact of agriculture on the development of soil and surface water quality due to agricultural use

As stated previously, the impact of intensive animal husbandry cannot be detected from current copper levels in soil (Figure 4 and Figure 6a). In Figure 6a the soil copper content is expressed as a percentage of the Agricultural Advisory Levels for copper used in the

Netherlands (Römkens *et al.*, 2007). These advisory levels (for soil) are derived from quality standards of agricultural products (meat, milk) and to prevent health effects for animals due to the consumption of grass and soil. In general, soil copper levels throughout the country are still below these limits (except in some of the peat areas in the western peat area), but the net positive copper balance at the field level will result in ongoing accumulation. Effects of this ongoing accumulation will become obvious within 30 to 100 years as is shown in Figures 6b and 6c. Based on the present copper loads, the changes in the soil copper content were calculated according to the model described earlier.

Figures 6a/c clearly illustrate the future impact of intensive animal husbandry and use of manure on soil quality in the eastern sandy soils (especially obvious in the provinces of Noord-Brabant and Overijssel). Ultimately, copper levels can reach 10 times the recommended maximum levels based on the protection of animal health and product quality. Obviously the impact as expressed by an area where specific soil quality standards are exceeded can be smaller or larger depending on the magnitude of the soil quality standard used. In general, the agricultural advisory level (ranging from 30 to 50 mg Cu kg^{-1}) is, for example stricter than the quality criteria based on ecological effects. Calculated PNEC levels (Predicted No Effect Concentration levels, defined as the 5% level of a species sensitivity curve) for copper range from 79 mg kg^{-1} in sandy soils to 238 mg kg^{-1} in peat soils (average 134 mg kg^{-1}, ECI, 2005). Obviously, the degree by which the soil copper content within 100 years will reach the PNEC level is (much) lower than that of the agricultural advisory level. Nevertheless, these PNEC levels will be exceeded albeit after a much longer period; ranging from approx. 200 years in sandy soils to almost 2000 years in areas where inputs are lower (Groenenberg *et al.*, 2006).

For surface waters, the model to calculate trends in loads to ground- and surface water is being developed at present and will be available soon but no dynamic calculations have been performed yet. Results in Figure 5 clearly indicate that in low-lying areas, critical limits (MTR) are already exceeded. It can be expected that under current conditions (continuation

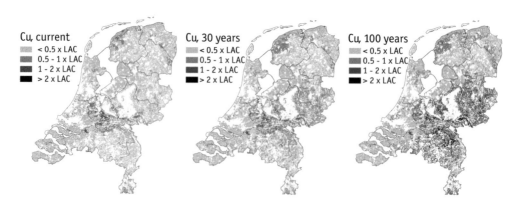

Figure 6. Impact of input of copper to agricultural soils (grassland and arable land): copper levels in soil relative to the agricultural advisory levels in 2005 (left), 2035 (middle) and 2105 (right).

of the load) no improvements can be expected. The impact of leaching and run off in the eastern sand district is, at present, still limited. For zinc, effects on soil quality are less when considering agricultural risks. However, zinc accumulation will result in an impact on soil regarding ecological thresholds (data not shown here, for more results see De Vries *et al.*, 2004) although times to reach these critical levels range from 100 to more than 2000 years. At present also the impact on the quality of surface waters is limited although it can be expected that, similar to copper, water quality criteria in low-lying areas will be exceeded on a larger scale in the future.

Copper and zinc balances from the perspective of nutritional requirements of cattle

The data of Hin and Boer (2003) were used to calculate the actual annual intake of copper and zinc per animal (cattle). A distinction was made between intake through roughage including grass, maize and imported fodder and additives (mineral mixes or otherwise as listed in the farm balance tables). Also a distinction was made between the summer and winter (stable) period, data on how many days the animals were kept outside and inside were available as well as the composition of their daily ration during both periods. On the other hand, the daily mineral allowance was calculated based on the actual recommendations. For copper a value of 10 mg kg^{-1} DM on the basis of the complete feed intake was used, for zinc this was 25 mg kg^{-1} DM, both values are based on the current Dutch advisory levels (COMV, 2005). In Table 4 the results for 8 individual farms are shown for copper and zinc. It should be kept in mind that the results of such calculations strongly depend on the criteria used. Both the daily allowances can be higher (or lower) and also the concentration and absorbability of metals in feed raw materials may differ substantially. The results shown in Table 4 are, therefore, not the 'absolute' truth but merely an indication of a shortage or excess of copper and zinc based on an animal-based balance. Results in Table 4 are based on recommendations as suggested by COMV (2005), that is, the sum of copper or zinc from feed raw materials together with those of additives has to equal the daily requirements. The excess is calculated as the actual supply minus the daily need.

For copper, the use of additional feed supplements seems to be necessary for farm number 1 and 2, whereas it can be reduced significantly for the other farms. In general, copper levels in roughage are below the recommended level so feed additives are still required in some cases. Based on the data in Table 4, an average level of 20 mg Cu kg^{-1} DM in additives seems to be sufficient to meet the requirements of animals. Present levels in additives are, on average, 30 mg kg^{-1} so a reduction of approx. 30% of the copper used in additives seems possible.

For zinc a totally different result was obtained. Based on the farm data and the nutritional requirements of 25 mg kg^{-1} DM in feed for cattle, no additives are required. All farms have an estimated excess of approx. 50%. The total nutritional need can be obtained from zinc in roughage. In this example no correction for the absorbability of copper and zinc in roughage has been made. It is clear that the composition of feed will affect the absorbability though. Nevertheless, it is also clear that the use of large amounts of feed additives for zinc are not necessary.

Table 4. Overview of recommended and actual intake by cattle for copper and zinc.

Farm no.	Farm balance	Cu in roughage	Cu in additives	Recommended amount	Actual supply	Difference	Excess
	g/ha	mg kg^{-1} DM	mg kg^{-1} DM	g/animal/yr	g/animal/yr	g/animal/yr	%
Copper							
1	77	8.3	16.1	65	63	-2	-3%
2	574	5.9	27.2	74	76	2	3%
3	59	6.4	30.3	76	91	15	16%
4	362	6.9	28.2	74	88	15	17%
5	17	6.0	29.4	75	90	15	17%
6	129	7.0	31.1	74	95	20	22%
7	227	5.7	41.2	78	132	54	41%
8	87	6.6	38.3	73	132	58	44%
Average	192	6.6	30.2	74	96	22	20%
Zinc							
1	363	51	52	162	329	167	51%
2	641	34	83	185	327	142	43%
3	-29	49	74	191	420	229	55%
4	1187	50	61	184	390	205	53%
5	42	42	126	188	476	288	61%
6	293	41	69	185	360	176	49%
7	498	41	72	196	398	203	51%
8	86	28	85	183	354	171	48%
Average	385	42	78	184	382	198	51%

Based on the actual nutritional requirement of an animal, and the resulting excess (or shortage) of supply, a comparison was made between the farm balance and the balance based on the difference between requirement and supply of the cattle. The data in Table 4 indicate that there is no relation for both elements between a farm balance and the excess load based on animal requirements. This puts the use of farm balances in a different perspective. Farm balances obviously are useful to determine whether or not inputs exceed outputs, but they cannot be used directly to determine to what extent inputs should be reduced. Data from farm number 2 (for copper) and 3 (for zinc) are very illustrative. Despite the net negative zinc balance in farm number 3, there appears to be an excess of approx. 50% based on animal requirements. Data from farm number 2 for copper show the opposite; despite a large positive copper balance, the requirement for copper of the animal is just met. One reason for this discrepancy is that in the farm balances also changes in stock supplies (e.g. the amount of mineral supplements that can be used in the coming years) are accounted for. These can vary per year depending on the stock present at the farm. Also the use of copper food baths or other sources like fertilizer affects the discrepancy between both approaches. On the long term, of course, these purchases that are not used within the same year will be part of the average farm balance (and the net excess load). This example is meant to illustrate that a direct comparison between an *annual* farm balance -where temporal changes can be large- and animal-based nutritional balances is difficult and may lead to erroneous conclusions.

Obviously, a lower degree of the absorbability of metals in feed raw materials, strongly affects the degree to which the daily metal supply exceeds the nutritional need. Reducing the absorbability of zinc by 50% in feed raw material for example reduces the average daily excess to 20%. For copper the results are even more pronounced; if the absorbability in roughage is 40% of the adopted value, the average excess for cattle is reduced to zero!

These examples nevertheless clearly show that mineral balances have to be used carefully. Proper estimates of a potential excess load should always include the analyses based on nutritional needs as is illustrated here. Farm balances alone can lead to wrong conclusions as far as reductions of inputs are concerned.

Options to reduce inputs on the farm level

Both the environmental assessment (farm and national level) as well as the mineral balance based on the nutritional need of animals indicate that input levels of copper and zinc can and should be reduced. In the Netherlands, awareness has grown that indeed action is needed to avoid unwanted effects of metals both in soils and surface waters. Recently a few studies have been performed to identify measures at the farm level to reduce the emission to soil (Kool *et al.*, 2006; den Boer *et al.*, 2007; Bussink *et al.*, 2007). Here a short overview of the major conclusions of these studies is presented.

Bussink *et al.* (2007) distinguish between two main routes for supplying animals with the trace elements they need:
• soil-crop route, using fertilizers for optimal supply of Cu and Zn in grass and maize (roughage);

- animal feed route, using feed concentrates and mineral supplements for adding Cu and Zn to the diet.

Because research and practice have not found the best way to optimize/combine these two routes yet, farmers end up using both routes for providing their animals with sufficient Cu and Zn: this means a 'double supply' both by fertilization and by feed suppletion. This can potentially result in adverse effects on animal health, on environmental (soil and water) quality and in unnecessary high costs for farmers.

Some farmers have used additional mineral fertilizer containing copper to increase the copper content in grass, because Cu in organic manure has a very low effectiveness. Field data on the composition of grass (copper content) show that the range in the copper content across a wide range of soil types and copper levels in soil is rather limited (Table 4). For example, the copper level in the farms studied by Boer and Hin (2003) indicate that the copper level in grass ranges generally from 6 to 9 mg kg^{-1} (dry matter) which is in close accordance with national data (www.blgg.nl), but values up to 13 mg kg^{-1} are sometimes found as well. To meet the animal requirements not only the total Cu content is important but also the Cu absorption coefficient. The latter is strongly influenced by the concentration of S and Mo in the diet. High S and Mo levels suppress the absorption of Cu by the animal (Jongbloed *et al.*, 2005). Grassland management should therefore not only focus on maintaining the Cu content of grass but also on preventing high levels of S and Mo. At present a fixed factor is used for the Cu absorption coefficient (based on the average concentration of S and Mo in grass products). More knowledge on the exact relationship between Cu, Mo and S could be useful to determine the 'true' absorption coefficient (Bussink *et al.*, 2007). In case of a low S and Mo content this could result in a higher absorption coefficient and thereby in a lower total Cu supply to the animal.

Bussink *et al.* (2007) investigated the pro's and con's of the possibilities to supply trace elements by the soil-crop route and by the feed (supplement) route. Their investigation shows a different need for each element by each animal category. The degree to which the amounts of Cu and Zn needed are covered, is largely determined by the variation in the composition of the main feed products that are available (both roughage and feed concentrates). The availability of trace elements in soil, whether animals are kept outside or inside and the absorption of elements by different animal categories are all very important determining factors in this respect.

Variation in, for example, grass composition determines whether or not specific elements are supplied sufficiently; dairy cows with mainly grass-feeding will not be supplied with sufficient Cu based on the average copper content of Dutch grass (concentrations available through www.blgg.nl). Zinc however will be supplied mostly in excess of the actual need in that case.

Bussink *et al.* (2007) showed that (based on Dutch recommendations) no extra supply of Zn through feed is needed and Zn may be left out of mineral supplements in almost all cases. These results are in close agreement with the data shown in Table 4 in this paper. Also

they point to the need for products that differentiate more regarding their trace element contents. One size clearly does not fit all: actual nutritional recommendations and actual supply should be tuned to one another. A balanced (fine-tuned) supply with Cu and Zn is necessary. In order to achieve such a balanced approach, some basic steps are needed:
1. determine the actual supply of the livestock (farmers do not know this in most instances);
2. determine the actual requirement of the livestock;
3. develop a strategy to change the actual supply to an optimal supply; either by fertilization or by feed (supplements) or (in rare occasions) through both routes.

These steps were carried out by Den Boer *et al.* (2007). In this study a practical tool was developed by NMI (Nutrient Management Institute) to calculate the trace element supply of dairy cows. This tool can be used as a scan for micronutrient supply (Zn, Se, Cu, Co), actual need and resulting recommendations for a more balanced feed management. In this way, feed management was fine-tuned for 8 dairy farms in the Kempen area. Also attention was given to using copper sulfate in foot baths. The project was initiated in the context of 'good agricultural water management'.

The input/load of Cu and Zn resulting from metal intake through the feed is shown for 4 situations (see Table 5):
- the actual feeding regime (current diet);
- the new feeding strategy based on the analysis of actual supply and actual need;
- the situation when no extra Zn is added to feed concentrates and mineral mixtures;
- the situation when animals are fed strictly according to the recommended allowances.

The results are then converted to fluxes in g ha^{-1} yr^{-1}.

Table 5. Average Zn and Cu input/load in g per ha per year through feed/manure.

	Actual situation	New strategy	No extra Zn added to feed concentrates and min. mixtures	Supply according to standard needs
Zinc (feed)	958	889	782	336
% to standard*	285	265	233	100
Copper (feed)	247	223	223	176
% to standard*	140	127	127	100
	Foot bath:			
Copper (CuSO$_4$)	275	147	0**	0**

*100% means supply according to standard needs.
**when no copper sulfate is used in foot baths.

In this pilot with 8 farms the supply of Zn proved to be 200-300% compared to the estimated requirement of the livestock. This results especially from extra Zn addition to feed concentrates and through mineral mixtures.

Adjusting the feed management and leaving out Zn additions from concentrates and mixtures results in a significant decrease in total Zn input and consequently losses. However, feeding livestock strictly according to their standard needs for trace metals is hardly ever possible because:
- different animal categories within the cattle herd differ with regard to their diet and their standard needs;
- Zn contents in roughage and raw materials of feed concentrates are already much higher than needed for covering the livestock's need.

For copper only a significant reduction of the inputs can be reached by a different strategy for hoof disinfection; so by using alternatives for copper sulfate in foot baths.

An important aspect in the discussion on the reduction of copper and zinc deals with the absorbability of metals in feed. The requirement for copper and zinc is based on estimates of the absorption coefficient of animals. Kool *et al.* (2006) pointed out that the safety margins that are used for absorption coefficient can be as high as 50%. Also in their study, Kool *et al.* (2006) concluded that addition of Zn to the diet through mineral mixtures is unnecessary for all cattle categories. Their results suggest that roughage and feed concentrates together suffice for Zn supply. Leaving out Zn from mineral mixtures would result in a decreased Zn load of about 60%. This again is in good agreement with the average reduction of the total input of 51% (Table 4) presented here. Additional input reduction can be reached by leaving Zn supplements out of the feed concentrates.

For copper the easiest solution to reduce inputs is to use alternatives for copper sulfate in foot baths (in the Netherlands used on approx. 40-50% of the farms). Another significant reduction (approx. 25%) can be reached by leaving out Cu of the mineral mixtures for lactating cows, when animal requirements are already met with the basic diet. Again, this reduction is close to the reported 20% reduction in inputs reported here (Table 4).

Contribution of other farming types to copper and zinc load

The results for copper and zinc shown in the previous section were based on the recommendations for cattle only (the results presenting the impact on soil and water are based on total emissions). Reduction rates are not based on those of other types of animals. To assess what the relevance of other sources of copper and zinc in manure from other livestock sectors is, we calculated the load of both metals from the 3 major farming types (cattle, pigs, poultry). Based on the total N-production in manure (data from the Dutch National Bureau of Statistics for 2005, www.statline.nl) and the average ratio of Zn or Cu to N in manure, the specific contribution of each sector can be calculated. In Table 6 the results from this exercise are shown. No distinction between liquid and solid manure or different types of animals within the specified categories were made. The metal to N ratios are based on data

Table 6. Total N-production in 2005 and calculated Cu and Zn load in different types of manure.

	Animal group				Total
	Cattle	Pigs	Poultry	Other[1]	
Total N produced (thousand tons)	289.4	98.3	57.9	14.8	460.4
Zn/N ratio (g Zn/g N)	0.0021	0.0066	0.0044	0.0021	-
Zn emission (tons)	608	649	255	31	1542
Percent of total	39%	42%	17%	2%	
Cu/N ratio (mg Cu/g N)	0.41	2.91	0.77	0.41	
Cu emission (tons)	119	286	45	6	455
Percent of total	26%	63%	10%	1%	

[1]Goat, sheep, rabbits, the conversion factor was assumed to equal that of cattle.

from Driessen and Roos (1996) but are scaled (down) to match the observed decrease in metal levels in animal feed that was enforced in 2000. More details on the Cu and Zn to N ratios can be found in Groenenberg *et al.* (2006) for copper and De Vries *et al.* (2004) for Zn.

The calculated total emission of Cu and Zn in manure is in line with other reported national values (www.statline.nl). Reported copper loads in manure in 2005 equalled 435 tons (here: 455) whereas that for zinc was equal to 1245 tons (here: 1542). The latter difference can be explained by the fact that for zinc even lower levels in feeds have been enforced during the last 3 years (EU, 2003). The metal to N ratios were scaled for the year 2000 after which a further reduction of Zn in manure has occurred (see also Figure 1).

For zinc the contribution of cattle and pig manure is approximately equal (each 40% of the total load in manure). For copper however, the contribution of pig manure is far greater than that of cattle manure. This is largely due to much higher levels of copper in the feeds of pigs. For young pigs up to 170 mg kg^{-1} is allowed, whereas levels for cattle range from 15 (non-ruminating calves) to 35 mg kg^{-1} (EU, 2003).

These data show that reductions in the level of copper in feeding regimes for cattle will reduce the total load but the effect is limited if no measures are taken to reduce the level in diets for pigs and poultry as well. Rough estimates indicate that a reduction by 50% of copper in diets of cattle will result in a net decrease of the total load of 10 to 15%. For zinc, a zero addition of additives as suggested by several studies, would result in an estimated decrease of the load by cattle manure of 50% (half of the present load is from additives and half comes from roughage), which would reduce the total manure zinc load by 20%.

Clearly the intake by pigs and to a lesser extent poultry has to be decreased to obtain a further substantial decrease in the total load of copper and zinc in manure. This further urges the need to evaluate the nutritional needs of animals for copper and zinc similar as was done for cattle.

Discussion and conclusions

Environmental impact of the use of animal manure at present loading rates

The results presented for the Netherlands clearly show that accumulation of both copper and zinc does occur. However, the fact that inputs exceed outputs as such is not very relevant when considering animal needs nor does it indicate where and when critical limits in soil are exceeded. The major issue to decide whether or not input reductions should be enforced is, therefore, whether or not the observed degree of accumulation leads, within relevant time frames (10?, 50?, 100? years), to unacceptable levels in the soil and or other compartments of the environment. As such different criteria can be used as guidelines. These include (1) agricultural advisory levels in soil based on crop and animal product quality criteria, (2) ecological criteria in soils, (3) criteria in ground- and surface waters. Other criteria currently under debate include the so called stand still principle which does not allow for any accumulation at all in soils. Obviously the latter one is a very desirable goal but considering present day agriculture not very realistic. Ultimately, the supply of metals in feeds and additives should be based on the requirement of animals.

This study clearly shows that using different criteria different degrees of 'risk' are eminent. For the Netherlands at present, soil quality criteria are hardly exceeded except in areas with a specific history of regional polluting activities. On the other hand, surface water criteria already are exceeded in some areas. It can be safely assumed that under present conditions, leaching losses to ground and surface waters will not decrease so if the water criteria will serve as the first indicator to decide whether or inputs have to be reduced, large reductions are needed in the low lying areas of the Netherlands.

Data from various studies on the actual intake by animals show that in general the supply of copper and especially zinc exceeds the demand according to present recommendations on animal needs. The degree by which inputs through feeds and additives can be reduced ranges from 20% for copper to 50% for zinc. The majority of the studies suggest that addition of zinc to diets for cattle is largely unnecessary due to the relatively high supply by roughage.

An interesting observation is also that the excess load at the farm level is not correlated at all to the excess supply for the animals. This indicates that farm balances alone are not a good indicator for the estimated reduction that can be achieved. Both animal-based analyses (what is the requirement of an animal) in combination with farm data have to be used to assess the real excess load. The data seem to suggest that a zero-balance at the farm level in that case is hard to achieve. Whether or not that is considered acceptable would require a detailed analysis as to how the animal-based approach affects soil and water quality within the next decades.

An estimate of the contribution of pigs and poultry manure to the total zinc and copper load in manure clearly shows that measures to reduce both metals in feeds have to be extended to these categories as well. Reducing copper and zinc in feed additives for cattle only would result in a maximum reduction between 10 (for Cu) and 20% (for Zn) of the total load.

Measures to reduce inputs and additional research needs to improve animal-based nutrition regimes for copper and zinc.

Based on the farm results several no regret measures (what can be done already, which approach is practical and cost-effective) can be identified:
- Reducing the amount of copper in foot baths or the use of alternatives (see for example: Kool *et al.*, 2006; Den Boer *et al.*, 2007).
- Reducing the amount and composition of feed concentrates (see for example: Den Boer *et al.*, 2007).
- Reducing the amount and composition of mineral supplements (mixtures). (see for example Bussink *et al.*, 2007; Den Boer *et al.*, 2007).
- Improve knowledge of the Cu and Zn availability in soils by using new soil analysis techniques and soil chemistry in order to improve/fine-tune fertilizer recommendations; these will generally become lower. (Bussink and Temminghoff, 2004).

At this moment, a mineral mixture without Cu and Zn is more expensive than the same product including these metals. This will not stimulate farmers to use these instead of cheaper alternatives.

Input reduction at the source can be reached by decreasing heavy metal loads through fine-tuning the feed composition (feed concentrates and mineral supplements) and by reduced use of copper baths.

Effect-oriented measures to reduce environmental impact of leaching of copper to ground- and surface water could consist of lowering the groundwater table and increasing the pH of the soil.

Grass and maize on sandy soils are often grown at low pH values (4 to 4.5) and leaching of heavy metals at these pH values can be significant. Managing groundwater levels at a regional scale, however, is difficult due to the fact that arable land and pasture are often present in the same area. Lowering the water table might be beneficial to reduce leaching (or better increase the time needed before metals reach the groundwater), but it is perhaps less positive for arable crops who might suffer from water shortage.

Animal requirements (based on the absorption coefficients of distinct animal categories) and soil type (availability of Cu and Zn for crop uptake; e.g. competition of Cu with Mo and S) largely determine how realistic it is to leave out the Cu and Zn additions in mineral supplements and feed concentrates. However, it seems that Zn can be completely left out of the mineral mixtures and feed concentrates for a very broad range of cattle farms in practice.

In order for farmers to be able to determine the supply of trace elements and to compare this with their animals' needs it is necessary to give them information on the heavy metal contents in their feed:

- Declaration of contents on labels of mineral mixtures and feed concentrates.
- Routine analysis of contents in roughage.

Aside from the actions needed to stimulate farmers to fine tune the requirement of animals regarding copper and zinc in diets, also the information regarding the requirement of metals by animals has to be improved.

Issues that need to be discussed include:

- Which uptake of Cu and Zn is required by several animal categories for a good metabolism? Are current recommendations for minimum uptake realistic? Is more fine-tuning possible for diverse animal categories?
- Which content in animal feed is needed to reach that uptake (what is the absorption coefficient)? Are current absorption coefficients realistic? Is more fine-tuning possible for diverse animal categories?
- Can uptake of trace elements from feed be optimized?
- How do current ways of feeding and farming relate to these minimum requirements?
- When the conclusion is reached that more fine-tuning is possible and that more 'tailor-made' feeding regimes can be used in daily farming regimes (considering the diversity in animal categories), are feed manufacturers willing to produce a wider range of products that are more in line with specific animal needs?

References

Bang, K.S., A.S. Familton and A.R. Sykes, 1990. Effect of copper oxide wire particle treatment on establishment of major gastrointestinal nematodes in lambs. Res. Vet. Sci. 49, 132-137.

Boer, M. and K.J. Hin, 2003. Heavy metals in Dutch dairy farming systems. Results and recommendations obtained in the project 'Koeien en Kansen' (In Dutch). Centrum voor landbouw en milieu, CLM report no. 587-2003.

Boer, den, D.J., S.W. Moolenaar and R.F. Bakker, 2007. Reduction of heavy metal supply through feed on dairy farms by optimizing supply with the trace element scan ('spoorwijzer'). NMI-report 1216.N.06 (In Dutch), Nutrient Management Institute, the Netherlands (the trace element scan is downloadable from www.nmi-agro.nl).

Bonten, L.T.C. and D.J. Brus, 2006. Belasting van het oppervlaktewater in het landelijk gebied door uitspoeling van zware metalen; Modelberekeningen t.b.v. emissieregistratie 2006 en invloed van redoxcondities. Alterra report 1340 (In Dutch), Wageningen University and Research Center, the Netherlands.

Bussink, D.W. and E.J.M. Temminghoff, 2004. Soil and tissue testing for micronutrient status. The International Fertiliser Society. Proceedings No: 548. York, United Kingdom.

Bussink, D.W., D.J. den Boer, G. van Duinkerken and R.L.G. Zom, 2007. Trace-element supply of dairy cows through the feed route or the soil-crop route. NMI report O 1139 (In Dutch). Co-production of the Nutrient Management Institute and the Animal Sciences Group (ASG) of Wageningen University and Research Center, the Netherlands.

COMV, 2005. Handleiding mineralenvoorziening rundvee, schapen, geiten. CVB, Lelystad.

De Vries, W., P.F.A.M. Römkens, J.J.M. van Bronswijk and T. van Leeuwen, 2002. Heavy metals. In: P. Haygarth and S. Jarvis (Eds): Agriculture, Hydrology and Water Quality: 107-132.

De Vries, W., P.F.A.M. Römkens and J.C.H. Voogd, 2004. Prediction of the long term accumulation and leaching of zinc in Dutch agricultural soils: a risk assessment study. Alterra report 1030, Wageningen University and Research Center, the Netherlands.

Delahaye, R., P.K.N. Fong, M.M. van Eerdt, K.W. van der Hoek and C.S.M. Olsthoorn, 2003. Emission of seven heavy metals to agricultural land (In Dutch). Central Bureau of Statistics, Voorburg, the Netherlands.

Driessen, J.J.M., and A.H. Roos, 1996. Heavy metals, organic micropollutants and nutrients in animal manure, compost, sewage sludge, soil and fertilizers. Rikilt-DLO report no. 96.14. Rikilt, Wageningen UR, the Netherlands.

ECI, 2005. European Union Risk Assessment Report copper, copper ii sulphate pentahydrate, copper(i)oxide, copper(ii)oxide, dicopper chloride trihydroxide. Voluntary Risk assessment European Copper Institute Draft May 2005 chapter 3, part 3.

Eckel, H., U. Roth, H. Döhler, F. Nicholson and R. Unwin (eds.), 2005. Assessment and reduction of heavy metal input into agro-ecosystems. Final report of the EU-Concerted Action AROMIS. KTBL publication no. 432, Darmstadt, Germany.

EU, 2000. Directive 2000/60/EC. Establishing a framework for Community action in the field of water policy. Official Journal of the European Communities L 327.

EU, 2003. Commission Regulation (EC) No 1334/2003 of 25 July 2003. Amending the conditions for authorisation of a number of additives in feedingstuffs belonging to the group of trace elements. Official Journal of the European Union L 187.

Groenenberg, J.E., P.F.A.M. Römkens and W. de Vries, 2006. Prediction of the long term accumulation and leaching of copper in Dutch agricultural soils: a risk assessment study, Alterra report 1278, Wageningen University and Research Center, the Netherlands.

Jongbloed, A.W., Tsikakis, P., Kogut, J., 2004. Quantification of the effects of copper, molybdenum and sulphur on the copper status of cattle and sheep and inventory of these mineral contents in roughages. Report 04/0000637 ASG Nutrition and Food.

Keller, A. and A. Desaules, 2003. The Swiss soil monitoring network: regular measurements of heavy metals in soil and field balances. In: Eckel, H., U. Roth, H. Döhler, F. Nicholson and R. Unwin (Eds.), Assessment and reduction of heavy metal input into agro-ecosystems. Final report of the EU-Concerted Action AROMIS. Darmstadt, Germany, KTBL publication no. 432:217-221.

Kool, A., A.W. Jongbloed, S.W. Moolenaar, G.J. Hilhorst and F.C. van der Schans, 2006. Tackling heavy metals at dairy farms. CLM report no. 640-2006 (In Dutch), Culemborg, the Netherlands.

Kroes, J.G., P.J.T. van Bakel, J. Huygen, T. Kroon en R. Pastoors, 2001. Actualisatie van de hydrologie voor STONE 2.0. Alterra, Wageningen. Reeks Milieuplanbureau 16 (In Dutch). Alterra rapport 298, Wageningen University and Research Center, the Netherlands.

Kroon, T., P. Finke, I. Peereboom en A. Beusen, 2001. Redesign STONE. De nieuwe schematisatie voor STONE: de ruimtelijke indeling en de toekenning van hydrologische en bodemchemische parameters. RIZA, Lelystad (In Dutch). RIZA rapport 2001.017.

Moolenaar, S.W., S.E.A.T.M. van der Zee and Th.M. Lexmond, 1997. Indicators of the sustainability of heavy-metal management in agro-ecosystems. Sci. Tot. Env. 201, 155-169.

Moolenaar, S.W., 1999. Heavy-Metal Balances, Part II. Management of Cadmium, Copper, Lead, and Zinc in European agro-ecosystems. J. Ind. Ecol. 3, 41-53.

Moolenaar, S.W. and Th.M. Lexmond, 1999. Heavy-Metal Balances, Part I. General aspects of Cadmium, Copper, Lead and Zinc balance studies in agro-ecosystems. J. Ind. Ecol. 2, 45-60.

National Research Council, 2005. Mineral tolerance of animals, 2nd revised edition. National Academic Press, Washington, D.C., USA.

Römkens, P.F.A.M., J.E. Groenenberg, L.T.C. Bonten, W. de Vries and J. Bril, 2004. Derivation of partition relationships to calculate Cd, Cu, Ni, Pb and Zn solubility and activity in soil solutions. Alterra report 305, Wageningen University and Research Center, the Netherlands.

Römkens, P.F.A.M., J.E. Groenenberg, R.P.J.J. Rietra and W. de Vries, 2007. Derivation of Dutch Agricultural Advisory levels and overview of soil to plant relationships to be used in risk assessment. Alterra report 1442 (In Dutch w. English summary). Alterra, Wageningen UR, the Netherlands.

Nutrition

Improvement of balance of trace elements in pig farming systems

J.Y. Dourmad and C. Jondreville

Keywords: copper, zinc, pig, environment, manure

Introduction

Although Cu and Zn are minor body components in pigs, they are involved in many metabolic functions (Jondreville *et al.*, 2003). Their provision in sufficient amount in pig feeding is therefore indispensable to ensure good performance and health. However, because they are used as growth promoters or because large safety margins are applied, Cu and Zn are often oversupplied in pig diets. Consequently, manure is highly concentrated in these elements, which accumulate in top soil and cause toxicity to plants and microorganisms, and become an environmental concern in areas of intensive pig farming. Moreover, when a treatment is applied to the slurry, Cu and Zn concentrate in the solid fraction and their concentration may exceed the maximal values allowed for the valorisation of these products as organic fertilisers.

Effect of Cu and Zn supply on their contents in manure

The amount of Cu and Zn excreted can be estimated by difference between intake and retention. Cu and Zn retention may depend on their dietary supplies. However, because supplies generally largely exceed the requirements, Cu and Zn retentions by pigs are maximized and represents about 1.1 and 21.8 mg/kg BW, respectively (Dourmad *et al.*, 2002). According to different scenarios of Cu and Zn dietary provision in a farrow-to-finish farm we calculated the concentration of Cu and Zn in manure, their annual application per ha, and the time needed for Cu and Zn to reach 50 and 150 mg/kg DM soil, respectively. These levels correspond to half of the maximum allowed concentration of Cu and Zn in sludge-treated soils in France. The effect of Cu and Zn on plant toxicity depends on plant species and type and pH of soil (Hartmans, 1978, Coppenet *et al.*, 1993) which also affect their leaching (Römkens *et al.* 2008). Soil microflora is more sensitive than plants to high soil Cu and Zn concentrations (McGrath *et al.* 1995, Morel, 1997). The upper limits of 50 and 150 mg/kg DM soil used for Cu and Zn, respectively, correspond to about half of the levels required to reach plant toxicity and are close to levels for which soil microflora is affected. However, for Cu the considered level exceeds the maximum level of 25-30 mg/kg DM soil recommended for sheep grazing (Hartmans, 1978). The annual uptake by crops is assumed to be 50 and 300 g/ha for Cu and Zn, respectively. The amount of soil is 3000 t/ha (20 cm deep ploughing) with an initial content of Cu and Zn of 15 and 20 mg/kg DM, respectively. No leaching of Cu and Zn is considered in the calculation. The nutritional scenarios were chosen in order to evaluate the effect of the reduction in 2003 of the maximal allowed levels of Cu and Zn in pig diets in EU and identify possible further improvements.

The incorporation of 150 to 250 ppm Cu in pig diets has been employed for a long time because of its growth promoting effect (Braude, 1980). This practice is currently authorized in EU for pigs up to 12 weeks, with diets containing a maximum of 170 ppm Cu. After 12 weeks of age, Cu is no more used as growth factor in EU, and the maximal level of incorporation is 25 ppm. Compared to the former regulation (scenario B), this new regulation (scenario C) results in a drastic reduction of Cu in manure, by almost 60% (Table 1). Nevertheless supplies remain higher than usual published requirements (less than 10 ppm) and average retention efficiency is still less than 1%. This efficiency remains low even when pigs are fed diets corresponding to actual requirements (about 4%).

Supplementing weaned piglets diets with 1500 to 3000 ppm Zn as ZnO was also reported to stimulate their growth (Poulsen, 1995). This practice is still allowed in many countries but should no more be used in EU. In fact, in 2003 (EC, 1334/2003), maximal level of Zn incorporation in all pig diets was reduced to 150 ppm, compared to 250 ppm before. This level is much closer to the published requirement which vary between 100 and 50 ppm according to growing stage and authors. Compared to a situation in which weaning pigs are fed a diet with 2500 ppm Zn from 8 to 15 kg BW and 250 ppm thereafter (scenario A) the present EU regulation (scenario C) results in 53% reduction of Zn excretion (Table 2).

With the present EU regulation, Cu and Zn contents in slurry DM (about 350 and 1250 mg/kg DM, respectively) are below the maximal concentration allowed in sewage sludge in France (1000 and 3000 mg/kg DM, respectively), but they exceed the concentration allowed for organic fertilizers (300 and 600 mg/kg DM, respectively). With the hypotheses that 170 kg

Table 1. Estimates of Cu balance according to different hypothesis of supply in pig feeding[1].

	B	C	D
Concentration in diets, ppm			
piglets 1	175	170	10
piglets 2	175	170	10
fattening pigs	175-120	25	10
sows	100	25	10
Cu balance, g/ pig (0-110 kg BW)			
intake	42.65	13.48	3.33
excreted	42.53	13.36	3.21
Slurry composition, mg/kg DM	1119	351	84
Application (kg.ha^{-1}.year^{-1})			
no slurry treatment	1.89	0.59	0.14
50% of N removed	3.78	1.19	0.29
Years to reach 50 mg/kg soil DM			
no slurry treatment	47	160	941
50% of N removed	23	77	371

[1]Scenario B: former EU regulation, scenario C: actual EU regulation, scenario D: perspectives.

Table 2. Estimates of Zn balance according to different hypothesis of supply in pig feeding[1].

	A	B	C	D
Concentration in diets, ppm				
piglets 1	2500	250	150	70
piglets 2	250	250	150	50
fattening pigs	250	250	150	30
sows	250	250	150	70
Zn balance, pig (0-110 kg BW)				
intake	84.1	68.3	41.7	9.0
excreted	81.7	65.9	39.3	6.7
Slurry composition, mg/kg DM	2542	2128	1269	284
Application (kg.ha^{-1}.year^{-1})				
no slurry treatment	4.29	3.59	2.14	0.48
50% of N removed	8.59	7.19	4.29	0.96
Years to reach 150 mg/kg soil DM				
no slurry treatment	79	95	167	1160
50% of N removed	39	46	79	427

[1]Scenario A: former EU regulation, with the use of Zn as growth promoter; scenario B: former EU regulation; scenario C: actual EU regulation; scenario D: perspectives.

N/ha are spread each year, it will take 80 to 160 years for the soil to reach the maximal limit fixed. This is much longer than with the previous regulation (25 to 50 years).

The time required to reach the upper limits of Cu and Zn in the soil is highly dependent on the hypotheses used for calculation and the choice of these upper limits. The export by crops was calculated in the case of wheat production with the export of both grain (8 t/ha) and straw (5 t/ha). In the situation of decreased export due to lower yield, or the use of crops with lower Cu or Zn content, the delay will be shortened. However this effect is limited. For instance, this delay is shortened by only 10 years (compared to about 160 years), in the case of a 50% reduction of crop export of Cu and Zn. In our calculations we assumed that only a marginal leaching and runoff of Cu and Zn occurred. This may not be the case especially for sandy soils for which a significant leaching of Zn has been measured which might lead to an underestimation of the time needed to reach the upper limit in the soil (Römkens *et al.*, 2007). However leaching and runoff of Cu and Zn may result in an accumulation of these elements in rivers and sea sediments (Arzul and Mager, 1990) with possible adverse effects on flora and fauna.

Another important point to consider refers to the criteria used for the determination of the upper limit of Cu and Zn in soils. A level close the level required for plant toxicity is often used in such estimations (Jondreville *et al.*, 2003). In our calculations we considered 50% of this level which is close to the level for which some effects are measured on soil microflora. This results in a much shorter delay than with the previous assumption. But the sustainability of such practices may still be discussed.

In the future, further reductions in Cu and Zn excretion should be possible (scenario D, Tables 1 and 2) resulting in a more equilibrated balance between spreading and export by plants. For instance, Revy *et al.* (2006) recommended a supply of Zn for weaned piglets of 90 mg/kg diet with a possible reduction by around 35 mg when the diet is supplemented with microbial phytase. However this will require a more precise evaluation of the requirements for all types of pigs and a better understanding of the factors that affect Cu and Zn availability.

References

Arzul, G., and J.F. Maguer, 1990. Influence of pig farming on the copper content of estuarine sediments in Brittany, France. Marine Pollution Bull. 21, 91-107.

Braude, R., 1980. Twenty five years of widespread use of copper as an additive to diets of growing pigs. In: P. L'Hermite and J. Dehandtschutter (Eds.), Copper in Animal Wastes and Sewage Sludge, pp. 3-15

Coppenet, M., J. Golven, J.C. Simon and M. Le Roy, 1993. Evolution chimique des sols en exploitations d'élevage intensif: exemple du Finistère. Agronomie 13, 77-83.

Dourmad, J.Y., C. Pomar and D. Massé, 2002. Modélisation du flux de composés à risque pour l'environnement. Journ. Rech. PorcineFr. 34, 183-194.

Hartmans, J., 1978. Identifying the priority contaminants toxicological aspects of animal effluents. In : W.R. Kelly (Ed.), Animal and Human Health Hazards associated with the utilization of animal efflents, EEC Publication, 21-23 Novembre 1978, Dublin, Ireland, pp 35-56.

Jondreville, C., P.S. Revy and J.Y. Dourmad, 2003. Dietary means to better control the environmental impact of Cu and Zn by pigs from weaning to slaughter. Livestock Production Science, 84, 147-156.

McGrath, S.P., A.M. Chaudri and K.E. Giller, 1995. Long term effects of metals in sewage sludge on soils, microorganisms and plants. J. Ind. Microbiol. 14, 94-104.

Morel, J.L., 1997. Bioavaibility of trace elements to terrestrial plants- Chapter 6. Tarradellas, In: J., Bitton and G. Rossel (Eds.). Soil Ecotoxicology, Lewis Publihers, CRC Press, Boca Raton, Fl, 141-176.

Poulsen, H.D., 1995. Zinc oxide for weanling piglets. Acta. Agric. Scand. 59, 159-167.

Römkens, P.F.A.M., S.W. Moolenaar, J.E. Groenenberg, L.T.C. Bonten and W. de Vries, 2008. Copper and Zinc in feed (additives): an essential burden? In: P. Schlegel, S. Durosoy and A.W. Jongbloed (Eds.), Trace elements in animal production systems, Wageningen Academic Publishers, Wageningen, the Netherlands, pp.115-136.

Reduction in trace element excretion in swine and poultry in the United States

G.M. Hill

Keywords: copper, zinc, iron, swine, poultry

Introduction

An over-supplementation pattern of trace elements has developed in the United States because marginal or deficient dietary concentrations of minerals were well known to reduce productivity, and mineral costs were low compared to the cost of other nutrients. Like pigs in all phases of production, many birds in the broiler and turkey industries are fed dietary concentrations beyond those recommended by the National Research Council (NRC, 1998).

In the United States, swine producers have utilized the performance benefits of pharmacological zinc (Zn) and copper (Cu) especially in nursery diets for the last 10 to 15 years. However, due to concerns about the faecal excretion of these nutrients and other minerals, scientists have begun to look for alternative sources and synergistic interactions to maintain enhanced growth and reduce threats to the environment.

Influence of genetics

Swine producers have known and utilized in feed formulations that different genotypes have differing nutrient needs during their life cycle due to their unique growth patterns relative to muscle and fat accretion. However, the influence of genetics is usually not considered in studying trace element excretion patterns. Pigs from sows of traditional maternal breeds (Large White X Landrace), sires of terminal breeds (Hampshire X Duroc) and their F_1 progeny were utilized to investigate the effect of genetics and sex on trace elements in swine excreta (Crocker and Robison, 2002). Faecal and urinary output were collected for three days and analyzed for trace elements. When daily output per kilogram of pig weight was adjusted for feed disappearance, Zn and Cu excretion was less for pigs from the maternal line than for those of the terminal line or generated by the F_1 progeny. Even within the Duroc breed, they found that pigs generated from a low testosterone line had lower Cu in their excreta than those generated from a high testosterone line. After the adjustment for weight differences, they also observed that gilts excreted less Zn, Cu and iron (Fe) than barrows. While this could be due to differing tissue accretion patterns, it could also be influenced by nutritional needs as they approach sexual maturity. There are limited reports in the literature about the influence of genetics on the excretion of minerals. Genetic strain, gender and their interaction was reported to have a limited affect on body concentrations of Zn, Fe, Cu and Mn in the chicken (Mohanna and Nys, 1998). However, younger birds were reported to have higher concentrations of these elements than older chickens.

Manganese

As previously noted (Hill and Spears, 2000) the requirements and dietary factors that affect manganese (Mn) requirements during the pig's life cycle are not well defined. The need for Mn is well documented in the chick due to its role in bone health. However, seeds and their products vary in Mn content depending on the plant species, soil pH and portion of the seed utilized (Underwood and Suttle, 1999). The Mn concentration in corn, sorghum and barley is less than wheat and oats. Utilizing cannulas in the bile duct, duodenum, portal vein, ileocolic vein and jugular vein, young pigs (20 to 40 kg) were found to have a true Mn absorption of 0.5% (Finley *et al.*, 1997). This resulted in apparent absorption of approximately 1.7% in pigs fed a corn and soybean meal diet due to the large amount of Mn that is absorbed and then excreted via the bile. In poultry, low dietary Fe concentrations will increase and high dietary phosphorus (P) and calcium (Ca) will decrease Mn absorption. Because of the incomplete information on Mn bioavailabilities in swine and the known need for Mn in bone structure in poultry species, most non-ruminant species' diets in the United States have higher Mn concentrations than required by the animal. Another example of an interaction that may affect bioavailability of Mn in turkeys, and thus ultimately affect excretion, was observed when dietary Mn, Cu and Zn concentrations were similar, but protein concentration differed. Male turkeys (32 to 35 week old Nicholas) with femoral fractures were found to have higher femoral concentrations of Mn and lower concentrations of Cu than male turkeys without femoral fractures. While the birds were fed similar concentrations of Ca and P, 14% protein was fed to the birds with no fractures and 8% to those with femoral fractures (Crespo *et al.*, 2002). While this influence of protein on bone health is not surprising, the differing concentrations of Cu and Mn in bone with similar dietary concentrations illustrates why (1) bone adequacy can not be predicted from dietary concentrations and (2) concentrations of minerals in bone can not be used as an indication of bioavailability.

Utilizing the limited data on the Mn/Fe interaction, Creech *et al.* (2004) hypothesized that if dietary Fe and Mn concentrations were reduced, the requirements for Zn and Cu would also be reduced for the pig. Gilts (18 to 21 days of age at weaning) were fed one of three dietary treatments from weaning until completion of the third parity (Creech *et al.*, 2004; Flowers *et al.*, 2001). The dietary treatments were: (1) a control diet with Cu, Zn, Fe and Mn concentrations typical of the swine industry in the United States, (2) reduced inorganic and (3) reduced chelated trace minerals. The control diet minerals were provided in the sulfate form, and the concentrations in the nursery diet were 25 ppm Cu, 150 ppm Zn, 180 ppm Fe and 60 ppm Mn and 15, 100, 100 and 40 ppm, respectively, in the grower, gilt developing, gestation and lactation diets. The reduced mineral diets were supplemented 5, 25, 25, and 10 ppm Cu, Zn, Fe and Mn, respectively in all dietary phases. The reduced inorganic minerals were provided in the sulfate form, and the reduced chelate treatment contained 50% of the Cu, Zn, Fe and Mn as proteinates and the remainder as sulfates. Faecal concentrations of Cu, Zn and Mn were lower when the reduced concentrations were fed regardless of source compared to concentrations typically fed in the United States. The Mn superoxide dismutase (SOD) activity in the kidney was significantly higher in gilts after 136 days of dietary intervention and in the third parity sows when fed the higher Mn concentrations compared to the activity from pigs fed the reduced inorganic minerals. Cardiac Mn concentration was also higher in first parity offspring from sows fed the industry standard diet compared to offspring from sows fed the two reduced mineral diets.

To determine if Mn SOD gene expression in cardiac tissue could be used to determine Mn source bioavailabilities, Luo *et al.* (2007) fed day-old broiler chicks an unsupplemented Mn diet compared to diets supplemented with 120 ppm from Mn sulfate, Mn methionine and two different Mn amino acid sources representing weak, moderate and strong chelation strength. Results indicate that after only seven days, differences in mRNA level due to Mn source could be detected. Thus, this molecular technique could be used to predict Mn bioavailability and lead to the reduction in the feeding of excess Mn.

Copper

While Cu has been fed at pharmacological concentrations for over 60 years in the swine industry to improve growth performance (Hill and Spears, 2000), only in the last 15 years have producers in the United States become concerned about potential environmental damage which appears to be mediated by soil pH. The Creech *et al.* (2004) study provided the first data showing the benefit of reduced mineral intake (5 ppm organic or inorganic) to improve feed efficiency during the nursery phase compared to when Cu was fed at the typical industry concentration (25 ppm). As expected, faecal Cu was reduced by the lower dietary concentrations, but perhaps more importantly an even greater reduction was observed when organic Cu was fed compared to the same dietary concentration of Cu provided as sulfate (71 vs. 108 ppm) during the grower and gilt-developer phases. The Cu/Zn SOD activity in the red blood cell was less in the third parity pigs of sows fed the reduced concentration of Cu. Also, the Cu concentration in the kidney of first parity pigs at 59 days of age and in third parity pigs at 136 days of age was lower when dietary Cu was reduced (Hill *et al.*, 2000).

When 200 ppm of Cu was added as either sulfate or lysine complex to a basal diet with 36 ppm Cu, the absorption and retention of Cu did not differ (Apgar and Kornegay, 1996). With similar growth performance, pigs fed 100 ppm Cu as a proteinate absorbed and retained more Cu but excreted less Cu than pigs fed 250 ppm as sulfate (Veum *et al.*, 2004). In a large commercial setting, nursery pigs fed 125 or 250 ppm Cu as sulfate or 125 ppm as citrate had similar daily gain, feed intake and feed efficiency, but faecal Cu was decreased when pigs consumed 125 ppm Cu citrate compared to 250 ppm Cu sulfate (Armstrong *et al.*, 2004). Thus, if gut microbiota can be altered by Cu sulfate as indicated by polymerase chain reaction and denaturing gradient gel electrophoresis with Cu sulfate (Namkung *et al.*, 2006), pharmacological concentrations with growth benefits may be utilized with reduced faecal Cu excretion when some organic forms of Cu are fed.

Similar to the swine industry, 125 to 250 ppm Cu from Cu sulfate is usually fed in the poultry industry as a growth promoter. With many hypothetical explanations for this response and the known importance of pH in the gastrointestinal tract, Pang and Applegate (2007) studied the effect of Cu source on digesta pH, mineral solubility, soluble complex size in the duodenum and P retention to determine if all forms were effective. To a basal corn-soybean meal diet, they added 250 ppm Cu as sulfate, lysinate or tribasic Cu chloride and found that source did not affect pH of the gizzard, duodenum or jejunum. While the Cu sulfate was more soluble at the pH of 2.5, 5.5 and 6.5 than lysinate with lysinate more soluble than tribasic chloride, it also decreased the least when the three sources were compared. The solubility of supplemental Cu in digesta was approximately 60% and was not affected by source. Only Cu lysinate decreased the solubility of Zn, but all sources increased the Zn associated with large complexes (≥100,000 MW) suggesting the known antagonism between Cu and Zn.

When trace mineral premixes were removed from the diet of nursery pigs, the Cu concentrations in the bone, loin and kidney were reduced (Shelton *et al.*, 2005). Similarly, when vitamin and trace mineral premixes were removed from the diet for either the last 6 or 12 weeks in finisher diets, the Cu concentration in the ham muscle was reduced (Edmonds and Arentson, 2001). However, Shaw *et al.* (2002) did not find a decrease in Cu, Zn or Fe concentrations or Cu/Zn SOD in the longissimus muscle when vitamins and minerals were withdrawn 28 days prior to slaughter. Analyzed dietary concentrations with supplementation were 11.5 ppm Cu, 246.7 ppm Fe, 20.3 ppm Mn and 146.6 ppm Zn. Thus, like the Creech *et al.* (2004) industry control diet, all minerals were supplied at concentrations beyond those suggested by NRC (1998). Perhaps more importantly, Cu, Zn, Fe and Mn faecal concentrations were reduced from 50 to 75% by the removal of trace minerals and vitamins 28 days prior to slaughter. These reductions in faecal trace mineral concentrations were reflective of the changed analyzed dietary content that occurred due to the removal.

Zinc

For approximately 20 years, producers in the United States have been adding pharmacological Zn to nursery diets (2,000 to 3,000 ppm Zn from oxide) to enhance performance (Hill and Spears, 2000). Various modes of action have been proposed such as control of *E. coli,* improvement of gut morphology (Carlson *et al.*, 1998) and microbiota and stimulation of metallothionein (Mt) production (Carlson *et al.*, 1998). Even pigs challenged with gastrointestinal disease such as transmissible gastroenteritis (TGE) have been found to have improved gut health and growth when fed 3,000 ppm Zn as Zn oxide (Stanger *et al.*, 1998). A multi-state research project revealed that with varying genetics and management strategies, 2,000 ppm Zn from Zn oxide was as effective for growth promotion as 3,000 ppm Zn (Hill *et al.*, 2001). Feeding pharmacological Zn for 10 to 14 days has been shown to be the minimum time necessary to stimulate growth and improve feed efficiency (Carlson *et al.*, 1999). By weaning pigs directly to metabolism cages, feeding 2,000 ppm Zn as oxide or methionine and collecting all feces and urine daily for 14 days, Rincker *et al.* (2005a) were able to show that for approximately 10 days faecal Zn is similar to that of pigs fed 150 ppm Zn. It appears that during this time, the body Zn loads and then faecal Zn is increased. Thus, to increase growth and minimize faecal Zn excretion, most swine producers in the United States feed pharmacological Zn as Zn oxide in the first two dietary phases after weaning. To further reduce the amount of Zn excreted with this management practice, researchers have studied various strategies such as using phytase to increase Zn bioavailability (Martinez *et al.*, 2004). Intestinal mucosa Mt protein and renal and intestinal mucosa relative Mt mRNA abundance were greater in pigs fed 2,000 ppm Zn plus 500 FTU/kg phytase compared to pigs fed 150, 1,000 or 2,000 ppm Zn as Zn oxide. To determine if organic Zn sources would result in decreased faecal excretion, nursery pigs were fed 2,000 pm Zn as oxide or methionine. While urinary Zn was increased with Zn methionine, Zn retention and faecal excretion were similar (Rincker *et al.*, 2002). Whole body Zn concentrations were also similar (Rincker *et al.*, 2003). In comparing the excretion of pigs fed 100, 1,000 or 4,000 ppm Zn from Zn oxide, pigs fed 4,000 ppm had increased urinary and decreased faecal Cu when the treatments were fed for 21 days compared to pigs fed the lower concentrations of Zn (Martinez *et al.*, 2005). The Mt concentration in the duodenum was similar for pigs fed either 1,000 or 4,000 ppm Zn, but the jejunum Mt was similar for pigs fed 100 and 1,000 ppm Zn, but significantly higher when 4,000 ppm Zn was fed as an oxide. Certainly this data indicates that Mt controls of Zn metabolism are different throughout the Zn absorptive area of the gastrointestinal tract giving researchers another opportunity to reduce faecal Zn when pharmacological Zn is fed to the nursery pig.

Organic Zn fed as a polysaccharide or a proteinate with concentrations from 125 to 800 ppm in a basal diet with 165 ppm Zn as sulfate did not improve growth performance when compared to 2,000 ppm Zn as Zn oxide. As expected, faecal Zn was decreased at these lower inclusion rates of organic minerals. Percent absorbed did not differ by treatment, but the amount absorbed per day (mg/d) was significantly higher for 250, 375 and 500 ppm Zn polysaccharide compared to 2,000 ppm Zn as Zn oxide (Carlson *et al.*, 2004). When feeding 300 ppm Zn as Zn polysaccharide, Zn excretion decreased by 76% compared to feeding 2,000 ppm Zn as Zn oxide (Buff *et al.*, 2005).

In the research of Creech *et al.* (2004) with reduced dietary mineral concentrations, faecal Zn was reduced when organic chelated Zn (25 ppm) was fed compared to 100 or 150 ppm Zn supplied as Zn sulfate. Offspring at 59 days of age from the first parity had higher Zn concentrations in the liver, kidney and muscle when the gilts and their offspring were fed 100 and 150 ppm of Zn as sulfate compared to those fed the reduced Zn diets (25 ppm) in the sulfate or chelate forms. At 136 days of age, pigs from the third parity still retained these differences in hepatic and renal tissue (Hill *et al.*, 2000). This research was the first to report that when reduced concentrations of Cu, Zn, Fe and Mn were fed throughout the life cycle, there was no reduction in maternal performance through three parities, and that a greater proportion of sows fed the reduced mineral treatments (inorganic and organic) farrowed in parity three than parity two compared to those fed the industry standard diet. Litter weaning weights were also higher for the sows on the reduced dietary trace element diets compared to those fed the higher concentration of minerals. These findings have many implications for the swine industry including: (1) reduction of feed ingredient costs, (2) reduced faecal mineral concentrations and (3) improved productivity in sows beyond the first parity.

Crop scientists are striving to provide grains with lower phytate content to improve the availability of many minerals required by non-ruminant animals. Recently, broiler chicks have been shown to utilize dietary Zn and P when low phytate barley is fed compared to wild type barley even if the endogenous phytase in the barley has been destroyed (Linares *et al.*, 2007). Performance was not affected by barley type. Tibia and toe Zn concentrations were not increased by the addition of 10 or 20 ppm Zn in the low phytate barley, but the percent retained was decreased. However, tibia and toe Zn concentrations increased with 10 or 20 ppm Zn in the diets containing the wild type barley as percent retained also increased. Thus, the bioavailability of Zn in barley appears to be related to more than the endogenous phytate and phytase content and once again indicates that the Zn content of bone is not a good indicator of Zn bioavailability.

Iron

Indicating that organic Fe is more available to the young pig than inorganic Fe (sulfate), haemoglobin concentrations were significantly higher at the end of the nursery portion of the study of Creech *et al.* (2004) when the same concentration of Fe was fed (25 ppm) and was similar to the haemoglobin concentration of pigs fed 180 ppm Fe. The ability to control the absorption of Fe is essential for homeostasis. Thus, it is interesting to find that sows after the completion of the third parity had higher hepatic Fe concentrations when fed 25 ppm Fe from sulfate since weaning than sows fed 180 (nursery) and 100 ppm for the remaining phases. It is important to note that the NRC recommendation ranges from 80 to 40 ppm Fe as the age of the animal advances. While many feed ingredients contain Fe that are used in swine diets, the bioavailability is usually very limited if of plant or mineral source origin (Rincker *et al.*, 2005b).

Iron regulatory proteins (IRP) have recently been shown to be influenced by the dietary Fe source (Rincker *et al.*, 2005a) and may be useful in determining the changing Fe requirement of the pig and to reduce faecal Fe excretion. Increasing dietary Fe from 0 to 150 ppm resulted in an increase in faecal Fe while there was a linear increase in haemoglobin, hematocrit and plasma Fe (Rincker *et al.*, 2005b) and average daily gain responded in a linear increase from day 7 to 21 post weaning (Rincker *et al.*, 2004). This data indicates that nursery pigs raised without access to soil and with a rapid muscle accretion pattern have a greater need for bioavailable Fe than pigs raised 30 to 40 years ago from which the 100 ppm Fe NRC requirement was established. However, it is extremely important that organic Fe sources be fed to a similar genetic population as used in production in the United States to determine if faecal Fe excretion can be reduced as Fe homeostasis needs are met.

Conclusion

The trace elements Cu, Zn, Fe and Mn are essential for health and productivity of non-ruminant animals yet it is essential that we prevent the contamination of our environment. These four elements interact with each other so it is appropriate that all be considered when one or more elements are being increased or decreased in the diet. New and traditional laboratory techniques have enabled researchers to further investigate the biological utilization of these elements for the maintenance of homeostasis in meeting the animal's needs. Changes in the parameters measured can not always be deemed positive or negative since the animal is able to utilize these elements over a fairly wide range of dietary concentrations. Additionally, there is limited data available for the interpretation of data resulting from the use of newer laboratory techniques. It is clear that improved villus height and crypt depth gained from supplementation of pharmacological Zn are important for the health of the animal. Recent data indicates that some organic mineral sources may be more available to the animal, and thus may provide the opportunity to reduce environmental contamination and increase productivity.

References

Apgar, G.A. and E.T. Kornegay, 1996. Mineral balance of finishing pigs fed copper sulfate or a copper-lysine complex at growth-stimulating levels. J. Anim. Sci. 74, 1594-1600.

Armstrong, T.A., D.R. Cook, M.M. Ward, C.M. Williams and J.W. Spears, 2004. Effect of dietary copper source (cupric citrate and cupric sulfate) and concentration on growth performance and fecal copper excretion in weanling pigs. J. Anim. Sci. 82, 1234-1240.

Buff, C.E., D.W. Bollinger, M.R. Ellersieck, W.A. Brommelsiek and T.L. Veum, 2005. Comparison of growth performance and zinc absorption, retention, and excretion in weanling pigs fed diets supplemented with zinc-polysaccharide or zinc oxide. J. Anim. Sci. 83, 2380-2386.

Carlson, M.S., C.A. Boren, C. Wu, C.E. Huntington, D.W. Bollinger and T.L. Veum, 2004. Evaluation of various inclusion rates of organic zinc either as polysaccharide or proteinate complex on the growth performance, plasma, and excretion of nursery pigs. J. Anim. Sci. 82, 1359-1366.

Carlson, M.S., G.M. Hill and J.E. Link, 1999. Early- and traditionally weaned nursery pigs benefit from phase-feeding pharmacological concentrations of zinc oxide: effect on metallothionein and mineral concentrations. J. Anim. Sci. 77, 1199-1207.

Carlson, M.S., S.L. Hoover, G.M. Hill, J.E. Link and J.R. Turk, 1998. Effect of pharmacological zinc on intestinal metallothionein concentration and morphology in nursery pig. J. Anim. Sci. 76 (Suppl. 2), 53 (Abstr.).

Creech, B.L., J.W. Spears, W.L. Flowers, G.M. Hill, K.E. Lloyd, T.A. Armstrong and T.E. Engle, 2004. Effect of dietary trace mineral concentration and source (inorganic vs. chelated) on performance, mineral status, and fecal mineral excretion in pigs from weaning through finishing. J. Anim. Sci. 82, 2140-2147.

Crespo, R., S.M. Stover, H.L. Shivaprasad and R.P. Chin, 2002. Microstructure and mineral content of femora in male turkeys with and without fractures. Poult. Sci. 81, 1184-1190.

Crocker, A.W. and O.W. Robison, 2002. Genetic and nutritional effects on swine excreta. J. Anim. Sci. 80, 2809-2816.

Edmonds, M.S. and B.E. Arentson, 2001. Effect of supplemental vitamins and trace minerals on performance and carcass quality in finishing pigs. J. Anim. Sci. 79, 141-147.

Finley, J.W., J.S. Caton, Z. Zhou and K.L. Davison, 1997. A surgical model for determination of true absorption and biliary excretion of manganese in conscious swine fed commercial diets. J. Nutr. 127, 2334-2341.

Flowers, W.L., J.W. Spears and G.M. Hill, 2001. Effect of reduced dietary Cu, Zn, Fe, and Mn on reproduction performance of sows. J. Anim. Sci. 79 (Suppl. 2), 61 (Abstr.).

Hill, G.M., J.E. Link, J.W. Spears and W.L. Flowers, 2000. Impact of reduced dietary trace minerals on mineral and anti-oxidant status in swine. J. Anim. Sci. 78 (Suppl. 1), 175 (Abstr.).

Hill, G.M., D.C. Mahan, S.D. Carter, G.L. Cromwell, R.C. Ewan, R.L. Harrold, A.J. Lewis, P.S. Miller, G.C. Shurson and T.L. Veum, 2001. Effect of pharmacological concentrations of zinc oxide with or without the inclusion of an antibacterial agent on nursery pig performance. J. Anim. Sci. 79, 934-941.

Hill, G.M. and J.W. Spears, 2000. Trace and ultratrace elements in swine nutrition. In: L.L. Southern and A.J. Lewis (Eds.), Swine Nutrition. CRC Press, Boca Raton, 229-261.

Linares, L.B., J.N. Broomhead, E.A. Guaiume, D.R. Ledoux, T.L. Veum and V. Raboy, 2007. Effects of low phytate barley (Hordeum vulgare L.) on zinc utilization in young broiler chicks. Poult. Sci. 86, 299-308.

Luo, X.G., S.F. Li, L. Lu, B. Liu, X. Kuang, G.Z. Shao and S.X. Yu, 2007. Gene expression of manganese-containing superoxide dismutase as a biomarker of manganese bioavailability for manganese sources in broilers. Poult. Sci. 86, 888-894.

Martinez, M.M., G.M. Hill, J.E. Link, N.E. Raney, R.J. Tempelman and C.W. Ernst, 2004. Pharmacological zinc and phytase supplementation enhance metallothionein mRNA abundance and protein concentration in newly weaned pigs. J. Nutr. 134, 538-544.

Martinez, M.M., J.E. Link and G.M. Hill, 2005. Dietary pharmacological or excess zinc and phytase effects on tissue mineral concentrations, metallothionein, and apparent mineral retention in the newly weaned pig. Biol. Trace Elem. Res. 105, 97-115.

Mohanna, C. and Y. Nys, 1998. Influence of age, sex and cross on body concentrations of trace elements (zinc, iron, copper and manganese) in chickens. Br. Poult. Sci. 39, 536-543.

Namkung, H., J. Gong, H. Yu and C.F.M. de Lange, 2006. Effect of pharmacological intakes of zinc and copper on growth performance, circulating cytokines and gut microbiota of newly weaned piglets challenged with coliform lipopolysaccharides. Can. J. Anim. Sci. 86, 511-522.

NRC, 1998. Nutrient Requirements of Swine. Natl. Acad. Press, Washington, DC.

Pang, Y. and T.J. Applegate, 2007. Effects of dietary copper supplementation and copper source on digesta pH, calcium, zinc, and copper complex size in the gastrointestinal tract of the broiler chicken. Poult. Sci. 86, 531-537.

Rincker, M., G.M. Hill, J.E. Link, J.E. Rowntree, D.M. Dvoracek-Driksna and J.G. Green, 2002. Comparison of organic vs. inorganic sources of Zn supplementation on zinc retention in young pigs. Presented at the 11th International Symposium on Trace Elements in Man and Animals (TEMA).

Rincker, M.J., S.L. Clarke, R.S. Eisenstein, J.E. Link and G.M. Hill, 2005a. Effects of iron supplementation on binding activity of iron regulatory proteins and the subsequent effect on growth performance and indices of hematological and mineral status of young pigs. J. Anim. Sci. 83, 2137-2145.

Rincker, M.J., G.M. Hill, J.E. Link, A.M. Meyer and J.E. Rowntree, 2005b. Effects of dietary zinc and iron supplementation on mineral excretion, body composition, and mineral status of nursery pigs. J. Anim. Sci. 83, 2762-2774.

Rincker, M.J., G.M. Hill, J.E. Link and J.E. Rowntree, 2004. Effects of dietary iron supplementation on growth performance, hematological status, and whole-body mineral concentrations of nursery pigs. J. Anim. Sci. 82, 3189-3197.

Rincker, M.J., G.M. Hill, J.E. Link, J.E. Rowntree, J.G. Green and D.M. Dvoracek-Driksna, 2003. Effects of organic vs. inorganic sources of zinc supplementation on the whole body mineral composition (Cu, Zn, Fe, Mn, P, and N) of nursery pigs. J. Anim. Sci. 81 (Suppl. 2), 85 (Abstr.).

Shaw, D.T., D.W. Rozeboom, G.M. Hill, A.M. Booren and J.E. Link, 2002. Impact of vitamin and mineral supplement withdrawal and wheat middling inclusion on finishing pig growth performance, fecal mineral concentration, carcass characteristics, and the nutrient content and oxidative stability of pork. J. Anim. Sci. 80, 2920-2930.

Shelton, J.L., F.M. LeMieux, L.L. Southern and T.D. Bidner, 2005. Effect of microbial phytase addition with or without the trace mineral premix in nursery, growing, and finishing pig diets. J. Anim. Sci. 83, 376-385.

Stanger, B.R., G.M. Hill, J.E. Link, J.R. Turk, M.S. Carlson and D.W. Rozeboom, 1998. Effect of high Zn diets on TGE-challenged early-weaned pigs. J. Anim. Sci. 76 (Suppl. 2).

Underwood, E.J. and N.F. Suttle, 1999. The Mineral Nutrition of Livestock. CABI Publishing, New York.

Veum, T.L., M.S. Carlson, C.W. Wu, D.W. Bollinger and M.R. Ellersieck, 2004. Copper proteinate in weanling pig diets for enhancing growth performance and reducing fecal copper excretion compared with copper sulfate. J. Anim. Sci. 82, 1062-1070.

Zinc and copper for piglets – how do high dietary levels of these minerals function?

H.D. Poulsen and D. Carlson

Keywords: zinc, copper, piglets, diarrhoea, performance, interaction, environmental impact

Introduction

Zinc (Zn) and copper (Cu) are essential trace elements for animals affecting health and productivity. Zinc deficiency results in e.g. growth retardation, anaemia, skin diseases and impaired wound healing (Cousins, 1985; McDowell, 2003). Furthermore, hypozincemia has been reported in a variety of diseases including diarrhoea, anorexia, skin diseases and impaired wound healing (McDowell, 2003). Copper is involved in the processes concerning e.g. iron utilization and the synthesis of connective tissue (McDowell, 2003). Newly weaned piglets have a low feed intake and have to adapt to a lot of changes at weaning (nutritional, environmental, loss of intestinal immune protection provided by the sows' milk, etc.). These changes may result in alterations in the morphology and function of the gastrointestinal tract and in the intestinal microbial balance providing an opportunity for pathogens to colonize and cause diseases, like diarrhoea, that may result in poor growth and ultimately in death. Furthermore, the change from milk to solid feed (mainly plant-based) imposes an urgent need for quick intestinal adaptations to the new feed components in order to maintain a sufficient uptake of the plant-borne nutrients.

This may call for special attention to the need for Zn and Cu and a better understanding of their functions in piglets around weaning (Poulsen, 1995). The Zn body stores of piglets are very small resulting in a limited labile Zn pool, whereas the liver Cu stores are high (McDowell, 2003). Overall, it is important to ensure a daily intake of sufficient Zn to meet the physiological demand as insufficient intake of Zn is reported to result in growth retardation, bacterial infections and diarrhoea.

Effects on microflora

Many studies over the last two to three decades have shown that Zn has a preventive effect on diarrhoea when piglets are fed high dietary concentrations of Zn (e.g. Poulsen, 1989, 1995; Hahn and Baker, 1993; Carlson *et al.*, 1999; Hill *et al.*, 2001). Studies have also shown that high dosages of Zn stimulate feed intake and growth of piglets whether they were suffering from diarrhoea or not. Although some studies have revealed no or small effects of Zn, the overall picture is that high dietary levels of Zn fed as zinc oxide (ZnO) may reduce the problems around weaning in young piglets. However, the mechanisms behind the beneficial effects of Zn are still not well understood although several hypotheses have been proposed and addressed. Katouli *et al.* (1999) reported that high dietary Zn supports a large diversity of coliforms in weaned piglets, and Mores *et al.* (1998) found that Zn reduced the pigs´ susceptibility to *E. coli* infections. This was supported by an *in vitro* study indicating that ZnO protects intestinal cells from *E. coli* infections by inhibiting the adhesion and internalization of these bacteria, rather than having a direct antibacterial effect (Roselli *et al.*, 2003). However, Jensen-Waern *et al.* (1998) found that feeding high dietary Zn temporally reduced the faecal counts of enterococci three days after weaning without affecting the number of *E. coli*.

For many years it has been common practice to include extra Cu in diets for weaned piglets as a lot of studies have shown that Cu exerts a growth stimulating effect (Braude, 1967). Many studies have focused on either the growth stimulating effect of Cu on daily gain/FCR or on the effects of Cu on the microbial ecosystem in the gut mainly by studying the effects on the faecal microflora. As such, Fuller *et al.* (1960) reported that high dietary Cu (250 mg/kg as CuSO$_4$) specifically reduced the number of streptococci in faeces. The number of lactobacilli has also been shown to decrease in pigs fed high Cu diets (170-250 mg Cu/kg, as CuSO$_4$) (Jensen, 1998; Kellogg *et al.*, 1964), whereas the coliforms remained unaffected or indeed tended to increase (Fuller *et al.*, 1960; Jensen, 1998). Shurson *et al.* (1990) found positive effects of high Cu diets on growth rate in conventionally-reared pigs and tendencies for negative effects of the same dietary Cu concentration in germ-free pigs. However, no data has so far definitely linked the growth stimulating effect of Cu to the effects on the microflora. Despite this, it is generally accepted that the action of Cu is attributed to its antimicrobial activity.

Recently, a study compared the interactive effects of high and low dietary Zn and Cu levels in a factorial design (Højberg *et al.*, 2005). They found that high dietary Zn (2500 mg Zn/kg, as ZnO) reduced the bacterial activity in the digesta from the gastrointestinal tract in newly weaned piglets compared with piglets fed low Zn (100 mg Zn/kg). High Zn resulted in a reduced number of lactic acid bacteria and lactobacilli whereas the coliforms and enterococci were more numerous in animals fed high Zn compared with low dietary Zn. It was stated that the influence of Zn on the gastrointestinal microflora resembles the working mechanisms suggested for some growth-promoting antibiotics by suppression of gram-positive rather than potential pathogenic gram-negative organisms (Gaskins *et al.*, 2002). Therefore, the reduced fermentation of digestible nutrients in the gastrointestinal tract may render more energy available for the piglet and contribute to the growth stimulating effect of high Zn doses. The same study found that high doses of Cu inhibited the coliforms but the overall observed effect of Cu was limited compared with that of Zn (Højberg *et al.*, 2005). However, it is generally accepted that lactic acid producing bacteria are a part of the natural defence, and consequently the observed Zn effects may not fully support the hypothesis of an antimicrobial effect affecting the diarrhoea inducing microorganisms.

Physiological effects

The first study on high dietary inclusion of ZnO (1000, 2500 or 4000 mg Zn/kg) reported that the increased performance and reduced frequency of diarrhoea was accompanied with an increase in plasma Zn concentration (Poulsen, 1995). This relationship was confirmed in later studies (e.g. Carlson *et al.*, 1999, 2007b). On the contrary, high dietary Cu intake post-weaning (175 to 250 mg Cu/kg) does not to the same extent increase plasma Cu concentrations

(Veum *et al.*, 2004; Carlson *et al.*, 2007b). These observations led to the working hypothesis suggesting that high dietary Zn has a physiological effect in the young weaned piglet and that Zn exerts its effect via a general improvement in the Zn status of the piglets, whereas the main effect of high dietary Cu may be explained by a more local effect in the lumen of the intestinal tract, e.g. an antimicrobial effect.

Zinc is a component of more than 200 metalloenzymes and in addition, Zn increases the activity of a great number of other enzymes (Prasad, 1969; Cousins, 1985). In pigs, improved Zn status has often been measured by increased activity of the Zn-dependent enzyme alkaline phosphatase (AP) in serum (Prasad *et al.*, 1971; Swinkels *et al.*, 1996). In piglets fed high dietary Zn (2500 mg/kg) concentrations post-weaning, plasma AP activity increased in a number of studies (Poulsen, 1989, 1995; Feng *et al.*, 2006; Carlson *et al.*, 2007b), which may indicate improved Zn status of these piglets. However, plasma AP is also influenced by e.g. bone health and animal growth and as such it is not a specific marker for Zn status.

It has been hypothesized that supplementing pig diets with high dietary concentrations of Zn and Cu would stimulate the synthesis of digestive enzymes, resulting in a better digestion and absorption of nutrients (Hedemann *et al.*, 2006). However, the study with weaned piglets fed the high dietary Zn and Cu treatments for 14 days after weaning did not consistently support this hypothesis. It was found that 2500 mg Zn/kg significantly increased the amylase, carboxypeptidase A, chymotrypsin, trypsin and lipase activity in pancreatic tissue (Hedemann *et al.*, 2006). However, the activity of the same enzymes in the small intestinal content was either reduced or unaffected by the dietary Zn treatment. The mucosal enzyme activity was not affected by dietary treatments except maltase that was significantly reduced due to the Zn treatment (Hedemann *et al.*, 2006).

Cell type specific localization of metallothionein (MT) in the small intestine of rats indicates a function of Zn and MT in intestinal mucosal turnover (Szczurek *et al.*, 2001). Tran *et al.* (1999) and Carlson *et al.* (1999) suggested that MT provides Zn for mucosal cell function, differentiation, growth and repair. Consequently, the positive effect of dietary Zn on post-weaning piglets may be a secondary effect of improved mucosal growth and cell function, because of increased Zn held in the intestinal cells by MT. Recently, it was found that MT mRNA abundances increased in intestinal epithelium of piglets fed 2500 mg Zn/kg from ZnO compared to 100 mg/kg for 5-7 days after weaning (Carlson *et al.*, 2007b), whereas MT protein was only numerically increased after a similar feeding period in another study (Feng *et al.*, 2006). But if high Zn concentrations are fed for 1 or 2 weeks after weaning the MT protein concentrations in intestinal epithelium have been found to increase (Carlson *et al.*, 1999; Martinez *et al*, 2004). These discrepancies may confirm that MT gene expression does not always imply the presence of the MT protein (Vasconcelos *et al.*, 2002). In addition, the results may indicate that piglets should be fed a diet with 2500 mg Zn/kg for more than 5 to 6 d to increase epithelial MT protein concentrations (Carlson *et al.*, 2007b). The liver MT concentration has been found to increase 3 to 4 times when pigs are fed 2500-3000 mg Zn/kg for 5-7 day (Carlson *et al.*, 1999; Feng *et al.*, 2006).

Insulin-like growth factor-I (IGF-I) is a central hormone in growth regulation and an association between high dietary Zn and increased serum IGF I levels was found at 5 and 14 days after weaning (Carlson *et al.*, 2004). In the same study there was no effect of diets with a high Cu supplementation (175 mg/kg) on serum IGF-I concentrations. A recent study found increased protein and mRNA levels for IGF-I in the small intestinal mucosa in Zn-supplemented piglets 2 weeks after weaning (Li *et al.*, 2006). The intestinal mucosa from the same piglets had increased villous heights, perhaps indicating a positive relationship between high dietary Zn concentrations, IGF-I and intestinal morphology. However, conflicting results exist regarding the effect of high dietary Zn concentrations on intestinal morphology. Li *et al.* (2001) found that villous height increased and crypt depth decreased throughout the small intestine at 11 days after weaning in pigs fed 3000 mg Zn/kg as ZnO. Li *et al.*, (2006) found increased villous height but unchanged crypt depth in piglets fed 3000 mg Zn/kg (ZnO) 14 days after weaning, whereas Hedemann *et al.* (2006) found no effect of 2500 mg Zn/kg (ZnO) on any of these parameters 14 days after weaning. Of these studies, only Li *et al.* (2006) found significant effects on growth rate and feed intake. Copper supplementation (200 mg/kg) reduced crypt depth in the duodenum of piglets 10 days after weaning and tended to increase villous height in jejunum (Zhao *et al.*, 2007). In the latter study, growth rate and feed intake were also increased due to the dietary Cu treatment. Because increased feed intake improves intestinal morphology (Pluske *et al.*, 1997) after weaning, it may be suggested that the effect of Cu on intestinal morphology found by Zhao *et al.* (2007) was an indirect effect of increased feed intake rather than a direct effect of Cu per se. Furthermore, as villous height normally reaches pre-weaning values at 9 days post-weaning (Hedemann *et al.*, 2003) it is possible that the lack of response to dietary Zn at 14 days after weaning was a result of measurements being made after intestinal conditions had stabilized (Hedemann *et al.*, 2006).

The underlying mechanisms behind the preventive effect of Zn supplementation on post-weaning diarrhoea have until recently been largely unknown. But during the last years a few studies have been made to address this issue and some possible mechanisms have come into view. A possible mechanism could be that Zn may have an effect on intestinal permeability. In a study with guinea pigs, a low protein diet was associated with increased intestinal permeability to electrolytes and small molecules, but high doses of Zn were able to prevent this increase in permeability (Rodriguez *et al.*, 1996). The effect of weaning on intestinal permeability in pigs is ambiguous as in some studies the intestinal permeability was reduced (Carlson *et al.*, 2004), whereas in others the permeability was increased after weaning (Spreeuwenberg *et al.*, 2001). However, in accordance with the guinea pig studies (Rodriguez *et al.*, 1996) *in vitro* studies indicate that adding Zn to the bathing media reduces intestinal permeability across the intestinal epithelium in piglets fed 100 or 2500 mg Zn/kg diet the first 5-7 days after weaning (Carlson *et al.*, 2006; Feng *et al.*, 2006).

In addition to a possible effect on intestinal permeability Zn may affect the severity of pathogenic induced water and ion secretion. When pathogenic bacteria secrete toxins in the intestinal lumen the neurotransmitter 5-HT is released from special cells located in the intestinal epithelium and several receptors at the epithelial cells are activated resulting in water and chloride secretion (Skadhauge *et al.*, 1997). Furthermore, 5-HT induces the release of other neurotransmitters (e.g. vasoactive intestinal peptide, VIP) from enteric nerve endings which elevates intracellular messengers (e.g. cAMP) followed by water and chloride secretion (Barrett and Keely, 2000). An *in vitro* study (Ussing chambers) with small intestinal epithelium from piglets fed high or low Zn and/or high or low Cu revealed that weaning resulted in increased ileal glucose absorption as well as increased neuroendocrine-regulated and cAMP-dependent chloride secretion that may result into diarrhoea. However, high Zn supplementation (2500 mg Zn/kg as ZnO) reduced the responses to these secretagogues (5-HT and theophylline), but the study did not reveal any influence of Cu (175 mg Cu/kg as $CuSO_4$) (Carlson *et al.*, 2004). As such, the positive effect of high dietary Zn may be due to Zn reducing the intestinal mucosal susceptibility to secretagogues that activate chloride and water secretion. Another *in vitro* study suggests that dietary Zn reduces diarrhoea directly through a regulatory role of serosal Zn on chloride secretion and indirectly by improving the nutritional status which may stabilize the function of the intestinal epithelium (Carlson *et al.*, 2007a). A factor involved in stabilising the intestinal epithelium may be the mast cells of the intestinal epithelium. Mast cells contribute to diarrhoea through the production and release of e.g. histamine and 5-HT and the number of mast cells was found to be reduced in piglets fed 3000 mg Zn/kg for 10 days post-weaning (Ou *et al.*, 2007). Consequently, the intestinal immune system may also be altered due to high dietary Zn concentrations post weaning. However, going further into the effect of Zn on the immune system will be beyond the scope of this paper.

Zinc seems to exert its attenuating effect on secretion from the basolateral side of the intestinal epithelium (Carlson *et al.*, 2007a), which may indicate that dietary zinc needs to be absorbed and circulated in the blood before it is able to reduce secretion. *In vitro* studies with rat intestinal epithelium by Hoque *et al.* (2005) also found that zinc at the serosal side reduces the c-AMP-dependent chloride secretion caused by forskolin. Furthermore, Feng *et al.* (2006) concluded that Zn attenuates ion secretion through an effect on the epithelial cells rather than an effect on the enteric nervous system situated in the epithelial submucosa. Carlson *et al.* (2006) found that Zn only had an inhibitory effect on secretory responses to secretagogues that act directly through receptors situated at the basolateral membrane of the epithelial cells. From the above mentioned findings, a theoretical model of the mode of action of Zn on intestinal secretion is suggested and illustrated in Figure 1.

These new findings emphasise that the effects of Zn on post-weaning diarrhoea are partially exerted from the serosal side of the epithelial cells of the intestinal wall which means that it is a physiological effect achieved when the Zn status of the piglets is ensured by a sufficient absorption of Zn. These findings indicate that the physiological Zn demand may, due to the limited feed intake, solely be fulfilled if the Zn concentration in the diet is high. This is also supported by the improved weight gain of piglets receiving high dietary Zn.

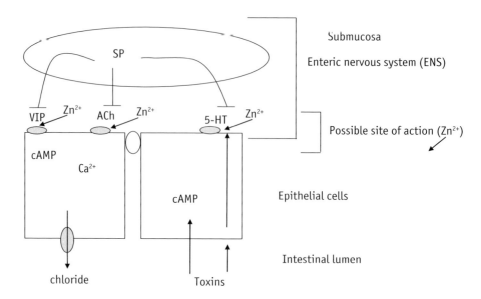

Figure 1. Hypothesized model of the suggested mechanisms behind the secretory effect of the secretagogues serotonin (5-HT), Substance P (SP), Acetylcholone (Ach) and Vasoactive Intestinal Peptide (VIP) and the possible site for the inhibitory effect of zinc on these processes. Toxins from intestinal pathogenic bacteria trigger the 5-HT release, which activates chloride secretion through a variety of epithelial receptors and mediators, including the release of acetylcholine and VIP from enteric nerve endings. VIP and Ach bind to their receptors, at the basolateral membrane and intracellular cAMP and Ca^{2+} concentrations increase, respectively. Ca^{2+} and cAMP activates chloride channels resulting in chloride secretion. SP mediates its effects through secondary cell types. The Zn^{2+} with arrows illustrate that the inhibitory mechanism of zinc ions may take place at the receptors situated at the basolateral membrane.

Environmental impacts

The specialization within modern pig production may result in large production units which may not have sufficient cropland to make use of all the nutrients in the manure without accumulation in the soil. This may result in leaching and thus pollution of e.g. lakes and streams. In the past, attention was paid to nitrogen and phosphorus but now heavy metals like Zn and Cu are also included in this environmental concern. As such, feeding high dietary concentrations of Zn and Cu may be in disharmony with the environment. Simulating different feeding scenarios have shown that the accumulation of Zn and Cu in the soil may reach critical levels within 75 to 100 years if high Zn (2,500 mg Zn/kg as ZnO) is included in the diets for weaned piglets for two weeks after weaning and about 110 years when applying high Cu (175 mg Cu/kg) in intensive pig producing areas in Denmark (Poulsen, 1998).

Due to these environmental considerations, emphasis has been put on strategies for lowering the excretion of Zn and Cu. This includes assessment of the requirement of Zn and Cu in weaned piglets in relation to diet composition and phytase activity (Revy *et al.*, 2006).

Another approach is to use organic Zn and Cu sources instead of inorganic sources that are supposed to have a lower bioavailabiliLy. These studies have shown contradictory results but some studies have found that lower dietary Zn levels are needed when organic Zn sources are used (e.g. Case and Carlson, 2002). However, the conflicting results may imply that the effects may not solely be related to differences in Zn bioavailability as the availability may not differ much between organic and inorganic Zn sources when expressed in absolute and not in relative terms (Revy *et al.*, 2002; Poulsen and Larsen, 1995). As such the availability of organic and inorganic Zn sources was almost similar (approx. 25%). In general, the nutrients should be available in appropriate amounts at the right time and the right position of the GI-tract so that the net absorbed amount is sufficient to meet the requirement of the piglet.

Concluding remarks

Fundamentally, the Zn demand in young fast-growing animals is high, and the labile Zn pool in the body is very limited so that a daily uptake is needed. As such, piglets have a daily requirement for Zn which should be provided by the diet. This means that weaning diets must contain sufficient available Zn to fulfil this daily requirement in order to maintain the physiological processes properly to ensure growth, health and wellbeing of the young fast-growing piglets. Consequently, nutritionists should – when defining the necessary Zn concentration in the diets – adjust the dietary Zn concentration to (1) the limited feed intake of the piglets during the first days after weaning, (2) the bioavailability of Zn in the feedstuffs, (3) the bioavailability of Zn in the supplementing sources, and (4) interactive effects of e.g. anti-nutritional components in the diets like phytate that may interfere with the availability and uptake of Zn. Due to these factors, the concentration of Zn in diets for newly weaned piglets may be rather high the first days after weaning in order to compensate for the low daily feed intake and a low bioavailability of Zn in feedstuffs as well as in the Zn-supplying sources. It could be speculated that many of the health problems like diarrhoea associated with the weaning of piglets in modern piggeries are caused by (marginal) Zn deficiency of the piglets and that high dietary Zn has to be supplied to alleviate these problems which was previously controlled via the use of antibiotic growth promoters (AGPs) (until 2006 within EU). This statement is supported by the recent findings that the effects of Zn on the microflora resemble the effects seen when AGPs are amended to pig diets. So far, the effects of high Cu in piglets could not be definitely related to either an effect on the microflora or a physiological effect.

In conclusion, there is a need for a balanced approach taking proper care of a correct physiological feeding of piglets around weaning without medication and a proper environmental carefulness. Zinc and Cu feeding strategies and sources should be defined so that it will be possible at the same time to (1) meet the physiological requirements to ensure proper health and performance and (2) minimize the excretions and the soil accumulations. The overall goal is to ensure a high performing healthy and ethical sound pig production (without AGPs) in balance with environmental concerns.

References

Barrett, K.E. and S.J. Keely, 2000. Chloride secretion by the intestinal epithelium: Molecular basis and regulatory aspects. Annu. Rev. Physiol. 62, 535-572.

Braude, R., 1967. Copper as a stimulant in pig feeding (*cuprom pro pecunia*). World. Rev. Anim. Prod. 3, 69-82.

Carlson, D., H.D.Poulsen and M.Vestergaard, 2004. Additional dietary zinc for weaning piglets is associated with elevated concentrations of serum IGF-I. J. Anim. Physiol. a. Anim. Nutr. 88, 332-339.

Carlson, D., J. Sehested, Z. Feng and H.D. Poulsen, 2007a. Zinc is involved in regulation of secretion from intestinal epithelium in weaned piglets. Livest. Sci.. 108, 45-48.

Carlson, D., J.H.Beattie and H.D. Poulsen, 2007b. Assessment of zinc and copper status in weaned piglets in relation to dietary zinc and copper supply. J. Anim. Physiol. and Anim. Nutr. 91, 19-28.

Carlson, D., H.D. Poulsen and J. Sehested, 2004. Influence of weaning and effect of post weaning dietary zinc and copper on electrophysiological response to glucose, theophylline and 5-HT in piglet small intestinal mucosa. Comp. Biochem. Physiol. A. 137, 757-765.

Carlson, D., J. Sehested and H.D. Poulsen, 2006. Zinc reduces the electrophysiological responses *in vitro* to basolateral receptor mediated secretagogues in piglet small intestinal epithelium. Comp. Biochem. Physiol. 144, 514-519.

Carlson, M.S., G.M. Hill and J.E. Link, 1999. Early- and traditionally weaned nursery pigs benefit from phase-feeding pharmacological concentrations of zinc oxide: effect on metallothionein and mineral concentrations. J. Anim. Sci. 77, 1199-1207.

Case, C.L. and M.S. Carlson, 2002. Effect of feeding organic and inorganic sources of additional zinc on growth performance and zinc balance in nursery pigs. J. Anim. Sci. 80, 1917-1924.

Cousins, R.J., 1985. Absorption, transport, and hepatic metabolism of copper and zinc: special reference to metallothionein and ceruloplasmin. Physiol. Rev. 65, 238-309.

Feng, Z., D.Carlson and H.D. Poulsen, 2006. Zinc attenuates forskolin-stimulated electrolyte secretion without involvement of the enteric nervous system in small intestinal epithelium from weaned piglets. Comp. Biochem. Physiol. A. 145, 328-333.

Fuller, R., L.G.M. Newland, C.A.E. Briggs, R. Braude and K.G. Mitchell, 1960. The normal intestinal flora of the pig. IV. The effect of dietary supplements of penicillin, chlortetracycline or copper sulfate on the feacal flora. J. Appl. Bacterol., 23, 195-205.

Gaskins, H.R., C.T. Collier and D.B. Anderson, 2002. Antibiotics as growth promotants: mode of actions. Anim. Biotechnol. 13, 29-42.

Hahn, J.D. and D.H. Baker, 1993. Growth and plasma zinc responses of young pigs fed pharmacological levels of zinc. J. Anim. Sci. 71, 3020-3024.

Hedemann, M.S., S. Højsgaard and B.B. Jensen, 2003. Small intestinal morphology and activity of intestinal peptidases in piglets around weaning. J. Anim. Physiol. a. Anim. Nutr. 87, 32-41.

Hedemann, M.S., B.B. Jensen, and H.D. Poulsen, 2006. Influence of dietary zinc and copper on digestive enzyme activity and intestinal morphology in weaned pigs. J. Anim. Sci. 84, 3310-3320.

Hill, G.M., D.C. Mahan, S.D. Carter, G.L.Cromwell, R.C. Ewan, R.L. Harrold, A.J. Lewis, P.S. Miller, G.C. Shurson, and T.L. Veum, 2001: Effect of pharmacological concentrations of zinc oxide with or without the inclusion of an antibacterial agent on nursery pig performance. J. Anim. Sci. *79*, 934-941.

Hoque, K.M., V.M. Rajendran, and H.J. Binder, 2005. Zinc inhibits cAMP-stimulated Cl secretion via basolateral K-channel blockade in rat ileum. Amer. J. Physiol. -Gastrointest. L. 288, G956-G963.

Højberg, O., N. Canibe, H.D. Poulsen, M.S. Hedemann, and B.B. Jensen, 2005. Influence of dietary zinc oxide and copper sulfate on the gastrointestinal ecosystem in newly weaned piglets. Appl. Environ. Microbiol. 71, 2267-2277.

Jensen-Waern, M., L. Melin, R. Lindberg, A. Johannisson, L. Petersson, and P. Wallgren, 1998. Dietary zinc oxide in weaned pigs – effect on performance, tissue concentrations, morphology, neutrophil functions and faecal microflora. Res. Vet. Sci. 64, 225-231.

Jensen, B.B., 1998. The impact of feed additives on the microbial ecology of the gut in young pigs. J. Anim.Feed Sci. 7, 45-64.

Katouli, M., L. Melin, M. Jensen-Waern, M. Jensen-Waern, P. Allgren, and Ollby, R., 1999. The effect of zinc oxide supplementation on the stability of the intestinal flora with special reference to composition of coliforms in weaned pigs. J. Appl. Microbiol.. 87, 564-573.

Kellogg, T.F., V.W. Hays, D.V. Catron, L.Y. Quinn, and V.C. Speer, 1964. Effect of dietary chemotherapeutics on the performance and fecal flora of baby pigs. J. Anim. Sci. 25, 1102-1106.

Li, B.T., A.G. Van Kessel, W.R. Caine, S.X. Huang, and R.N. Kirkwood, 2001. Small intestinal morphology and bacterial populations in ileal digesta and feces of newly weaned pigs receiving a high dietary level of zinc oxide. Can. J. Anim. Sci. 81, 511-516.

Li, X., J.Yin, D. Li, X. Chen, J. Zang, and X. Zhou, 2006. Dietary supplementation with zinc oxide increases IGF-I and IGF-I receptor gene expression in the small intestine of weaning piglets. J. Anim. Sci. 84, 74-75.

Martinez, M.M., G.M. Hill, J.E. Link, N.E. Raney, R.J. Tempelman and C.W. Ernst, 2004. Pharmacological zinc and phytase supplementation enhance metallothionein mRNA abundance and protein concentration in newly weaned pigs. J. Nutr. 134, 538-544.

McDowell, L.R., 2003., Minerals in Animal and Human Nutrition. Elsevier, 644 pp.

Mores, N., J. Cristani, I.A. Piffer, W. Barioni Jr, and G.M.M. Lima, 1998. Effect of zinc oxide on postweaning diarrhea control in pigs experimentally infected with E.coli. Arquivo Brasileiro de Medicina Veterinaria e Zootecnia. 50, 513-523.

Ou, D., D. Li, Y. Cao, X. Li, J. Yin, S. Qiao and G. Wu, 2007. Dietary supplementation with zinc oxide decreases expression of the stem cell factor in the small intestine of weanling pigs. J. Nutr. Biochem. in press, doi:10.1016/j.jnutbio.2006.12.022.

Pluske, J.R., D.J. Hampson, and I.H. Williams, 1997. Factors influencing the structure and function of the small intestine in the weaned pig: a review. Livest. Prod. Sci. 51, 215-236.

Poulsen, H.D. 1989. Zinc oxide for weaned pigs. In: anonymous, (Eds.) 40th Annual Meeting of the European Association for Animal Production, vol. 2. EAAP publications, Dublin, Ireland, pp. 265-266.

Poulsen, H.D., 1995. Zinc oxide for weanling piglets. Acta. Agric. Scand. 45, 159-167.

Poulsen, H.D., 1998. Zinc and copper as feed additives, growth factors or unwanted environmental factors. J. Anim. Feed. Sci. 7, 135-142.

Poulsen, H.D. and T. Larsen, 1995. Zinc excretion and retention in growing pigs fed increasing levels of zinc oxide. Livest. Prod. Sci. 43, 235-242.

Prasad, A.S., D. Oberleas, E.R. Miller and R.W. Luecke, 1971. Biochemical effects of zinc deficiency: changes in activities of zinc-dependent enzymes and ribonucleic acid and deoxyribonucleic acid content of tissues. J. Lab. Clin. Med. 77, 144-152.

Prasad, A.S., D. Oberleas, P. Wolf, J.P. Horwitz, E.R. Miller and R.W. Luecke, 1969. Changes in trace elements and enzyme activities in tissues of zinc-deficient pigs. Am. J. Clin. Nutr. 22, 628-637.

Revy, P.S., C. Jondreville, J.Y. Dourmad, F. Guinotte and Y. Nys, 2002. Bioavailability of two sources of zinc in weanling pigs. Animal Research. 51, 315-326.

Revy, P.S., C. Jondreville, J.Y. Dourmad, and Y. Nys, 2006. Assessment of dietary zinc requirement of weaned piglets fed diets with or without microbial phytase. J. Anim. Physiol. a. Anim. Nutr. 90, 50-59.

Rodriguez, P., N. Darmon, P. Chappuis, C. Candalh, M.A. Blaton, C. Bouchaud and M. Heyman, 1996. Intestinal paracellular permeability during malnutrition in guinea pigs: effect of high dietary zinc. Gut. 39, 416-422.

Roselli, M., A. Finamore, I. Garaguso, M.S. Britti and E. Mengheri, 2003. Zinc oxide protects cultured enterocytes from the damage induced by Escherichia coli. J.Nutr. 133, 4077-4082.

Shurson, G.C., P.K. Ku, G.L. Waxler, M.T. Yokoyama and E.R. Miller, E.R., 1990. Physiological relationships between microbiological status and dietary copper levels in the pig. J. Anim. Sci. 68, 1061-1071.

Skadhauge, E., M.L.Grøndahl and M.B. Hansen, M.B. 1997. Pathophysiology and symptomatic treatment of secretory and osmotic diarrhoea. In: Laplace, J.P., Février, C., Barbeau, A., (Eds.) Proceedings of the VIIth International symposium on 'Digestive physiology in pigs', vol. 88. EAAP publications, Saint Malo, France, pp. 241-254.

Spreeuwenberg, M.A.M., J.M.A.J.Verdonk, H.R.Gaskins and M.W.A.Verstegen, 2001. Small intestine epithelial barrier function is compromised in pigs with low feed intake at weaning. J. Nutr. 131, 1520-1527.

Swinkels, J.W.G.M., E.T. Kornegay, W. Zhou, M.D. Lindemann, K.E. Webb and M.W.A Verstegen, 1996. Effectiveness of a zinc amino acid chelate and zinc sulfate in restoring serum and soft tissue zinc concentrations when fed to zinc depleted pigs. J. Anim. Sci. 74, 2420-2430.

Szczurek, E.I., C.S. Bjornsson and C.G. Taylor, 2001. Dietary zinc deficiency and repletion modulate metallothionein immunolocalization and concentration in small intestine and liver of rats. J. Nutr. 131, 2132-2138.

Tran, C.D., R.N. Butler, G.S. Howarth, J.C. Philcox, A.M. Rofe and P. Coyle, 1999. Regional distribution and localization of zinc and metallothionein in the intestine of rats fed diets differing in zinc content. Scand. J. Gastroenterol. 34, 689-695.

Vasconcelos, M.H., S.C. Tam, J.E. Hesketh, M. Reid, M. and J.H. Beattie, 2002. Metal- and tissue-dependent relationship between metallothionein mRNA and protein. Toxicol. Appl. Pharmacol. 182, 91-97.

Veum, T.L., M.S. Carlson, C.W. Wu, D.W. Bollinger and M.R. Ellersieck, 2004. Copper proteinate in weanling pig diets for enhancing growth performance and reducing fecal copper excretion compared with copper sulfate. J. Anim. Sci. 82, 1062-1070.

Zhao, J., A.F. Harper, M.J. Estienne, K.E. Webb, A.P. McElroy and D.M. Denbow, 2007. Growth performance and intestinal morphology responses in early weaned pigs to supplementation of antibiotic-free diets with an organic copper complex and spray-dried plasma protein in sanitary and nonsanitary environments. J. Anim. Sci. 85, 1302-1310.

Bioavailability criteria for trace minerals in monogastrics and ruminants

J.W. Spears and S.L. Hansen

Keywords: bioavailability, trace minerals, zinc, copper, manganese, selenium, iron, cobalt

Introduction

A number of different definitions of bioavailability can be found in scientific literature. In regard to minerals, bioavailability has been defined as the proportion of the element consumed that is utilized for a biochemical or physiologic function (O'Dell, 1997). Other researchers have also stressed utilization when describing bioavailability (Fairweather-Tait, 1992). Absorption is clearly a key component of trace mineral utilization and is generally the major factor that limits bioavailability. However, in addition to absorption, utilization of trace minerals also consists of transport to the site of action, cellular uptake and incorporation into a biochemically active form (O'Dell, 1997). Less strenuous definitions of trace mineral bioavailability do not emphasize utilization but rather the degree to which an ingested trace mineral is absorbed in a form that can be utilized in metabolism (Ammerman et al., 1995). Metabolism may reflect use for a biochemical function, storage in the body or excretion from the body via urine, bile, etc.

Bioavailability of trace minerals is usually expressed relative to a standard source. It is desirable to evaluate at least three concentrations of the standard and test trace mineral source within a range of concentrations that will give a linear response in the bioavailability criteria measured. The ratio between the slopes of the dose-response curves can then be used to estimate bioavailability of the test source relative to the standard source. With large animals, such as cattle, it may be difficult to evaluate multiple levels of the mineral because of the cost involved. In such instances, relative bioavailability estimates may be based on a single dietary concentration rather than multiple points. Criteria that are most appropriate for estimating bioavailability will vary depending on the mineral and animal species in question. This paper will discuss criteria that may be used to assess bioavailability of the essential trace elements, zinc (Zn), copper (Cu), iron (Fe), manganese (Mn), selenium (Se), and cobalt (Co).

Approaches to assessing bioavailability

Two basic experimental approaches have been used to estimate relative bioavailability of trace mineral sources in animals. The first involves feeding the mineral at concentrations below the animal's requirement for a specific function. This approach allows for measurement of actual utilization of the particular mineral in the body. The second experimental protocol involves feeding the mineral at dietary concentrations considerably above the animal's requirement and measuring accumulation in blood or tissues. Supplementation of trace minerals at concentrations well above dietary requirements is also referred to as plethoric or pharmacologic supplementation. Use of plethoric concentrations of a trace mineral in bioavailability studies does not allow for measurement of utilization but instead measures the accumulation of the mineral in one or more tissues that serve at least as a temporary sink for excess absorbed mineral. The tissue mineral concentration measured under these conditions represents a balance between the amount of mineral absorbed and that excreted from the body, via urine or faecal endogenous origin (bile, pancreatic, etc.). As will be discussed under sections on specific minerals, relative bioavailability estimates obtained using high dietary mineral concentrations may or may not agree with estimates obtained using diets deficient in the mineral, depending on a number of factors including animal species and the mineral in question.

The advantages of using tissue accumulation in animals fed plethoric concentrations to assess bioavailability have been discussed (Ammerman, 1995) and include: (1) use of practical feedstuffs that allow for optimal growth, (2) greater statistical sensitivity (fewer animals may be needed to detect differences among sources), and (3) studies can be of fairly short duration. A major disadvantage of using plethoric concentrations to determine bioavailability is that normal physiological processes involved in regulating mineral metabolism are overwhelmed; thus, allowing the mineral to accumulate in certain tissues. It has long been recognized that homeostatic control mechanisms alter absorption and/or excretion of essential trace minerals, resulting in tissue concentrations of most minerals being maintained within a fairly narrow range (Miller, 1975; Windisch and Ettle, 2008). Using modern biochemical and molecular techniques specific transporters have been characterized for zinc (Cousins *et al.*, 2006), copper (Linder, 2002), and iron (Conrad and Umbreit, 2000) that are involved in cellular transport and assimilation of the metals by intestine and other tissues. Many of these transporters are either up or down regulated in rodent models when trace minerals are fed in excess of dietary requirements (Conrad and Umbreit, 2000; Cousins *et al.*, 2006). This suggests that absorptive and post-absorptive transport mechanisms used when plethoric concentrations of some trace minerals are fed may differ from those used when more physiological concentrations of the mineral are fed. Figure 1 shows a schematic view of changes in the absorptive mechanism of the intestine when trace minerals are supplemented at plethoric levels rather than within the physiological range, using Zn and Fe as examples.

Trace minerals are supplemented to diets to provide bioavailable mineral in situations where natural feedstuffs may not provide sufficient bioavailable quantities of the mineral to meet the animal's requirement. Therefore, bioavailability estimates for trace minerals should be most meaningful when assessed at dietary concentrations below the animal's requirement. This approach not only allows for measuring trace mineral utilization but also should result in absorptive and excretory pathways more similar to those observed under practical conditions. However, assessing bioavailability of certain trace minerals at concentrations below their requirement is not without potential drawbacks. It is essential that the control diet is deficient in the mineral in question if a biochemical function of the mineral is to respond in a dose manner to supplementation. Formulating diets, using practical feedstuffs that are consistently deficient in certain minerals can present a problem. With certain trace minerals, it may be necessary to use semi-purified diets to formulate a control that is deficient in the mineral. Extrapolation of bioavailability estimates obtained in animals fed

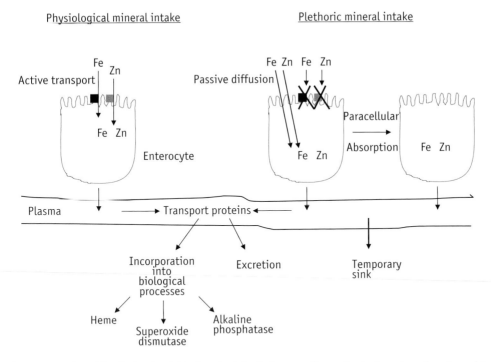

Figure 1. At mineral intake levels within the physiological range of an animal minerals such as Fe and Zn are absorbed via active transport by importers such as Divalent Metal Transporter I (Fe, ■) and the ZIP4 (Zn, ■) before being bound to transport proteins in plasma and eventually incorporated into biological processes. At high levels of mineral intake these transport mechanisms are often down regulated and absorption occurs by methods of passive diffusion. If an animal is not deficient in the mineral of interest much of this absorbed mineral will either be excreted or stored in a temporary sink that is of little or no use to the animal.

semi-purified diets to practical conditions can raise questions. Growth rate of animals fed semi-purified diets is generally lower compared with animals fed practical diets. Secondly, antagonists, such as phytate, that are known to reduce trace mineral absorption may be low or even absent in semi-purified diets. Thus, bioavailability estimates achieved under such conditions may not truly represent the natural conditions under which the mineral would normally be fed.

A number of methods have been used to predict bioavailability of trace minerals (Ammerman, 1995; O'Dell, 1997). These include: (1) apparent absorption, (2) true absorption, (3) mineral balance or retention, (4) plasma or tissue mineral concentrations, (5) growth, (6) biologically active forms of the mineral (i.e. haemoglobin, enzymes) and (7) prevention of pathologic deficiency signs. Although many of these methods do not measure utilization of the mineral directly, they provide information that can be useful in estimating bioavailability.

Zinc

Zinc absorption is affected by Zn status and bioavailability of dietary Zn. With low or marginal intakes of Zn relative to requirements, the efficiency of Zn absorption increases assuming that dietary Zn is present in a bioavailable form. However, when Zn is fed in excess of requirements, absorption of Zn is greatly reduced (Miller, 1975). In addition to changes in absorption, faecal endogenous secretion of Zn also plays an important role in Zn homeostasis (Miller, 1975).

In weanling pigs and young chicks, maize-soybean meal-based diets can be used to assess bioavailability of Zn when supplemented at physiological concentrations (Wedekind *et al.*, 1992; Mohanna and Nys, 1999; Revy *et al.*, 2006). Depending on the Zn content of the basal diet, response criteria that may respond to physiological concentrations of Zn include growth, feed intake, bone Zn and plasma Zn concentrations. Plasma alkaline phosphatase activity, a Zn metalloenzyme, and liver Zn also respond linearly to physiological Zn concentrations in young pigs (Revy *et al.*, 2006) but not in poultry (Dewar and Downie, 1984; Mohanna and Nys, 1999). These response criteria plateau at different dietary concentrations of Zn, with maximal growth reaching a plateau at lower dietary Zn concentrations than bone and plasma Zn concentrations. Edwards and Baker (1999) reported that gain and tibia Zn concentrations gave similar estimates of Zn bioavailability in chicks fed a semi-purified diet low in Zn. Zinc absorption and retention using the balance technique in pigs (Poulsen and Larsen, 1995; Revy *et al.*, 2002) as well as whole body Zn retention in chicks (Mohanna and Nys, 1999) also respond in a dose manner to physiological concentrations of Zn. Intestinal mucosa metallothionein in chicks also appears to be a suitable criterion for assessing bioavailability of Zn (Cao *et al.*, 2002). In older pigs (grow-finish) additions of low levels of Zn to a maize-soybean meal-based diet did not increase growth but resulted in a linear increase in plasma and bone Zn concentrations (Wedekind *et al.*, 1994). Bone Zn was less variable than plasma Zn in assessing bioavailability based on a lower coefficient of variation and R^2. Similarly in weanling pigs, the R^2 for regression of bone Zn on Zn intake was higher than for liver and plasma Zn, or plasma alkaline phosphatase activity (Revy *et al.*, 2006).

In lambs, Zn addition at low concentrations to semi-purified diets deficient in Zn increases growth, plasma Zn and plasma alkaline phosphatase activity (Spears, 1989; Droke and Spears, 1993). These response variables, as well as apparent absorption and retention of Zn, have been used to assess relative bioavailability of Zn sources in ruminants fed low dietary Zn concentrations (Spears, 1989; Spears *et al.*, 2004a). It is difficult to formulate practical ruminant diets that are deficient enough in Zn to observe consistent responses to Zn supplementation.

Bioavailability of Zn also has been estimated using high dietary Zn concentrations that exceed requirements. Sandoval *et al.* (1997a) fed chicks up to 1,200 mg Zn/kg diet and measured Zn accumulation in bone, liver, kidney, and pancreas. Bone was the only tissue examined that had sufficient sensitivity to be used as a criterion for Zn bioavailability. However, Wedekind *et al.* (1992) evaluated supplemental concentrations up to 750 mg Zn/kg in chicks and found that Zn in bone and other tissues examined did not consistently increase beyond their first dose increment (250 mg Zn/kg). Plasma Zn concentrations (Hahn and Baker, 1993) and concentrations in liver, kidney, and bone (Zhang and Guo, 2007) respond to plethoric supplementation of Zn in pigs, and these criteria have been used to assess Zn bioavailability.

In ruminants, bioavailability of Zn also has been assessed using plethoric concentrations of Zn. Zinc concentrations in liver, kidney, and pancreas, and liver metallothionein have been used to estimate Zn bioavailability in lambs (Sandoval *et al.*, 1997b; Cao *et al.*, 2000). In calves liver, kidney, and plasma Zn concentrations, and liver metallothionein respond to plethoric Zn supplementation (Kincaid *et al.*, 1997; Wright and Spears, 2004). In contrast to research with chicks and pigs, bone Zn does not increase greatly in response to high dietary Zn in ruminants (Rojas *et al.*, 1995; Wright and Spears, 2004).

Copper

Copper metabolism differs greatly among nonruminants and ruminants (NRC, 2005). In nonruminants, Cu is fairly well absorbed from the small intestine. Absorbed Cu is transported to the liver where Cu can be excreted via the bile, stored, or used for synthesis of ceruloplasmin or other Cu metalloenzymes. Biliary excretion is the major route involved in control of Cu homeostasis. Liver Cu concentrations are well regulated in nonruminants due largely to biliary Cu excretion. However, liver Cu concentrations increase when high dietary Cu is fed because biliary excretion of Cu becomes saturated. Copper is poorly absorbed (<1.0 to 10.0%) in ruminants with a functional rumen (Underwood and Suttle, 1999). The lower absorption of Cu in ruminants compared with nonruminants is primarily due to complex interactions that occur among Cu, sulfur, and molybdenum in the ruminal environment. Biliary Cu excretion is less effective in ruminants in regulating liver Cu concentration. As a result, normal liver Cu concentrations are considerably higher in ruminants than in nonruminants.

The growth stimulating effects of feeding pharmacological concentrations of Cu to swine and poultry are well documented (NRC, 2005). However, nutritional requirements for Cu, based on biochemical functions of Cu, are low in nonruminants (Underwood and Suttle, 1999). In fact, the Cu content of practical diets is high enough that it is difficult to measure responses to physiological additions of Cu. It is generally necessary to feed a semi-purified diet in order to measure responses to physiological concentrations of Cu. Aoyagi and Baker (1993a) fed chicks a semi-purified diet containing 0.56 mg Cu/kg for 7 days followed by Cu supplementation at 0 to 15 mg/kg for 14 days. Liver and gall bladder Cu concentrations increased linearly between 0 and 2 mg Cu/kg, and then appeared to plateau at higher Cu levels. In subsequent studies, bile Cu concentration was used to assess Cu bioavailability in chicks fed a Cu-deficient diet supplemented with 0, 0.5, or 1.0 mg Cu/kg (Aoyagi and Baker, 1993a, b). Bile Cu excretion reflects the amount of Cu absorbed but does not measure utilization of Cu for a biochemical function. Liver Cu concentrations also have been used to assess Cu bioavailability in chicks fed plethoric concentrations (100 to 450 mg Cu/kg) of Cu (Ledoux *et al.*, 1991; Aoyagi and Baker, 1993b; Guo *et al.* 2001). Aoyagi and Baker (1993b) reported that relative Cu bioavailability estimates were similar using bile Cu concentrations in chicks fed Cu-deficient diets and liver Cu concentrations in animals fed high concentrations of Cu. In weanling pigs, increasing supplemental Cu from 10 to 66 mg/kg did not affect liver or bile Cu concentrations (Armstrong *et al.*, 2000). Both liver and bile Cu concentrations were greatly increased when supplemental Cu was increased to 225 mg/kg. These authors concluded that neither liver nor bile Cu concentrations were good indicators of Cu bioavailability in pigs fed adequate to pharmacological Cu concentrations (Armstrong *et al.*, 2000).

In ruminants, Cu requirements are higher than in nonruminants. The higher requirements in ruminants are largely due to the ruminal environment and dietary antagonists that reduce Cu bioavailability. Two approaches have been used to estimate Cu bioavailability in ruminants fed physiological concentrations of Cu. The first involves feeding a Cu antagonist(s) and measuring the ability of different Cu sources to prevent a decline in Cu status over time. Copper antagonists that have been used to decrease Cu bioavailability include a combination of molybdenum and sulfur (Ward *et al.*, 1996; Spears *et al.*, 2004b) and high dietary Fe (Mullis *et al.*, 2003). The second approach is to assess Cu bioavailability by measuring the rate of repletion of Cu-deficient animals following supplementation with different Cu levels and sources (Kegley and Spears, 1994; Yost *et al.*, 2002; Spears *et al.*, 2004b). Criteria that have been used in these studies to assess bioavailability include plasma and liver Cu concentrations, and plasma ceruloplasmin activity. The Cu-dependent enzyme, diamine oxidase, may also be a suitable criterion for assessing Cu bioavailability. In rats, plasma diamine oxidase was a more sensitive biomarker of marginal Cu deficiency than ceruloplasmin or other Cu enzymes (Kehoe *et al.*, 2000). Plasma diamine oxidase was recently shown to be greatly reduced by Cu deficiency in cattle (Legleiter and Spears, 2007).

Copper bioavailability has been assessed in lambs fed plethoric concentrations (60 to 180 mg/kg) of Cu for 10 days (Pott *et al.*, 1995; Ledoux *et al.*, 1995). Accumulation of Cu in the liver was the criterion used in these studies to estimate bioavailability.

Iron

Iron homeostasis in animals is controlled by altering the efficiency of absorption from the small intestine (Underwood and Suttle, 1999). The body has a limited ability to excrete absorbed Fe. Therefore, bioavailability of Fe is primarily dependent on intestinal absorption.

Bioavailability studies with Fe must be conducted using diets deficient in Fe, due to the rigorous control of absorption once Fe requirements are met. Haemoglobin regeneration is the preferred criterion for measuring Fe bioavailability. This procedure involves depletion of Fe followed by repletion with different concentrations of Fe. Haemoglobin concentrations following a period of repletion (Boling *et al.*, 1998) or the increase in haemoglobin levels compared to initial values (Miller *et al.*, 1981) are used to estimate relative bioavailability. Growth (Boling *et al.*, 1998), and Fe balance (Miller *et al.* 1981) have been used to assess Fe bioavailability but they are less sensitive than haemoglobin regeneration. Weaning pigs can be used to assess Fe bioavailability without prior depletion of haemoglobin, because of their high Fe requirement. Growth, haemoglobin concentration, plasma Fe, and Fe retention have been found to respond linearly to increased dietary Fe in weanling pigs (Rincker *et al.*, 2005; Feng *et al.*, 2007).

Manganese

The percentage of dietary Mn absorbed from the small intestine is generally low. True absorption of Mn from practical diets is usually less than 2% of Mn intake (Halpin *et al.*, 1986; Finley *et al.*, 1997). Absorbed Mn is excreted primarily in bile and the liver is very efficient in removing excess Mn from the body via biliary excretion. In addition to biliary excretion, changes in the efficiency of Mn absorption also may be involved in homeostasis (Underwood and Suttle, 1999).

Growth, incidence of leg abnormalities, bone and heart Mn concentrations, and heart Mn-dependent superoxide dismutase (MnSOD) activity and MnSOD mRNA have been used to access Mn bioavailability in chicks fed physiological concentrations of dietary Mn (Watson *et al.*, 1970; 1971; Baker and Oduho, 1994; Li *et al.*, 2004). Growth responses to increasing dietary Mn only occur at concentrations well below the dietary Mn requirement of chicks, necessitating the use of semi-purified diets (Watson *et al.*, 1970; 1971; Baker and Oduho, 1994). In contrast, addition of 60-180 mg Mn/kg to a maize-soybean meal-based diet, containing approximately 20 mg Mn/kg, resulted in linear increases in bone and heart Mn concentrations, heart MnSOD activity and MnSOD mRNA (Li *et al.*, 2004; 2005). Of these indices, they reported that MnSOD mRNA was the most sensitive criterion for assessing differences in bioavailability among Mn sources. Manganese-dependent SOD is found largely in the mitochondria and is important in preventing cellular damage from superoxide radicals.

Manganese bioavailability also has been estimated in chicks using supplemental Mn concentrations (500 to 4,000 mg/kg) well above dietary requirements (Black *et al.*, 1984; Wedekind and Baker, 1990). Response criterions that have been used to assess Mn bioavailability following plethoric supplementation include Mn concentrations in bone, liver, kidney, and bile. Accumulation of Mn in bone appears to be the most sensitive tissue for estimating Mn bioavailability in poultry fed high dietary Mn. Evidence in chicks suggests that bioavailability estimates for Mn, based on bone Mn concentrations, are similar in chicks fed physiological (0, 50, and 100 mg Mn/kg) and high (0, 500 and 1,000 mg Mn/kg) dietary Mn concentrations (Wedekind and Baker, 1990).

Requirements for Mn in ruminants are much lower than in poultry (Underwood and Suttle, 1999). Studies have not been conducted comparing the relative bioavailability of Mn sources in ruminants supplemented with physiological concentrations of Mn. Based on studies investigating the effect of dietary Mn on various criteria, it will likely be difficult to assess relative Mn bioavailability in ruminants using practical diets and normal supplemental concentrations of Mn. Masters *et al.* (1988) fed ram lambs diets containing 13, 19, 30 or 45 mg Mn/kg for 84 days. Heart and lung Mn concentrations and heart MnSOD activity at the end of the study responded in a linear manner to dietary Mn. Plasma Mn concentrations showed a linear relationship with dietary Mn on only one of the three sampling days. Growth, MnSOD activity, and Mn concentrations in other tissues examined were not affected by dietary Mn. In contrast to results observed in chicks, bone Mn did not respond to physiological additions of dietary Mn in lambs (Masters *et al.*, 1988). Long term supplementation of physiological concentrations of Mn to diets of growing heifers (Hansen *et al.*, 2006; basal diet contained

15.8 mg Mn/kg) and growing and finishing steers (Legleiter *et al.*, 2005; basal growing and finishing diets contained 29 and 8 mg Mn/kg, respectively) did not affect growth or plasma Mn concentrations. Although liver Mn concentrations were increased by dietary Mn in both studies, the magnitude of the increases were probably too small to accurately assess relative bioavailability of different Mn sources.

High concentrations of dietary Mn have been used to estimate relative bioavailability of Mn from different sources in lambs (Wong-Valle *et al.*, 1989; Henry *et al.*, 1992). Accumulation of Mn in bone, liver and kidney was used as response criterion to assess bioavailability. Since relative bioavailability of Mn sources has not been evaluated using physiological concentrations of Mn in ruminants, it is unclear if bioavailability estimates using plethoric Mn concentrations can be related to practical feeding conditions.

Selenium

Selenium metabolism is affected by the form of dietary Se (Sunde, 1997). Selenium that occurs naturally in feeds is largely found as seleno-amino acids, with selenomethionine being the most predominant form. Selenomethionine also is the major form of Se in selenized yeast. Inorganic Se is generally supplemented to animal diets as selenite. Selenite is absorbed from the small intestine by simple diffusion, while selenomethionine is actively absorbed by the same amino acid transport system as methionine (Sunde, 1997). Both forms of Se are well absorbed in nonruminants. Absorption of Se is lower in ruminants and the lower absorption may relate to the reduction of dietary Se to insoluble forms in the rumen environment (Spears, 2003). Absorption of Se is not regulated by dietary Se concentration or Se status. Selenium homeostasis is primarily regulated by urinary excretion of Se.

Following absorption, selenite and selenomethionine also are metabolized differently (Sunde, 1997). Selenite is reduced to selenide which can be used for synthesis of selenocysteine, or methylated and excreted in urine. Selenocysteine is the form of Se present in selenoenyzmes such as glutathione peroxidase (GSH-Px). Selenomethionine can be incorporated into proteins in place of methionine, or be converted to selenocysteine. Dietary methionine level will affect the extent to which selenomethionine is incorporated into general proteins versus conversion to selenocysteine for use in Se specific enzymes (Butler *et al.*, 1989)

Criteria that have been used to assess Se bioavailability include GSH-Px activity (Gabrielsen and Opstvedt, 1980), tissue Se concentrations (Osman and Latshaw, 1976), and prevention of Se deficiency signs (Cantor *et al.*, 1975a, b). Bioavailability estimates for Se sources (especially selenomethionine relative to selenite) varies greatly depending on the criterion used. Feeding selenomethionine or selenized yeast increases Se concentrations in blood (Ortman and Pehrson, 1999) and muscle compared with selenite (Osman and Latshaw, 1976; Mahan *et al.*, 1999). However, bioavailability estimates based on GSH-Px activity have usually been lower for selenomethionine relative to selenite when evaluated at low Se concentrations in chicks (Cantor *et al.*, 1975a; Gabrielsen and Opstvedt, 1980; Laws *et al.*, 1986) and pigs (Mahan *et al.*, 1999). Glutathione peroxidase activity is the preferred criterion for assessing Se bioavailability and measures utilization of Se when animals are fed diets deficient in Se. Activity of GSH-Px in plasma, red blood cells, and a number of organs respond in a dose manner to dietary Se concentrations below the requirement (Oh *et al.*, 1976). Clearly Se incorporation into non-Se specific proteins does not represent utilization of Se for a specific biochemical function. Although selenomethionine incorporated into non-Se requiring proteins is not utilized immediately for production of Se enzymes, it does appear to represent a potential storage source of Se. When chicks are fed selenium-deficient diets after receiving supplemental Se from either selenite or selenomethionine, whole blood (Moksnes and Norheim, 1986) and plasma GSH-Px (Payne and Southern, 2005) decline more rapidly in birds that previously received selenite. This suggests that at least a portion of the selenomethionine in non-specific proteins is released during normal protein turnover and amino acid catabolism, and used in the synthesis of GSH-Px.

Cobalt

Cobalt functions as a component of vitamin B_{12}. Nonruminants are unable to synthesize vitamin B_{12} from dietary Co and generally require a dietary source of vitamin B_{12}. Ruminants are not dependent on a dietary source of vitamin B_{12} because ruminal microorganisms are capable of synthesizing vitamin B_{12} from Co.

A number of criteria can be used to assess Co bioavailability in ruminants. At deficient concentrations of Co, body weight gain and feed intake respond to increasing dietary Co (Schwarz *et al.*, 2000; Tiffany and Spears, 2005). Vitamin B_{12} concentrations in ruminal fluid, plasma, and liver also are responsive to increased dietary Co (Stangl *et al.*, 2000; Tiffany *et al.*, 2003; Tiffany and Spears, 2005). However, ruminal bacteria produce a number of vitamin B_{12} analogues that are active in bacteria but inactive in animal tissue (Bigger *et al.*, 1976). Plasma vitamin B_{12} concentrations increase to a much greater extent than liver in response to high dietary Co in ruminants fed high concentrate diets (Tiffany *et al.*, 2003). These authors suggested that under these conditions much of the circulating vitamin B_{12} was excreted in bile or urine rather than being stored in the liver. In mammals, vitamin B_{12} is an important cofactor for the enzymes methylmalonyl-CoA mutase and methionine synthase. In vitamin B_{12} deficiency substrates (methylmalonate and homocysteine) for these two enzymes accumulate in plasma, reflecting a metabolic deficiency of vitamin B_{12}. Thus, concentrations of methylmalonate and homocysteine in plasma have been used to define Co requirements (Stangl *et al.*, 2000; Tiffany *et al.*, 2003), and should be useful criteria for assessing bioavailability of Co.

Conclusions

Trace mineral requirements in animals are greatly affected by bioavailability of dietary trace minerals. Good estimates of trace mineral bioavailability from supplemental sources, as well as common feedstuffs, should allow for formulation of diets that not only maximize animal performance and health, but also minimizes excretion of trace minerals in manure. Reducing excretion of minerals in animal waste is already a concern in some countries where intensive animal production occurs, because of environmental issues related to long term accumulation of certain trace minerals in soils following manure application.

Experiments should be properly designed and appropriate criteria must be measured in order for bioavailability estimates to be most meaningful. In pigs and poultry, Cu and Zn are sometimes supplemented at pharmacological concentrations during certain production stages (i.e. weanling pigs) to achieve improved performance and/or health. However, in most instances, trace minerals are supplemented at physiological concentrations with the goal of providing sufficient bioavailable concentrations of the mineral to meet animal requirements. Ideally, bioavailability of trace mineral sources should be assessed at dietary concentrations below the animal's requirement, because essential trace minerals are under homeostatic control. This ensures that absorptive and excretory pathways are operating to maximize trace mineral utilization, and that homeostatic mechanisms regulating absorption or excretion of the mineral are not overwhelmed. Determining bioavailability in animals fed diets low in a given mineral also is consistent with the goal of providing adequate bioavailable trace minerals to meet nutritional requirements. Accumulation of trace minerals in tissues following supplementation at high concentrations may in some instances provide useful estimates of bioavailability, especially with Mn where it is difficult to formulate practical diets deficient in this mineral for some animals. However, assessing bioavailability in animals fed plethoric mineral concentrations does not allow for measurement of trace mineral utilization, or normal homeostatic regulatory processes that occur under physiological conditions.

Appropriate response criteria for estimating bioavailability of various trace minerals when fed at concentrations below dietary requirements are summarized in Table 1. It is important to recognize that different response criteria for a given mineral may plateau at different dietary concentrations. Criteria suitable for assessing bioavailability also will vary with animal species and age of the animal.

Table 1. Suitable criteria for assessing trace mineral bioavailability at concentrations below dietary requirements.

Mineral	Species	Criterion
Zinc	Pig	Growth, bone Zn plasma Zn, plasma alkaline phosphatase
	Chick	Growth, bone Zn
	Ruminants	Growth, plasma Zn, plasma alkaline phosphatase
Copper	Pig	Plasma ceruloplasmin, plasma Cu, liver Cu, erythrocyte SOD
	Chick	Bile Cu, liver Cu
	Ruminants	Liver Cu, plasma Cu, plasma ceruloplasmin, plasma diamine oxidase?
Iron	Pigs	Haemoglobin concentrations or regeneration, growth
	Poultry	Haemoglobin regeneration, growth
Manganese	Pig	?
	Poultry	Bone Mn, heart SOD, heart SOD mRNA
	Ruminants	Heart SOD
Selenium	Pig	GSH-Px
	Poultry	GSH-Px
	Ruminants	GSH-Px
Cobalt	Ruminants	Liver vitamin B_{12}, plasma methylmalonate, plasma homocysteine

References

Ammerman, C.B. 1995. Methods for estimation of mineral bioavailability. In: Ammerman, C.B., D.H. Baker and A.J. Lewis (Eds.), Bioavailability of Nutrients for Animals. Academic Press, San Diego, CA. pp 83-94.

Ammerman, C.B., D.H. Baker and A.J. Lewis, 1995. Bioavailability of Nutrients for Animals. Academic Press, San Diego, CA.

Aoyagi, S. and D.H. Baker, 1993a. Bioavailability of copper in analytical-grade and feed-grade inorganic copper sources when fed to provide copper at levels below the chick's requirement. Poultry Sci. 72, 1075-1083.

Aoyagi, S. and D.H. Baker, 1993b. Nutritional evaluation of a copper-methionine complex for chicks. Poultry Sci. 72, 2309-2315.

Armstrong, T.A., J.W. Spears, E.van Heugten, T.E. Engle and C.L. Wright, 2000. Effect of copper source (cupric citrate vs cupric sulfate) and level on growth performance and copper metabolism in pigs. Asian-Aus. J. Anim. Sci. 13, 1154-1161.

Baker, D.H. and G.W. Oduho, 1994. Manganese utilization in the chick: Effects of excess phosphorus on chicks fed manganese-deficient diets. Poultry Sci. 73, 1162-1165.

Bigger, G.W., J.M. Elliot and T.R. Richards, 1976. Estimated ruminal production of pseudovitamin B_{12}, factor A and factor B in sheep. J. Anim. Sci. 43, 1077-1081.

Black, J.R., C.B. Ammerman, P.R. Henry and R.D. Miles, 1984. Biological availability of manganese sources and levels of high dietary manganese on tissue mineral composition of broiler-type chicks. Poultry Sci. 63, 1999-2006.

Boling, S.D., H.M. Edwards, J.L. Emmert, R.R. Biehl and D.H. Baker, 1998. Bioavailability of iron in cottonseed meal, ferric sulfate, and two ferrous sulfate by-products of the galvanizing industry. Poultry Sci. 77, 1388-1392.

Butler, J.A., M.A. Beilstein and P.D. Whanger, 1989. Influence of dietary methionine on the metabolism of selenomethionine in rats. J. Nutr. 119, 1001-1009.

Cantor, A.H., M.L. Scott and T. Noguchi, 1975a. Biological availability of selenium in feedstuffs and selenium compounds for prevention of exudative diathesis in chicks. J. Nutr. 105, 96-105.

Cantor, A.H., M.L. Langevin, T. Noguchi and M.L. Scott, 1975b. Efficacy of selenium in selenium compounds and feedstuffs for prevention of pancreatic fibrosis in chicks. J. Nutr. 105, 106-111.

Cao, J., P.R. Henry, R. Guo, R.A. Holwerda, J.P. Toth, R.C. Littell, R.D. Miles and C.B. Ammerman, 2000. Chemical characteristics and relative bioavailability of supplemental organic zinc sources for poultry and ruminants. J. Anim. Sci. 78, 2039-2054.

Cao, J.H., P.R. Henry, S.R. Davis, R.J. Cousins, R.D. Miles, R.C. Littell and C.B. Ammerman, 2002. Relative bioavailability of organic zinc sources based on tissue zinc and metallothionein in chicks fed conventional dietary zinc concentrations. Anim. Feed Sci. Technol. 101, 161-170.

Conrad, M.E. and J.N. Umbreit, 2000. Iron absorption and transport – an update. Amer. J. Hematology 64, 287-298.

Cousins, R.J., J.P. Liuzzi and L.A. Lichten, 2006. Mammalian zinc transport, trafficking, and signals. J. Biol. Chem. 281, 24085-24089.

Dewar, W.A. and J.N. Downie, 1984. The zinc requirement of broiler chicks and turkey poults fed on purified diets. Br. J. Nutr. 51, 467-477.

Droke, E.A. and J.W. Spears, 1993. In vitro and in vivo immunological measurements in growing lambs fed diets deficient, marginal or adequate in zinc. J. Nutr. Immunol. 2, 71-90.

Edwards, H.M. and D.H. Baker, 1999. Bioavailability of zinc in several sources of zinc oxide, zinc sulfate, and zinc metal. J. Anim. Sci. 77, 2730-2735.

Fairweather-Tait, S.J., 1992. Bioavailability of trace elements. Food. Chem. 43, 213-217.

Feng, J., W.Q. Ma, Z.R. Xu, Y.Z. Wang and J.X. Liu, 2007. Effects of iron glycine chelate on growth, haematological and immunological characteristics in weanling pigs. Anim. Feed Sci. Technol. 134, 261-272.

Finley, J.W., J.S. Caton, Z. Zhou and K.L. Davison, 1997. A surgical model for determination of true absorption and biliary excretion of manganese in conscious swine fed commercial diets. J. Nutr. 127, 2334-2341.

Gabrielsen, B.O. and J. Opstvedt, 1980. Availability of selenium in fish meal in comparison with soybean meal, corn gluten meal and selenomethionine relative to selenium in sodium selenite for restoring glutathione peroxidase activity in selenium-depleted chicks. J. Nutr. 110, 1096-1100.

Guo, R., P.R. Henry, R.A. Holwerda, J. Cao, R.C. Littell, R.D. Miles and C.B. Ammerman, 2001. Chemical characteristics and relative bioavailability of supplemental organic copper sources for poultry. J. Anim. Sci. 79, 1132-1141.

Hahn, J.D. and D.H. Baker, 1993. Growth and plasma zinc responses of young pigs fed pharmacologic levels of zinc. J. Anim. Sci. 71, 3020-3024.

Halpin, K.M., D.G. Chausow and D.H. Baker, 1986. Efficiency of manganese absorption in chicks fed corn-soy and casein diets. J. Nutr. 116, 1747-1751.

Hansen, S.L., J.W. Spears, K.E. Lloyd and C.S. Whisnant, 2006. Growth, reproductive performance, and manganese status of heifers fed varying concentrations of manganese. J. Anim. Sci. 84, 3375-3380.

Henry, P.,R., C.B. Ammerman and R.C. Littell, 1992. Relative bioavailability of manganese from a manganese methionine complex and inorganic sources for ruminants. J. Dairy Sci. 75, 3473-3478.

Kegley, E.B. and J.W. Spears, 1994. Bioavailability of feed-grade copper sources (oxide, sulfate or lysine) in growing cattle. J. Anim. Sci. 72, 2728-2734.

Kehoe, C.A., M.S. Faughnan, W.S. Gilmore, J.S. Coulter, A.N. Howard and J.J. Strain, 2000. Plasma diamine oxidase activity is greater in copper-adequate than copper-marginal or copper-deficient rats. J. Nutr. 130, 30-33.

Kincaid, R.L., B.P. Chew and J.D. Cronrath, 1997. Zinc oxide and amino acids as sources of dietary zinc for calves: Effects on uptake and immunity. J. Dairy Sci. 80, 1381-1388.

Laws, J.E., J.D. Latshaw and M. Biggert, 1986. Selenium bioavailability in foods and feeds. Nutr. Reports Int. 33, 13-24.

Ledoux, D.R., P.R. Henry, C.B. Ammerman, P.V. Rao and R.D. Miles, 1991. Estimation of the relative bioavailability of inorganic copper sources for chicks using tissue uptake of copper. J. Anim. Sci. 69, 215-222.

Ledoux, D. R., E.B. Pott, P.R. Henry, C.B. Ammerman, A.M. Merritt and J.B. Madison, 1995. Estimation of the relative bioavailability of inorganic copper sources for sheep. Nutr. Res. 15, 1803-1813.

Legleiter, L.R. and J.W. Spears, 2007. Plasma diamine oxidase: A biomarker of copper deficiency in the bovine. J. Anim. Sci. 85, 2198-2204.

Legleiter, L.R., J.W. Spears and K.E. Lloyd, 2005. Influence of dietary manganese on performance, lipid metabolism, and carcass composition of growing and finishing steers. J. Anim. Sci. 83, 2434-2439.

Li, S., X. Luo, B. Liu, T.D. Crenshaw, X. Kuang, G. Shao and S. Yu, 2004. Use of chemical characteristics to predict the relative bioavailability of supplemental organic manganese sources for broilers. J. Anim. Sci. 82, 2352-2363.

Li, S.F., X.G. Luo, L. Lu, T.D. Crenshaw, Y.Q. Bu, B. Liu, X. Kuang, G.Z. Shao and S.X. Yu, 2005. Bioavailability of organic manganese sources in broilers fed high dietary calcium. Anim. Feed Sci. Technol. 123-124, 703-715.

Linder, M.C., 2002. Biochemistry and molecular biology of copper in mammals. In: Massaro, E. J. (Ed.), Handbook of Copper Pharmacology and Toxicity. Human Press Inc., Totowa, N. J. pp 3-32.

Mahan, D.C., T.R. Cline and B. Rickert, 1999. Effects of dietary levels of selenium-enriched yeast and sodium selenite as selenium sources fed to growing-finishing pigs on performance, tissue selenium, serum glutathione peroxidase activity, carcass characteristics, and loin quality. J. Anim. Sci. 77, 2172-2179.

Masters, D.G., D.I. Paynter, J. Briegel, S.K. Baker and D.P. Burser, 1988. Influence of manganese intake on body, wool and testicular growth of young rams and on the concentration of manganese and the activity of manganese enzymes in tissues. Aust. J. Agric. Res. 39, 517-524.

Miller, E.R., M.J. Parsons, D.E. Ullrey and P.K. Ku, 1981. Bioavailability of iron from ferric choline citrate and a ferric copper cobalt choline citrate complex for young pigs. J. Anim. Sci. 52, 783-787.

Miller, W.J., 1975. New concepts and developments in metabolism and homeostasis of inorganic elements in dairy cattle. A review. J. Dairy Sci. 58, 1549-1560.

Mohanna, C. and Y. Nys, 1999. Effect of dietary zinc content and sources on the growth, body zinc deposition and retention, zinc excretion and immune response in chickens. Br. Poultry Sci. 40, 108-114.

Moksnes, K. and G. Norheim, 1986. A comparison of selenomethionine and sodium selenite as a supplement in chicken feed. Acta Vet. Scand. 27, 103-114.

Mullis, L.A., J.W. Spears and R.L. McCraw, 2002. Effects of breed (Angus vs Simmental) and copper and zinc source on mineral status of steers fed high dietary iron. J. Anim. Sci. 81, 318-322.

NRC, 2005. Mineral Tolerance of Animals. The National Academics Press, Washington, D.C.

O'Dell, B.L., 1997. Mineral-ion interaction as assessed by bioavailability and ion channel function. In: O'Dell, B.L. and R.A. Sunde (Eds.), Handbook of Nutritionally Essential Elements. Marcel Dekker, Inc., New York, NY. pp 641-659.

Oh, S.H., R.A. Sunde, A.L. Pope and W.G. Hoekstra, 1976. Glutathione peroxidase response to selenium intake in lambs fed a torula yeast-based, artificial milk. J. Anim. Sci. 42, 977-983.

Okonkwo, A.C., P.K. Ku, E.R. Miller, K.K. Keahey and D.E. Ullrey, 1979. Copper requirement of baby pigs fed purified diets. J. Nutr. 109, 939-948.

Ortman, K. and B. Pehrson, 1999. Effect of selenite as a feed supplement to dairy cows in comparison to selenite and selenium yeast. J. Anim. Sci. 77, 3365-3370.

Osman, M. and J.D. Latshaw, 1976. Biological potency of selenium from sodium selenite, selenomethionine, and selenocystine in the chick. Poultry Sci. 55, 987-994.

Payne, R.L. and L.L. Southern, 2005. Changes in glutathione peroxidase and tissue selenium concentrations of broilers after consuming a diet adequate in selenium. Poultry Sci. 84, 1268-1276.

Pott, E.B., P.R. Henry, C.B. Ammerman, A.M. Merritt, J.B. Madison and R.D. Miles, 1994. Relative bioavailability of copper in a copper-lysine complex for chicks and lambs. Anim. Feed Sci. Technol. 45, 193-203.

Poulsen, H.D. and T. Larsen, 1995. Zinc excretion and retention in growing pigs fed increasing levels of zinc oxide. Livest. Prod. Sci. 43, 235-242.

Revy, P. S., C. Jondreville, J.Y. Dourmad, F. Guinotte and Y. Nys, 2002. Bioavailability of two sources of zinc in weanling pigs. Anim. Res. 51, 315-326.

Revy, P.S., C. Jondreville, J.Y. Dourmad and Y. Nys, 2006. Assessment of dietary zinc requirement of weanling piglets fed diets with or without microbial phytase. J. Anim. Physiol. a. Anim. Nutr. 90, 50-59.

Rincker, M.J., G.M. Hill, J.E. Link, A.M. Meyer and J.E. Rountree, 2005. Effects of dietary zinc and iron supplementation on mineral excretion, body composition, and mineral status of nursery pigs. J. Anim. Sci. 83, 2762-3774.

Rojas, L.X., L.R. McDowell, R.J. Cousins, F.G. Martin, N.S. Wilkinson, A.B. Johnson and J.B. Velasquez, 1995. Relative bioavailability of two organic and two inorganic zinc sources fed to sheep. J. Anim. Sci. 73, 1202-1207.

Sandoval, M., P.R. Henry, C.B. Ammerman, R.D. Miles and R.C. Littell, 1997a. Relative bioavailability of supplemental inorganic zinc sources for chicks. J. Anim. Sci. 75, 3195-3205.

Sandoval, M., P.R. Henry, R.C. Littell, R.J. Cousins and C.B. Ammerman, 1997b. Estimation of the relative bioavailability of zinc from inorganic zinc sources for sheep. Anim. Feed Sci. Technol. 66, 223-235.

Schwarz, F.J., M. Kirchgessner and G.I. Stangl, 2000. Cobalt requirement of beef cattle-feed intake and growth at different levels of cobalt supply. J. Anim. Physiol. a. Anim. Nutr. 83, 121-131.

Spears, J.W., 1989. Zinc methionine for ruminants: Relative bioavailability of zinc in lambs and effects on growth and performance of growing heifers. J. Anim. Sci. 67, 835-843.

Spears, J.W., 2003. Trace mineral bioavailability in ruminants. J. Nutr. 133, 1506S-1509S.

Spears, J.W., P. Schlegel, M.C. Seal and K.E. Lloyd, 2004a. Bioavailability of zinc from zinc sulfate and different organic zinc sources and their effects on ruminal volatile fatty acid proportions. Livest. Prod. Sci. 90, 211-217.

Spears, J.W., E.B. Kegley and L.A. Mullis, 2004b. Bioavailability of copper from tribasic copper chloride and copper sulfate in growing cattle. Anim. Feed Sci. Technol. 116, 1-13.

Stangl, G.I., F.J. Schwarz, H. Muller and M. Kirchgessner, 2000. Evaluation of the cobalt requirement of beef cattle based on vitamin B_{12}, folate, homocysteine and methylmalonic acid. Br. J. Nutr. 84, 645-653.

Sunde, R.A., 1999. Selenium. In: O'Dell, B.L. and R.A. Sunde (Eds.), Handbook of Nutritionally Essential Elements. Marcel Dekker, Inc., New York, NY. pp 493-556.

Tiffany, M.E., J.W. Spears, L. Xi and J. Horton, 2003. Influence of dietary cobalt source and concentration on performance, vitamin B_{12} status and ruminal and plasma metabolites in growing and finishing steers. J. Anim. Sci. 81, 3151-3159.

Tiffany, M.E. and J.W. Spears, 2005. Differential responses to dietary cobalt in finishing steers fed corn-versus barley-based diets. J. Anim. Sci. 83, 2580-2589.

Underwood, E.J. and N F. Suttle, 1999. The Mineral Nutritionof Livestock, 3rd edition. CABI Publishing, New York, NY.

Ward, J.D., J.W. Spears and E.B. Kegley, 1996. Bioavailability of copper proteinate and copper carbonate relative to copper sulfate in cattle. J. Dairy Sci. 79, 127-132.

Watson, L.T., C.B. Ammerman, S.M. Miller and R.H. Harms, 1970. Biological assay of inorganic manganese for chicks. Poultry Sci. 49, 1548-1554.

Watson, L.T., C.B. Ammerman, S.M. Miller and R.H. Harms, 1971. Biological availability to chicks of manganese from different inorganic sources. Poultry Sci. 50, 1693-1700.

Wedekind, K.J. and D.H. Baker, 1990. Effects of varying calcium and phosphorus level on manganese utilization. Poultry Sci. 69, 1156-1164.

Wedekind, K.J., A.E. Hortin and D.H. Baker, 1992. Methodology for assessing zinc bioavailability: Efficacy estimates for zinc-methionine, zinc sulfate, and zinc oxide. J. Anim. Sci. 70, 178-187.

Wedekind, K.J., A.J. Lewis, M.A. Giesemann and P.S. Miller, 1994. Bioavailability of zinc from inorganic and organic sources for pigs fed corn-soybean meal diets. J. Anim. Sci. 72, 2681-2689.

Windisch, W. and T. Ettle, 2008. Limitations and possibilities for progress in defining trace mineral requirements of earth animals. In: P. Schlegel, S. Durosoy and A.W. Jongbloed (Eds.), Trace elements in animal production systems, Wageningen Academic Publishers, Wageningen, the Netherlands, pp.187-201.

Wong-Valle, J., P. R. Henry, C.B. Ammerman and P.V. Rao, 1989. Estimation of the relative bioavailability of manganese sources for sheep. J. Anim. Sci. 67, 2409-2414.

Wright, C.L. and J.W. Spears, 2004. Effect of zinc source and dietary level on zinc metabolism in Holstein calves. J. Dairy Sci. 87,1085-1091.

Yost, G.P., J.D. Arthington, L.R. McDowell, F.G. Martin, N.S. Wilkinson and C.K. Swenson, 2002. Effect of copper source and level on the rate and extent of copper repletion in Holstein heifers. J. Dairy Sci. 85, 3297-3303.

Zhang, B. and Y. Guo, 2007. Beneficial effects of tetrabasic zinc chloride for weanling piglets and the bioavailability of zinc in tetrabasis form relative to ZnO. Anim. Feed Sci. Technol. 135, 75-85.

Bioavailability of trace minerals sources in swine

K. Männer

Keywords: swine, piglets, bioavailability, trace minerals, performance

Introduction and overview of today's trace mineral recommendations to swine

Actually eighteen trace minerals are defined as essential in swine nutrition. However, only a few of these like iron, manganese, zinc, copper, selenium and iodine are relevant, because their native concentrations could be marginal or deficient in feed ingredients for pigs. Adequate essential trace mineral intake and absorption is important for structural, physiological catalytic and regulatory metabolic functions including immune response, reproduction and growth. Caused by the complexity of the metabolic functions, the specific net requirements are in general identified by dose response curves. Studies on trace element requirements are complicated because of differences in experimental and environmental conditions, in breed, in feed intake and performance level as well as in the measurements taken. Consequently scientific committees establishing recommendations for the relevant trace element requirements have included safety margins in their figures, as shown in Table 1.

However, despite this safety margin, requirements at certain critical periods and practical conditions may not be covered when feeding trace minerals as recommended. The presence of stressors and/or the presence of dietary antagonistic interactions between minerals and nutrients or within minerals will increase the animal's trace mineral requirement and/or reduce the trace mineral absorbability. Absorbability is therefore one of the key components in striving towards optimal production. The lack of knowledge on dietary trace mineral interactions makes recommendations under practical conditions difficult because the absorption rate and the corresponding intermediary efficiency modifies the total requirement of the trace elements. Consequently low or high absorption rates of

Table 1. Dietary recommendations of relevant trace minerals in swine (ranges are depending on age and performance).

Trace mineral	ARC (1981) mg/kg DM	GfE (2006) mg/kg DM	NRC (1998) mg/kg (88% DM)
Iron (Fe)	10 - 40	50 - 120	40 - 100
Manganese (Mn)	10	15 - 25	2 - 10
Zinc (Zn)	50	50 - 100	50 - 100
Copper (Cu)	4 - 15	4 - 10	3 - 6
Selenium (Se)	0 10 - 0.40	0.15 - 0.60	0.10 - 0.30
Iodine (I)	0.46 - 0.60	0.15 - 0.60	0.14

trace elements need a higher or lower total requirement. The apparent absorbability of the relevant essential trace elements varies between 0.5 and 45% depending on the metal, diet components and especially on the animal's homeostatic regulation influenced by the mineral supply level (Hennig, 1973). Absorption and different intermediary factors are characterizing the bioavailability of trace minerals. The bioavailability of trace elements, defined by Ammermann *et al.* (1995) 'as the degree to which an ingested mineral is absorbed in a form that can be utilized in metabolism by the normal animal' has become highly important in swine nutrition, especially since the population's environmental concerns have risen and since the regulated upper limits (EC, 2003a,b) for the trace elements iron, manganese and zinc in compound feeds were reduced by 40% when compared to European legislation before year 2003. The newly limited concentrations are set above official recommendations (Table 1), but the safety margins are limited. This new situation may lead to critical situations when the diets is highly concentrated in mineral antagonists (such as earth or phytate) and in animal categories having high mineral requirements but low feed intake capacity like weaned piglets and pigs bread specifically for high feed efficiency (kg feed per kg body weight gain)

Inorganic (oxides, sulfates, chlorides, carbonates) or organic trace mineral (chelate, metal amino acid complex, metal proteinate or metal polysaccharide by the AAFCO (1998)) sources are commonly used for dietary supply. Their bioavailability, which is characterized not only by the apparent digestibility but also by retention in different organs, mineral specific enzyme activities (metalloenzymes), health and performance, varies relative to sulfates between 0 to 185% as shown in Table 2 (Ammermann *et al.* 1995). In conclusion, oxides are less available than sulfates, and inorganic sources are less available than organic sources, especially with regard to performance and health parameters in sows, piglets and growing or fattening pigs (Close, 1998; Lewis *et al.* 1999; Smits and Henman, 2000; Lee *et al.* 2001; Schenkel, 2000).

Influencing factors on the apparent trace mineral absorbability (Flachowsky, 1997) are mineral supplementation (source, solubility, dose level), feed (interactions: dietary components, fibre fraction, phytate, polyphenols, organic acids; feed processing) and animal factors (physiological status, mineral status, intestinal lumen: endogenous ligands, microbiota, transit time, ionic state). The binding of metallic ions to organic substances (ligands) through donor atoms (oxygen, nitrogen, sulfur) leads to increased bioavailability, which is probably caused by different absorption and intermediary transport to organs. The higher bioavailability of organic sources allows the replacement of inorganic sources at a lower concentration while performance and tissue level is maintained or enhanced. Consequently organic trace elements may have a potential for reducing environmental impact of pig production because the calculated mineral excretion may be reduced. In feeding practice either inorganic sources only, a combinations of inorganic and organic sources without or with lower supplementation rates or a total substitution of inorganic trace elements by organic sources at reduced concentrations are used. Organic iron and selenium have shown improved reproductive efficiency as measured by increased farrowing rate, reduced mortality, larger litter size and increased litter weight at birth and at weaning. Organic copper and zinc reveal the potential to reduce their inclusion level without compromising

Table 2. Relative bioavailability of different trace element sources (Ammermann et al., 1995).

Element	Bonding form	Bioavailability (%) sulfate sources = 100
Copper	Cu-lysine	100 - 105
	Cu-methionine	90 - 110
	Cu-proteinate	130
	Cu-carbonate	115
	Cu-chloride	110
	Cu-oxide	0 - 30
Iron	Fe-methionine	185
	Fe-proteinate	125
	Fe (II)-carbonate	85 - 95
	Fe (II)-chloride	100
	Fe (III)-chloride	60 - 91
Manganese	Mn-proteinate	110
	Mn-methionine	120 - 125
	Mn-carbonate	30 - 55
	Mn-chloride	100
Zinc	Zn-proteinate	100
	Zn-methionine	100 - 125
	Zn-carbonate	60 - 105
	Zn-oxide	50 - 100

animal performance and maintaining serum concentrations, and substantially reducing Cu and Zn excretion. Reasons for the inconsistent effects of organic trace minerals are the chemical and physical properties (ligands, stoechiometry, chemical structure, particle size distribution), the experimental design (depletion, repletion, parameters used for bioavailability, purified or conventional diets with or without feed additives, or animal and housing factors like physiological status (breed, age, health, performance stress, mineral status, climatic and hygienic conditions).

With regard to the abrupt change from high trace element absorbability during suckling to their relative low absorbability in compound feed after weaning the absorbability becomes a major limiting factor in mineral availability for weaned piglets. The goal of the following experiment was therefore to investigate the effect on mineral bioavailability (apparent absorbability, performance, haemoglobin and plasma concentrations) of three trace mineral sources (sulfates, chelates, glycinates) for iron, manganese, zinc and copper in restrictively fed weaned piglets (24th day of age) after a 14-day-depletion period.

Materials and methods

Forty early weaned piglets (24th day of age) were used from day 24 to 44 of age. The piglets were allocated pair wise in stainless steel pens, on body weight, gender and litter. For avoiding incomplete absorption all piglets were fed a basal diet (Table 3) with 14.1 MJ ME/kg and 192 g/kg of crude protein restrictively (0.77 MJ ME/kg$^{0.75}$) for the first 14 days after weaning which was formulated to meet the GfE requirements (2006) for piglets except iron, manganese, zinc and copper with concentrations of 48.0, 15.3, 24.6 and 3.8 mg/kg feed, respectively. In order to minimize possible interactions with native phytase activity, barley was heat treated (steam temperature 80 °C) prior mixing. Piglets had free access to drinking water containing 0.11, 0.008, 2.03 and 0.75 mg/l of iron, manganese, zinc and copper, respectively. After the depletion period pens were randomly allocated to 4 treatments: Control (no supplementation), sulfates, chelates (B-TRAXIM® TEC, Pancosma, Switzerland: based on enzymatically hydrolyzed amino acids from soya) and glycinates (B-TRAXIM® 2C, Pancosma, Switzerland: based on crystalline complexes with synthetically glycine). The three supplemented sources for iron, manganese, zinc and copper were included into the basal diet to reach 90% of NRC recommendations. Feed intake was similar across the treatments (0.77 MJ ME/kg$^{0.75}$).

The analysed trace elements in the diets are given in Table 4. The trace mineral concentrations in the depletion diet for iron, manganese, zinc and copper were 43.7, 23.5, 38.5 and 31.4% lower than the recommendations given by NRC (1998). In order to minimize homeostatic control the repletion period was limited to 8 days and the supplementation rates were with 12.1 (Fe), 25 (Zn) and 15.6% (Cu) still lower when compared to the NRC recommendations with exception of manganese (33.5% higher).

Table 3. Basal diet formulation.

Components	%
Barley	53.0
Skimmilkpowder	37.4
Corn starch	3.0
Calcium carbonate	1.5
Cellulose	1.4
Soy oil	1.3
Mono-sodiumphosphate	1.2
Premix[1]	1.2

[1]Contents per kg Premix: 400,000 IE Vit. A; 40,000 IE Vit. D$_3$; 4,200 mg Vit. E; 200 mg Vit. K$_3$; 200 mg Vit. B$_1$; 250 mg Vit. B$_2$; 3,500 mg niacine; 400 mg Vit. B$_6$; 3000 µg Vit. B$_{12}$; 20,000 µg biotin; 1,500 mg pantothenic acid; 150 mg folic acid; 80,000 mg choline chloride; 40 mg Co; 35 mg Se; 50 mg J ; 125 g Na; 50 g Mg.

Table 4. Analysed dietary trace element concentrations for iron, manganese, zinc and copper (mg/kg feed) during the depletion and repletion period.

Treatment	iron	manganese	zinc	copper
Depletion period (24th to 37th day of age)	45.0	15.3	24.6	3.82
Depletion period (37th to 44th day of age)				
control	45.0	15.3	24.6	3.82
sulfates	69.7	26.9	43.1	5.94
chelates	72.9	25.6	42.6	5.70
glycinates	68.2	27.5	44.4	6.18

The bioavailability was characterized by apparent digestibility measured by adding Cr_2O_3 (0.5%) in diets fed during the depletion period from 37 to 44 days of age. Faeces samples were collected by rectal sampling from each piglet at 24-h intervals during the last three days of the 8 day-repletion period. Additionally zootechnical performance (body weight gain, feed intake, kg feed per kg body weight gain) was measured during the 8 day-repletion period, as well as, haemoglobin and plasma concentrations for the supplemented trace minerals at the end of the 8 day-repletion period (44th day of age). Trace elements in feed, water, faeces, and plasma were analysed in accordance with the methods of VDLUFA (1988). All data were subjected to analyses of variance as completely randomized design. For treatment mean comparisons, Sheffe's test was used and significance level was set at $P < 0.05$.

Results and discussion

The apparent iron, manganese, zinc and copper absorbability is presented for each treatment in Table 5. In spite of the inhibition of native phytase and the relative high concentration of skimmilkpowder the mineral absorbability of trace minerals in piglets fed the control diet was relatively low, which underlines obviously the abrupt change from high to low absorbability after weaning. The average values for mineral sulfates were not any better than those naturally found in the unsupplemented control diet. The chelates improved the overall mineral absorbability relative to sulfates by 19.4%, whereas differences for manganese and copper were significant. The absorbability of the glycinate sources were highest with an overall improvement of 31.1% when compared to sulfates, being significant for iron, zinc and copper. Compared to chelates the absorbability of copper glycinate was significantly increased.

The efficiency of chelates and glycinates on improving the amount of dietary absorbable trace mineral concentration compared to respective concentrations of sulfates are presented in Figure 1. The quantity of absorbed trace minerals, using organic sources were up to 5 mg/kg feed higher than using sulfates.

Table 5. Apparent digestibility of different trace element sources in piglets at age 42 to 44 days (values in %).

Treatment	iron	manganese	zinc	copper
control	35.5 ± 7.7[a]	36.1 ± 6.9[ab]	38.8 ± 9.2[a]	24.1 ± 7.0[a]
sulfates	34.4 ± 5.5[a]	33.0 ± 7.5[a]	37.1 ± 6.7[a]	26.0 ± 5.1[a]
chelates	40.3 ± 4.9[ab]	39.9 ± 5.0[b]	42.2 ± 7.7[ab]	31.4 ± 7.1[b]
glycinates	41.8 ± 3.3[b]	41.2 ± 3.8[b]	46.0 ± 3.6[b]	38.7 ± 5.5[c]

[ab]Values with different superscript in the same column differed significantly (P < 0.05)

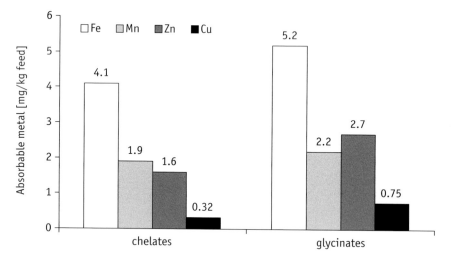

Figure 1. Efficiency of organic trace mineral sources on change in absorbable trace mineral concentration compared to dietary sulfates.

This also indicates that chelates and glycinates are able to provide identical quantities of absorbable Fe, Mn, Zn and Cu by reducing its dietary concentration when compared to sulfates. This potential is shown in Figure 2, whereas the relative dosage reduction ranged between 11 and 41%. Additionally, the data indicated that best effects were observed for glycinates.

Another approach is to calculate, based on the apparent absorbability values, the potential quantity of feed intake saving relative to the use of sulfates. The calculated data are presented in Figure 3.

The relative feed equivalents when supplying minerals with the tested organic sources were always positive and varied between 144 (zinc-chelate) and 490 g feed intake (copper-glycinate). Therefore the reduced feed intake, as it usually occurs following weaning, can

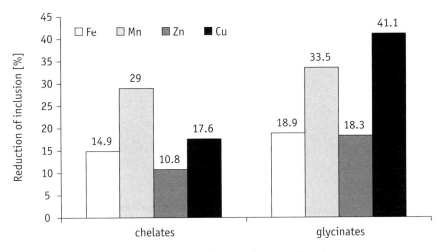

Figure 2. Potential reduction in mineral supply using organic trace mineral sources.

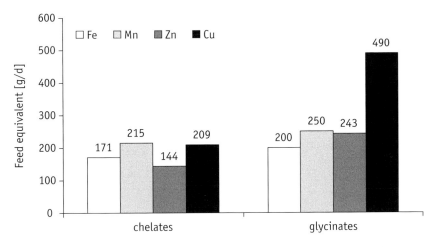

Figure 3. Potential feed intake replacement using organic trace mineral sources when compared to sulfates.

theoretically be reduced by a minimum of 144 g/d for chelates and by a minimum of 200 g/d for glycinates without harming the quantity of absorbable trace minerals, knowing that these piglets had an average daily feed intake of 300 g. In addition, the swine genetic evolution has in the past 37 years reduced feed conversion ratios from 3.5 for weaned piglets (5 to 20 kg), 4 for growing pigs (20 to 35 kg) and 4.5 for fattening pigs (35 to 115 kg) to 1.8, 2.1 and 2.5, respectively. Consequently the safety margins for absorbable trace mineral supply in modern genetics, related with high requirements would be comforted when using organic sources.

The haemoglobin status of piglets following a 8-day repletion period, is shown in Table 6. Piglets which were supplemented in trace minerals did not reach haemoglobin levels close to a normal range of 12 g/dl, but could however reach levels above 8 g/dl which is considered as anaemia. The reason for this is most probably the restricted feed intake and the limited time of the repletion period. Sulfates and chelates increased haemoglobin numerically by 13.8 and 19.8%, respectively. Glycinates increased the haemoglobin level significantly when compared to the other sources.

The average blood plasma manganese and copper concentrations were improved after trace mineral supplementation with an overall advantage (not significant) feeding organic sources (Table 7). However, with regards to plasma iron and plasma zinc organic sources did not indicate any benefits.

The performance of piglets during the 8 day-repletion period is presented in Table 8. The restricted feed intake for all treatments was in the range of 300 g per piglet and day. The use of dietary trace minerals tended to improved body weight gain and feed efficiency. Piglets supplied with organic trace minerals tended to perform even better than sulfates. However, with regard to the relative short supplementation period the results can be only interpreted as orientational.

Based on the method developed by Jongbloed *et al.* (2002) the overall bioavailability (ranking factors for absorbability: 4; haemoglobin: 2; feed efficiency: 1) for chelates and glycinates relative to sulfates was 115 and 124 respectively. The higher potential for glycinates when compared to chelates was especially due to the improved absorbability of Cu.

Table 6. Haemoglobin concentrations (g/dl) in piglets at age 44 days.

Treatment	control	sulfates	chelates	glycinates
haemoglobin	7.34 ± 2.06^a	8.35 ± 2.03^a	8.79 ± 1.21^{ab}	9.45 ± 1.13^b

[a,b]Values with different superscript differed significantly (P < 0.05).

Table 7. Trace mineral concentrations in blood plasma (µg/l) of piglets at age 44 days.

Treatment	iron	manganese	zinc	copper
control	720 ± 28^a	12 ± 10	1010 ± 300	1620 ± 270
sulfates	1290 ± 830^b	13 ± 10	1210 ± 240	1760 ± 240
chelates	900 ± 280^b	21 ± 20	1130 ± 340	1940 ± 290
glycinates	1210 ± 390^b	22 ± 11	1160 ± 220	1910 ± 410

[a,b]Values with different superscript in the same column differed significantly (P < 0.05).

Table 8. Piglet performance between 37 and 44 days of age.

Treatment	control	sulfates	chelates	glycinates
body weight (kg)				
- start	7.54 ± 0.81	7.62 ± 0.88	7.66 ± 1.33	7.58 ± 0.97
- end	9.54 ± 0.91	9.71 ± 0.90	9.80 ± 1.38	10.00 ± 0.94
body weight gain (kg)	1.90 ± 0.59	2.09 ± 0.31	2.14 ± 0.28	2.42 ± 0.05
feed efficiency[1]	1.383 ± 0.484	1.184 ± 0.193	1.149 ± 0.143	1.004 ± 0.022

[1]kg feed per kg body weight gain.

Conclusions

Supplementations of either chelates or glycinates into basal diets to reach 90% of NRC recommendations for Fe, Mn, Zn and Cu after a 14-day-depletion period increased trace mineral absorbability, haemoglobin concentrations, as well as, growth performance and reduced faecal excretion when compared to sulfates and the unsupplemented control. Therefore the two tested organic trace mineral sources, especially the crystalline glycinates (B-TRAXIM® 2C) would provide higher safety margins especially for animals with high mineral requirements but low feed intake capacity like early weaned piglets or growing and fattening pigs with genetically improved feed efficiency

Relative improvement of feed equivalency when supplying organic compared to inorganic sources is an additional way for characterizing the relative bioavailability of trace elements in swine. Caused by the fact that the bioavailability of the relevant trace elements can only partly be predicted for farming conditions it seems more realistic to present the recommendations at the level of net requirement and calculate the supplementation rate for the relevant trace minerals with regard to the bioavailability of the different defined trace mineral sources and the respective native concentrations.

References

AAFCO, 1998. Official publication of the association of american feed control officals incorporated. P. M. Bachman, (Ed.), pp. 237-238.

Ammermann, C.B, D.H Baker and A.J. Lewis, 1995. Bioavailability of nutrients for animals: Amino acids, minerals, vitamins. Academic Press, San Diego.

ARC,1981. Agriculture Research Council: The nutrient requirement of pigs. Commenwealth Agric. Bureaux, Farnham Royal, Slough.

Close, W.H., 1998. The role of trace mineral proteinates in pig nutrition. In: Biotechnology in the feed industry. T. P. Lyons and K. A. Jacques (Eds.), Nottingham, University Press. Nottingham, United Kingdom, pp. 469-483.

EC, 2003a. Verordnung (EG) Nr. 1334/2003 der Kommission vom 25. Juli 2003 zur Änderungen der Bedingungen für die Zulassung einer Reihe von zur Gruppe der Spurenelemente zählenden Futterzusatzstoffen. Amtsbl. Europ. Union L 187, 11-15.

EC, 2003b. Verordnung (EG) Nr. 2112/2003 der Kommission vom 1. Dezember 2003 zur Berichtigung der Verordnung (EG) Nr. 1334/2003 zur Änderungen der Bedingungen für die Zulassung einer Reihe von zur Gruppe der Spurenelemente zählenden Futterzusatzstoffen. Amtsbl. Europ. Union L 317, 22-23.

Flachowsky, G., 1997. Bewertung organischer Spurenelementverbindungen in der Tierernährung. 17. Arbeitstagung Mengen- und Spurenelemente, Jena, 599-619.

GfE, 2006. Empfehlungen zur Energie- und Nährstoffversorgung von Schweinen. Energie- und Nährstoffbedarf landwirtschaftlicher Nutztiere. Heft Nr. 10.

Hennig, A., 1973. Mineralstoffe, Vitamine, Ergotropika. VEB Deutscher Landwirtschafts-verlag, Berlin.

Jongbloed, A.W., P.A. Kemme, G. de Groote, M. Lippens and F. Meschy, 2002. Bioavailability of major and trace elements. EMFEMA, Brussels, Belgium.

Lee, S.H., S.C. Choi, B.J. Chae, S.P. Acda and Y.K. Han, 2001. Effect of feeding different chelated copper and zinc sources on growth performance and fecal exretions of weanling pigs. Asian-Aus. J. Anim. Sci., 14, 1616-1620.

Lewis, A.J. and P.S. Miller, 1999. Bioavailability of iron in iron methionine for weanling pigs. J. Anim. Sc., 73 (suppl. 1), 172.

NRC, 1998: National Research Council: Nutrient requirement of domestic animals. No.2. Nutrient requirement of swine, 10th ed., National Academy of Science, Washington D.C.

Schenkel, H., 2000. Einsatz organischer Spurenelementverbindungen zur Versorgung landwirtschaftlicher Nutztiere. Themen zur Tierernährung, 29-44.

Smits, R.J. and D.J. Henman, 2000. Practical experience with bioplexes in intensive pig production. In: Biotechnology in the feed industry. T. P. Lyons and K. A. Jacques (Eds.) Nottingham, University Press. Nottingham, United Kingdom, pp 293-300.

Limitations and possibilities for progress in defining trace mineral requirements of livestock

W. Windisch and T. Ettle

Keywords: trace minerals, requirement, recommendation, bioavailability, livestock

Introduction

Adequate supply of essential trace elements to earth animals gained in relevance during last decades, especially regarding agricultural livestock. The main reason was increasing productivity of livestock without adequate response in feed intake capacity. Accordingly, nutrient density including that of essential trace elements was increased in practical livestock feeding. Nutritionists tried to respond to this issue by exact description of the animals' requirement as well as a more precise estimation of the feeding value of dietary components. In case of essential trace elements, however, recommendations for adequate supply as well as feeding values are still based on a gross basis and not on an absorbable basis. The reason for this as well as limitations and possibilities for further improvements will be discussed in this chapter.

Experimental derivation of requirement

In principle, requirement of an essential nutrient is considered as the amount of dietary intake necessary to provide sufficient quantities to the metabolism to fulfil the specific physiological processes based on this nutrient. In this context, the physiological needs have to be distinguished from further benefits possibly arising from dietary supplies exceeding the threshold of requirement defined by biological parameter. In essential trace elements, an example for such benefits related to production traits is stabilization of intestinal health and hence stimulation of growth rates in weaned piglets treated with excessive amounts of zinc (Zn) or copper (Cu). Such effects cannot be accepted as part of requirement for Zn and Cu, as it needs dietary supplies far above the physiological requirements to become efficacious. Other traits might be considered more ambiguously, like possible effects of essential trace elements on product quality (e.g. with selenium (Se): Mahan *et al.*, 1999; Downs *et al.*, 2000) or immunity functions (e.g. Zn and mastitis: Dunkel *et al.*, 2004). It may be assumed that those production traits do not pose an extra requirement unless a specific physiological parameter indicates the contrary.

Experimental assessments of essential trace element requirements are usually done with dose-response studies from deficient to sufficient supplies and observation of a specific response parameter for characteristic changes in response behaviour. The typical change in response behaviour is lack of reaction to further increase of trace element supplies once the requirement is met (e.g. transition from retarded to normal growth). However, it has to be taken into account that the point of change in response behaviour may differ among response parameters, depending on the priority of the parameter within the entire organism and its interrelationship to specific processes of the respective trace element. For example, zootechnical performance traits (e.g. growth rate, milk yield) are of high priority to the organism and may turn to normal levels at marginal trace element intake. The same applies for absence of clinical symptoms of deficiency.

Physiological response parameter suitable to indicate transition from deficient to sufficient supplies are essential trace element concentrations in biomolecules directly involved into the trace mineral metabolism (e.g. haemoglobin) and activities of respective enzymes (e.g. alkaline phosphatase activity (Zn), ceruloplasmin (Cu), glutathione peroxidase (Se)). Also tissue concentrations of trace minerals may be suitable as soon as they are sensitive to supply status, like tissues involved in transport (blood) and storage (bone, liver, thyroid gland) (e.g. Kirchgessner *et al.*, 1999; Windisch, 2001, 2003; Ettle *et al.*, 2008).

Homeostatic regulation of trace element fluxes is the physiological reason for the typical transition of status parameter from a dose response relationship during deficiency towards lack of further response at the point of sufficient intake. The reason for this regulation is the fact that essential trace elements are also toxic to cellular metabolism when present at uncontrolled quantities. The homeostatic control of essential trace element flux through the body is mediated by down-regulation of absorption and/or stimulation of excretion (urine, faecal excretion from endogenous sources) (Kirchgessner, 1993). It becomes active once the dietary supply of essential trace elements turns from deficient to sufficient amounts and hence may be used to determine the requirement of essential trace minerals.

In Zn for example, homeostatic control acts already at the site of the intestine (absorption, endogenous faecal excretion). As soon as the requirement is met, net influx from the intestinal lumen towards the inside of the organism is down-regulated to precisely match metabolic needs and to maintain tissue Zn contents as well as Zn-dependent enzymes over a wide range of dietary supply (Windisch and Kirchgessner, 1995). Recent studies have demonstrated that this control of Zn influx is associated by respective modulation of mRNA expression rates of specific Zn transport proteins localized in intestinal tissues and kidney (Pfaffl and Windisch, 2003). Consequently, reactions of Zn contents in sensitive tissues (e.g. blood plasma, bone), activities of Zn metallo enzymes (e.g. alkaline phosphatase) as well as Zn contents in products (egg, milk) reflect absence or presence of homeostatic regulation and may thus indicate reliably the point of dietary supplies matching the metabolic requirement.

Homeostatic control at the site of the intestine is known also for iron (Fe) and Cu (Kirchgessner, 1993). However, Fe and Cu homeostasis is interlinked with each other (e.g. high Cu absorption at deficient Fe supply, Reichlmayr-Lais and Kirchgessner, 1992), which may cause mutual accumulations of Fe or Cu in storage tissues (e.g. liver) in case of deficiency of one of these elements. In dose-response studies designed to derive requirement of Fe or Cu, these tissues can be used only indirectly as response parameter by assuming reasonable tissue contents reflecting normal Fe and Cu supply on a long term base. Nevertheless, Fe- and Cu-dependent proteins and enzyme activities (e.g. haemoglobin, serum ferritin, serum transferrin saturation, ceruloplasmin) provide additional information to derive the Fe and Cu requirement in a reliable way.

Derivation of requirement is more difficult for essential trace elements whose homeostatic control is mediated mainly by urinary excretion, like Se, iodine (I) and cobalt (Co). These elements are absorbed from the intestinal tract at high rates without homeostatic control over a wide range of supply from deficient to considerable excess (Kirchgessner *et al.*, 1994, 1997, 1999). As homeostatic elimination from the organism becomes active after absorption, the respective elements intrude into the organism in a manner proportionally to intake. This explains why body fluids and tissue contents of these essential trace elements usually continue to raise at dietary supplies exceeding requirement (Kirchgessner *et al.*, 1994, 1997, 1999). Only I content of the thyroid gland seems to show some homeostatically mediated limits at sufficient dietary I intake (Kirchgessner *et al.*, 1999). Nevertheless, the onset of homeostasis in urinary excretion may be used as indicator of transition from deficient to sufficient supplies (e.g. Kirchgessner *et al.*, 1997) but experimental methodology for such measurements is extensive.

Another approach to determine requirement of Se, I and Co is measurement of related enzyme activities. Glutathione peroxidase activity has been accepted widely as such a parameter to Se. For Co, the plasma contents of homocysteine and methylmalonic acid activity were suggested as suitable response criteria (e.g. Stangl *et al.*, 2000), and for I, the content of protein-bound or butanol-soluble I in blood serum are generally accepted as indicators of supply status. In total, these parameters are considered to provide sufficient information to derive requirements of Se, I and Co in a dose-response study to a sufficient extent.

Chromium has been accepted to be an essential trace element to agricultural livestock but knowledge about metabolism is currently too small to derive requirement (NRC, 1989, 2001). Also molybdenum (Mo) is an essential trace element but relevance to agricultural livestock refers mainly to an excessive intake depressing Cu absorption in ruminants rather than deficiency of Mo per se (NRC, 2001).

In total, there is extensive experience available to derive essential trace element requirement by dose-response techniques for a given experimental situation (animal species and category of production, level of performance, dietary composition). However, the transfer of this knowledge into recommendations to trace element supply under common conditions of agricultural production remains difficult.

Formulation of recommendations to trace element supply in practice

Recommendations of essential trace element supply are *a priori* suggestions aiming to secure sufficient intake of these nutrients at common practical conditions derived from *ex ante* estimates of requirement done in well-described experimental situations. As common situation may considerably vary, a safety margin has to be included into recommendations in order to compensate for possible adverse conditions. Major reasons for inclusion of safety margins are actual trace element content in feedingstuffs being lower than expected levels (e.g. from feed tables) and a low availability of dietary trace elements actually used, either due to the nature of the respective compounds or due to unfavourable interactions with other dietary constituents. It has to be noticed that safety margins are designed to cover unfavourable variations occurring under normal conditions of agricultural practice, while extreme (and therefore rare) conditions may not be compensated completely. For the average feeding situation, however, the recommendations provide trace element intakes moderately exceeding the actual requirement, followed by excretion through homeostatic regulation without being used by the animal. As soon as this dietary excess contributes to feed costs and undesired ecological effects, the concept of high safety margins has to be revised. This is true e.g. for Zn and Cu, the return of them via dung onto agricultural areas being under recent discussion.

Knowledge of trace element contents of feedingstuffs actually used is an important criterion for application of recommendations in practice. Especially in feedingstuffs typical to ruminants (green fodder, silage, hay) variations may be quite high and may result in unperceived periods of marginal or deficient trace mineral supply due to either low contents of the respective trace element or high contents of other dietary factors possibly depressing its availability. For example, a large survey of nutrient contents of roughage produced in Austria revealed an overall average of 35±8 mg Zn and 759±469 mg Fe per kg of grass silage DM (1st cut, young state of vegetation), while means varied among major regions of production between 33±9 and 42±2 mg/kg DM for Zn, and 696±444 and 1104±732 mg/kg DM for Fe, respectively (Wiedner *et al.*, 2001). Also Se contents in feedingstuffs are known to vary considerably depending on their geographical origin. Consequently, any attempt to further raise precision and reliability of recommendations poses the precondition of best knowledge of actual trace element contents in feedingstuffs.

Another approach to improve recommendations is applying the concept of factorial derivation of requirement. The principle consists in estimation of net requirement for maintenance and production (growth, milk, egg, reproduction), and subsequent multiplication with factors accounting for gross utilization (absorption × metabolic utilization). For elements controlled homeostatically at the site of absorption (e.g. Zn), normal trace element contents in tissues and products are known in general and estimates of net requirements for maintenance and production should provide sufficient reliability. For trace elements controlled by renal excretion (e.g. Se, I), however, respective estimates are difficult as retention in tissues and products correlates with intake. Another bottleneck for the factorial approach is quantitative knowledge of the potential of dietary trace element sources to be absorbed and metabolically utilized. The information currently available, however, does not seem to be sufficient enough to provide reliable estimates on these factors of utilization, especially when addressing different trace mineral formulations. Calculation of gross requirement from net data has therefore to be done by a conversion factor set rather low in order to maintain security in adequate trace mineral supply even under adverse feeding conditions. This principle reflects again the concept of safety margins and hence the calculated 'requirement' may still be addressed to as 'recommendation'. Nevertheless, concepts based on factorial derivation of 'requirement' may be considered as significant progress since variations in trace element requirement caused by different levels of productivity are considered more precisely (e.g. Zn to dairy cows, NRC, 2001).

Factors disturbing bioavailability of trace mineral sources

Advanced nutritional systems for nutrients or energy are usually based on digestible, metabolizable, or net feed values in order to account for variations given by the nature of the individual feedingstuff. Also in essential trace elements, there are discussions to make use of such concepts, namely in view of better assessing bioavailability. In quantitative terms, bioavailability may be defined as product of true absorption and metabolic utilization determined at absence of homeostatic counter-regulation (i.e. at deficient supply status) (Kirchgessner *et al.*, 1993). It is often considered as a property given individually to single feedingstuffs or trace element formulations. However, an essential precondition to a nutritional concept based on digestibility, metabolizability, bioavailability, etc. is stability and additivity of individual feed values within the entire diet as well as absence of interactions with other dietary constituents. As will be demonstrated in the following, these preconditions do not sufficiently apply to different essential trace elements and hence tabulation of feed values other than gross contents is severely restricted.

In essential trace elements, quantitative measurements of absorption as major determinant of bioavailability need a high experimental effort (e.g. use of isotopes). This is due to the fact that true absorption of dietary trace elements from the gut lumen towards the inside of the organism has to be recorded separately from reflux of already absorbed (endogenous) quantities. In case of Zn, studies with rats fed purified diets revealed extremely high rates of true absorption close to 100% for both, inorganic as well as organic Zn sources (Table 1). This result is quite unexpected from the practical point of view as absorptive capacity of Zn in monogastrics is generally assumed to be considerably lower and to be more differentiated between inorganic and organic sources. But the diets used in these studies were based on highly purified components and were virtually free from any complexing agents (e.g. phytate). This reflects more or less an ideal situation being significantly different from common nutritional conditions. Obviously, absorbability and hence bioavailability of different dietary Zn sources cannot be described in terms of an independent property. It seems to result mainly from interactions with other dietary components during digestion and absorption. The same may be supposed to apply in principle also for Fe, Cu and manganese (Mn).

Phytic acid and respective salts (phytates) are the most relevant substances which may interact with essential trace elements during digestion and absorption in common monogastric livestock feeding. Phytates are the major storage compound of phosphorus to the plant embryo. It is abundant in all seeds and highly enriched in their by-products (e.g. bran, soya extracts) and is known to form insoluble complexes especially with Zn, Cu and Fe and to withdraw considerable amounts of these elements from absorption. In Zn for example, the highly depressive effect of phytate on Zn absorption has been demonstrated in a radiotracer study, where rats were exposed to rising additions of phytate to a highly purified diet (Windisch and Kirchgessner, 1999). The diet provided Zn from Zn sulfate at slight deficiency in order to avoid homeostatic counter-regulation of absorption and to assess the maximum possible absorptive capacity of the inorganic Zn source under the respective conditions. As shown in Figure 1, true absorption of Zn was close to 100% when extrapolated to a phytate-free diet which confirms results presented in Table 1. But with

Table 1. Maximum rates of true Zn absorption from different Zn sources measured in rats with purified diets at deficient Zn supply.

Dietary Zn source	True absorption (% of intake)	Reference
Zn sulfate	91-100	Weigand and Kirchgessner (1978)
		Weigand and Kirchgessner (1992)
		Windisch and Kirchgessner (1994)
		Windisch *et al.* (1996)
Zn chloride	95	Weigand and Kirchgessner (1997)
Zn fumarate	95	Weigand and Kirchgessner (1997)
Zn histidine	95	Weigand and Kirchgessner (1997)

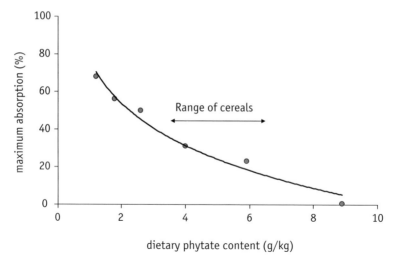

Figure 1. Effect of phytate additions to a purified diet on true Zn absorption from Zn sulfate in Zn deficient rats.

rising phytate additions, Zn absorption decreased to almost zero in a dose-related manner. At doses reflecting native phytate contents in cereals (about 4 - 6 g/kg (e.g. Lantzsch, 1990)), true absorption ranged at 30 to 40% which indeed seems to reflect the nutritional situation in practice. This example demonstrates that Zn absorption and hence bioavailability is widely subject to interactions with other dietary constituents. Due to chemical similarity of phytate complexes with Fe, Cu and Mn, interactions between these trace elements and phytate may be assumed to be of similar significance like with Zn.

In ruminants, interactions between Cu, Mo and sulfur (S) are quantitatively described and have been confirmed for their relevance by recent studies (e.g. Suttle, 1991; Spears, 2003; Hansen *et al.*, 2008). Regarding dietary chelators like phytate, ruminants are considered to be far less sensitive due degradation of these compounds in the forestomachs. Fe contents may be quite high in ruminant feedingstuffs (e.g. Wiedner *et al.*, 2001) and may impair availability especially of Zn (e.g. Standish *et al.*, 1969). Furthermore, changes in ruminal fermentation characteristics such as varying presence of fermentable energy may significantly affect the extent to which dietary Se (organic as well as inorganic) is transferred into bacterial protein and hence converted into a highly bioavailable organic form prior to absorption (Koenig *et al.*, 1997). Further ruminal interactions may arise from dietary S (for review cf. Spears, 2003).

Another type of interaction is competition for absorption. It is well known that Ca, Fe, Zn and Cu may potentially impair absorbability of each other (e.g. Windisch and Kirchgessner, 1994; Bremner *et al.*, 1987; Phillipo *et al.*, 1987; Prabowo *et al.*, 1988), but depression of Zn and Cu through excessive Ca and/or Fe is probably more relevant to practice. Following absorption, Fe, Zn, Cu and Mn may be considered to be highly available irrespective of their origin in feedingstuffs and supplements.

A further well known example of post-absorptive interactions is I metabolism as affected by goitrogens. It impairs I uptake by the thyroid gland, stimulates renal I excretion and seems to lower blood serum concentrations as a consequence (e.g. Schöne *et al.*, 2006).

In total, bioavailability of essential trace elements is subject to various significant interactions with other dietary compounds. This impairs massively attempts to get away from gross systems currently used, to improve nutritional systems to describe requirement and to tabulate the feed value more detailed. On the other hand it demonstrates that improvement of current recommendations can be achieved only through progress in quantitatively understanding these interactions affecting bioavailability of total dietary trace elements.

Improving bioavailability of total dietary trace elements

One way to improve bioavailability of total dietary trace elements is to eliminate factors responsible for negative interactions. For example, phytic acid in common diets to monogastric species may be at least partially converted into inactive compounds by adding commercial phytase formulations. This may exert a significant effect on trace mineral metabolism as has been shown in piglets fed a Zn deficient diet fortified with phytase (Figure 2) (Ettle *et al.*, 2005). Without added phytase, Zn absorption was negative (severe Zn deficiency), while degradation of phytate by the phytase supplementation increased apparent Zn absorption significantly to positive values and hence relieved the animals from severe Zn deficiency (rise in blood plasma Zn). Comparable results of added phytase formulations were observed also with Fe and Cu (Adeola *et al.*, 1995; Pallauf *et al.*, 1999; Jongbloed *et al.*, 2004; Kies *et al.*, 2006). Furthermore, dietary additions of e.g. ascorbic acid or citrate have been demonstrated to enhance absorption of trace elements (Fe) (for review see Beard and Dawson, 1997). These results further confirm the hypothesis that bioavailability of Fe, Zn, Cu

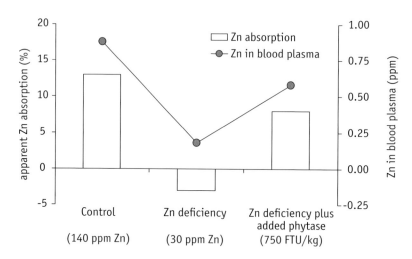

Figure 2. Effect of dietary phytase additions to a Zn deficient, cereal-based diet on apparent Zn absorption and blood plasma Zn in piglets.

and Mn in monogastrics is determined mainly by interactions with other dietary components rather than by the nature of the compound *per se*.

During recent years, organic trace mineral formulations have attracted significant interest as they are considered to be higher bioavailable than inorganic compounds (for review cf. e.g. Jongbloed *et al.*, 2002). However, it remains questionable whether superiority in bioavailability is based on the (bio)chemical nature of the compound alone. As shown above for Zn, absorption is determined mainly by interactions with other compounds of the meal (Figure 2), while general potential to be absorbed *per se* is quite high in both inorganic and organic Zn compounds (Table 1). In a recent radiotracer study, however, the maximum true absorption of inorganic and organic Zn compounds was determined when added to a purified diet enriched with phytate at amounts representing cereals (Schlegel and Windisch, 2006). As presented in Figure 3, Zn ingested as sulfate (control) was truly absorbable at 44%, which confirms the observations mentioned above (Figure 1) while Zn from a glycinate or protein complex revealed a significantly higher absorption potential. Obviously, Zn bound to organic formulations can resist the precipitating attack of phytate better than Zn from inorganic sources. Similar results were retrieved also for Zn from lipoate versus sulfate (Walter *et al.*, 2003).

Similarly, organically bound Fe is discussed to have a higher bioavailability compared to Fe from inorganic sources. This item is of special interest in piglets, in which the capacity to store Fe is low but the Fe requirement is relatively high due to rapid growth. Nevertheless, information about the use of organically-bound Fe in pig nutrition is scarce. In a study with newborn piglets, a single oral dose of 200mg Fe from FeMet or $FeSO_4$ given at an age of 3 days resulted in similar hematocrits and haemoglobin values at 7 days of age, but revealed higher

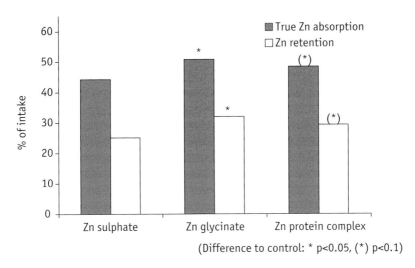

Figure 3. True Zn absorption and retention in Zn-deficient rats fed Zn from sulfate, glycinate or protein complex.

hematocrits and haemoglobin values in the FeMet group at 14 and 21 days of age (Kegley *et al.*, 2002). Both sources seemed to be less effective in maintaining high hematocrits or haemoglobin values when compared to a single Fe injection (200 mg Fe as gleptoferrin) within 12h after birth, but Fe from FeMet seemed to be more effective when compared to Fe from $FeSO_4$. Similarly, Yu *et al.* (2000) reported higher haemoglobin and packed cell volume values in weanling piglets fed a diet supplemented with 120 ppm Fe from a Fe chelate compared to feeding the same amount of Fe from $FeSO_4$. A recent study (Ettle *et al.*, 2008) revealed no effects of 30ppm Fe from organic sources or from $FeSO_4$ on blood parameters of weaned piglets. However, as shown in Figure 4, apparent absorption of Fe from an Fe-glycinate complex tended to be higher than Fe from $FeSO_4$, and Fe balance as calculated from Fe intake and Fe excretion via faeces and urine was significantly (P< 0.05) increased.

Also in ruminants, bioavailability of organic trace minerals is discussed to be superior to inorganic sources (e.g. Kincaid *et al.*, 1997; Spears, 2003). The major reason for this is probably different to that assumed in monogastrics because dietary chelators like phytate do not play a significant role on trace mineral absorption. Nevertheless, it has to be assumed that other dietary components may effectively depress absorption of trace minerals like Zn and Cu. One of these candidates might be a high content of Fe originating e.g. from contamination of ruminant feedingstuffs (green fodder, silage, hay) with soil.

Although high bioavailability of organic trace mineral formulations is widely accepted, the hypotheses on the mode of action are still controversial. Besides the above mentioned mechanism of preventing trace minerals from precipitating attack of strong dietary chelators, organic trace minerals might also enter the organism by pathways apart from regular transport systems. However, such a mode of action would overrule homeostatic

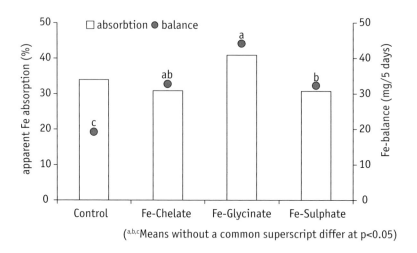

(a,b,cMeans without a common superscript differ at p<0.05)

Figure 4. Apparent absorbability of Fe and Fe-balance in piglets fed Fe from different sources.

regulation of absorption. In order to test this question for Zn, a model study with [65]Zn labelled rats was performed on basis of a purified diet enriched with phytate at amounts representing cereals (Windisch *et al.*, 2003). The control diet contained sufficient amounts of Zn (50 mg/kg from Zn sulfate) in order to stimulate homeostatic counter-regulation. On top of the basal Zn level, the diets were fortified with an extra Zn supplement of 50mg/kg from sulfate, gluconate, orotate, aspartate and histidine. As shown in Figure 5, intake of extra Zn was completely counterbalanced by a respective rise in faecal Zn excretion irrespective of the type of Zn formulation. Obviously, Zn homeostasis recognized any kind of dietary Zn (inorganic vs. organic) offered in this study and responded uniformly to Zn excess by down-regulation of absorption. Furthermore, the rate of [65]Zn exchange in whole body of labelled animals did not differ between added Zn sources after a short period of adaptation of Zn homeostasis (3-4 days; Windisch *et al.*, 2003). This strongly suggests that Zn from organic sources was not able to intrude into the organism to a higher degree than Zn from inorganic sources. Obviously, the major mode of action explaining better bioavailability of organic Zn sources is higher provision of absorbable Zn to the specific transporters at the intestinal mucosa due to prevention of negative interactions occurring within the lumen of the intestinal tract (e.g. precipitation by phytate). This gives further rise to the necessity to quantitatively understand interactions between essential trace minerals and other dietary compounds in order to further improve current recommendations.

Organic Se compounds are widely accepted to be highly bioavailable due to a higher retention in tissues compared to inorganic sources as demonstrated in numerous studies (e.g. Jang *et al.*, 2008). This effect is mediated mainly by selenomethionine (SeMet), the major Se compound in native Se of plant origin as well as in commercial organic Se formulations (e.g. selenized yeast). Data from a ^{75}Se radiotracer study comparing Se metabolism of Se from

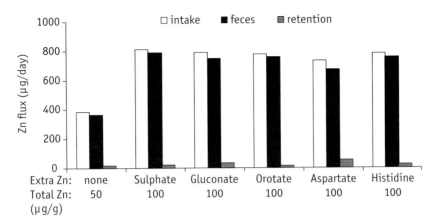

Figure 5. Reaction of faecal Zn excretion and retention on extra Zn additions of inorganic and organic origin in rats supplied with sufficient amounts of dietary Zn.

selenite and SeMet in rats fed Se close to requirement (150 µg/kg dietary Se) demonstrated almost complete true absorption of SeMet as well as a high rate of retention in tissues (Table 2) (Windisch *et al.*, 1997). These effects are known to arise from the fact that SeMet is recognized as regular methionine (Met), absorbed very efficiently along the Met pathway, incorporated unspecifically into protein synthesis and hence withdrawn from Se metabolism due to retention in tissue protein. As the pronounced drain of Se into protein retention reduces current amounts of directly available Se, the Se homeostasis responds with lower Se excretions along the endogenous faecal and especially the urinary path. Se ingested as selenite is absorbed also very efficiently, but is recognized by Se homeostasis and immediately excreted as far as it exceeds requirement. Consequently, the difference in 'bioavailability' between inorganic and organic Se is not only based on the chemical nature of the compounds *per se*, but also on interaction with regular Met ingested with the diet. The extent to which Se from SeMet is accumulated into protein may be supposed to rise with an overall increase of protein retention (high zootechnical performance) and with sinking supply of Met compared to requirement (high rate of utilization of absorbed Met for protein synthesis) (e.g. Butler *et al.*, 1989). Nevertheless, this interaction is beneficial to agricultural livestock production as it generates storage of Se within the animals' body protein which may be mobilized during episodes of deficient Se supply along with protein turnover and amino acid catabolism (Zuberbuehler *et al.*, 2006).

Table 2. Absorption, excretion and retention of Se from selenite or selenomethionine in rats fed restrictively a purified diet containing 150 µg/kg Se.

Se flux	Se from Selenite	Se from Selenomethionine
Se intake (ng/day)	1549	1550
Apparent Se absorption (% intake)	82[b]	92[a]
True Se absorption (% intake)	92[b]	99[a]
Endogenous faecal Se excretion (% intake)	10[a]	7[b]
Urinary Se excretion (% intake)	36[a]	25[b]
Se retention (% intake)	47[b]	67[a]

Conclusions

Experimental derivation of essential trace mineral requirement is well developed in terms of using reliable response parameters to estimate the transition from deficient to sufficient supply in individual dose-response studies. Transferring this information into recommendations, however, is still difficult. This is especially due to insufficient knowledge about availability of dietary trace element formulations, the more so as absorption and metabolic utilization may be severely affected by interactions with other dietary constituents. This forces nutritionists to continue with comparably large safety margins when expressing recommendations. Therefore, the bottleneck in improving precision and reliability of recommendations is quantitative understanding of these interactions. This refers also to the use of highly bioavailable trace element compounds (e.g. organic formulations) as differences in bioavailability originate mainly from variations in the extent of interaction with other dietary compounds (e.g. with strong chelators).

References

Adeola, O., 1995. Digestive utilization of minerals by weanling pigs fed copper- and phytase-supplemented diets. Can. J. Anim. Sci. 75, 603-610.

Beard, J.L. and H.D. Dawson, 1997. Iron. In: O´Dell BL, Sunde RA (Eds.) Handbook of nutritionally essential elements, Marcel Dekker Inc, New York, pp. 275-334.

Bremner, I., W.R. Humphries, M. Phillippo, M.J. Walker and P.C. Morrice, 1987. Iron-induced copper deficiency in calves: dose response relationships and interactions with molybdenum and sulphur. Anim. Prod. 45, 403-414.

Butler, J.A., M.A. Beilstein and P.D. Whanger, 1989. Influence of dietary methionine on the metabolism of selenomethionine in rats. J. Nutr. 119, 1001-1009.

Downs, K.M., J.B. Hess, and S.F. Bilgili, 2000. Selenium source effect on broiler carcass characteristics, meat quality and drip loss. J. Appl. Anim. Res. 18, 61-72.

Dunkel, S., H. Kluge, J. Spilke and K. Eder, 2004. Wirkungen von Aminosäure-Zink- und Manganchelaten auf Milchleistung und Tiergesundheit. Milchpraxis. 42, 180-181.

Ettle, T., W. Windisch and F.X. Roth, 2005. The effect of phytase on the bioavailability of zinc in piglets. In: Strain, J.J. *et al.* (Eds.): TEMA 12: 12th International Symposium on Trace Elements in Man and Animals, 19.-23.06.2005, University of Ulster, Coleraine, Northern Ireland, United Kingdom: 55.

Ettle, T., P. Schlegel and F.X. Roth, 2008. Investigations on iron bioavailability of different sources and supply levels in piglets. J. Anim. Phys. Anim. Nutr. In Press.

Hansen, S. L., P. Schlegel, K.E. Lloyd and J.W. Spears, 2008. Bioavailability of copper from copper glycinate in steers fed high dietary sulfur and molybdenum. In: P. Schlegel, S. Durosoy and A.W. Jongbloed (Eds.), Trace elements in animal production systems, Wageningen Academic Publishers, Wageningen, the Netherlands, pp.276-277.

Jang, Y.D., H.B. Choi, S. Durosoy, P. Schlegel, B.R. Choi and Y.Y.Kim, 2008. Efficacy of three selenium sources for growing finishing pigs. . In: P. Schlegel, S. Durosoy and A.W. Jongbloed (Eds.), Trace elements in animal production systems, Wageningen Academic Publishers, Wageningen, the Netherlands, pp.264-266.

Jongbloed, A.W., J.T.M. van Diepen, P.A. Kemme and J. Broz, 2004. Efficacy of microbial phytase on mineral digestibility in diets for gestating and lactating sows. Livest. Prod. Sci. 91, 143-155.

Jongbloed, A.W., P. Kemme, G. De Groote, M. Lippens and F. Meschy, 2002. Bioavailability of major and trace minerals. International Association of the European (EU) Manufacturers of Major, Trace and specific Fed Materials (EMFEMA), Brussels, Belgium.

Kies, A.K., P.A. Kemme, L.B.J. Sebek, J.T.M. van Diepen and A.W. Jongbloed, 2006. Effect of graded doses and a high dose of microbial phytase on the digestibility of various minerals in weaner pigs. J. Anim. Sci. 84, 1169-1175.

Kegley, E.B., Spears, J.W., Flowers, W.L., Schoenherr, W.D., 2002: Iron methionine as a source of iron for the neonatal pig, Nutr. Res. 22, 1209-1217.

Kies, A.K., Kemme, P.A., Sebek, L.B.J., van Diepen, J.T.M., Jongbloed, A.W., 2006: Effect of graded doses and a high dose of microbial phytase on the digestibility of various minerals in weaner pigs. J. Anim. Sci. 84, 1169-1175.

Kincaid, R.L., Chew, B.P., Cronrath, J.D., 1997. Zinc oxide and amino acids as sources of dietary zinc for calves: Effects on uptake and immunity. J. Dairy Sci. 80, 1381-1388.

Kirchgessner M., 1993. Homeostasis and homeorhesis in trace element metabolism. In: Trace Elements in Man and Animals – TEMA 8. Anke M. *et al.* (Ed.) Gerstorf Germany: Verlag Media Touristik (1993) pp.4-21.

Kirchgessner, M., W. Windisch and E. Weigand, 1993. True bioavailability of zinc and manganese by isotope dilution technique. In: Schlemmer, U. (Ed.): Bioavailability '93: Nutritional, chemical and food processing implication of nutrient availability, 09.-12-05.1993, Karlsruhe, D, 213-222.

Kirchgessner, M., S. Reuber and M. Kreuzer, 1994. Endogenous excretion and true absorption of cobalt as affected by the oral supply of cobalt. Biol. Trace Elem. Res. 41, 175-189.

Kirchgessner, M., S. Gabler and W. Windisch, 1997. Homeostatic adjustments of selenium metabolism and tissue selenium to widely varying selenium supply in ^{75}Se labelled rats. J. Anim. Physiol. a. Anim. Nutr. 78, 20-30.

Kirchgessner, M., J. He and W. Windisch, 1999. Homeostatic adjustments of iodine metabolism and tissue iodine to widely varying iodine supply in ^{125}I labelled rats. J. Anim. Physiol. a. Anim. Nutr. 82, 238-250.

Koenig, K.M., L.M. Rode, L.M. Coen and W.T. Bucklet, 1997. Effects of diet and chemical form of selenium on selenium metabolims in sheep. J. Anim. Sci. 75, 817-827.

Lantzsch, H.-J., 1990. Untersuchungen über ernährungsphysiologische Effekte des Phytats bei Monogastriern (Ratte, Schwein). Übers. Tierernährg. 18, 187-211.

Mahan, D.C., T.R. Cline and B. Richert, 1999. Effect of dietary levels of selenium enriched yeast and sodium selenite as selenium sources fed to growing-finishing pigs on performance, tissue selenium, serum glutathione peroxidase activity, carcass characteristics, and loin quality. J. Anim. Sci. 77, 2172-2179.

NRC, 1998. Nutrient requirements of swine. 10th ed. National Academy Press. Washington D.C., NW, USA.

NRC, 2001. Nutrient requirements of dairy cattle. 7th ed. National Academy Press. Washington D.C., NW, USA.

Pallauf, J., S. Pippig, E. Most and G. Rimbach, 1999. Supplemental sodium phytate and microbial phytase influence iron availability in growing rats. J. Trace Elem. Med. Bio. 13, 134-140.

Pfaffl, M.W. and W. Windisch, 2003. Influence of zinc deficiency on the mRNA expression of zinc transporters in adult rats. J. Trace Elem. Med. Biol. 17, 197-206.

Phillippo, M., W.R. Humphries and P.H. Garthwaite, 1987. The effect of dietary molybdenum and iron on copper status and growth in cattle. J. Agric. Sci. Camb. 109, 315-320.

Prabowo, A., J.W. Spears and L. Goode, 1988. Effects of dietary iron on performance and mineral utilization in lambs fed a forage-based diet. J. Anim. Sci. 66, 2028-2035.

Reichlmayr-Lais, A.M. and M. Kirchgessner, 1992. Effects of increasing alimentary Fe supply on the apparent digestibility of Fe, Cu, Zn, and Mn as well as on the contents of these elements in liver and carcass. J. Anim. Phys. Anim. Nutr. 67, 67-73.

Schlegel, P. and W. Windisch, 2006. Bioavailability of zinc glycinate in comparison to zinc sulfate in the presence of dietary phytate in an animal model with ^{65}Zn labelled rats. J. Anim. Physiol. a. Anim. Nutr. 90, 216-222.

Schöne, F., Chr. Zimmermann, G. Quanz, G. Richter and M. Leiterer, 2006. A high dietary iodine increases thyroid iodine stores and iodine concentration in blood serum but has little effect on muscle iodine content in pigs. Meat Science 72, 365-372.

Spears, J. W., 2003. Trace mineral bioavailability in ruminants. J. Nutr. 133, 1506S-1509S.

Spears, J. W., P. Schlegel, M.C. Seal and K.E. Lloyd, 2004. Bioavailability of zinc from zinc sulfate and different organic zinc sources and their effects on ruminal volatile fatty acid proportions. Lives. Prod. Sci. 90, 211-217.

Standish, J.F., C.B. Ammerman, C.F. Simpson, F.C. Neal and A.Z. Palmer, 1969. Influence of graded levels of dietary iron, as ferrous. J. Anim. Sci. 29, 496-503.

Stangl, G.I., F.J. Schwarz, H. Mueller and M. Kirchgessner, 2000. Evaluation of the cobalt requirement of beef cattle based on vitamin B12, folate, homocysteine and methylmalonic acid. Br. J. Nutr. 84, 645-653.

Suttle, N.F., 1991. The interactions between copper, molybdenum, and sulphur in ruminant nutrition. Ann. Rev. Nutr. 11, 121-140.

Walter, A., K. Kraemer, E. Most and J. Pallauf, 2003. Zinc bioavailability from zinc lipoate and zinc sulfate in growing rats. Trace Elem. Med. Biol. 16, 169-174.

Weigand, E. and M. Kirchgessner, 1978. Homeostatic adjustments in zinc digestion to widely varying dietary zinc intake. Nutr. Metab. 22, 101-112.

Weigand, E. and M. Kirchgessner, 1997. Zur Absorbierbarkeit von Zink aus unterschiedlichen Verbindungen. Z. Tierphys. Tierernährg. u. Futtermittelkde. 42, 137-146.

Weigand E. and M. Kirchgessner, 1992. Absorption, endogenous excretion, and balance of zinc in growing rats on diets with various sugars replacing starch. Biol. Trace Elem. Res. 34, 67-77.

Wiedner, G., T. Guggenberger and H. Fachberger, 2001. Futterwerttabelle der österreichischen Grundfuttermittel. Niederösterreichische Landeslandwirtchaftskammer, Bundesanstalt fuer alpenlaendische Landwirtschaft Gumpenstein (Eds.) Druckhaus Lahnsteiner GmbH, Wieselburg, Austria

Windisch, W., 2001. Homeostatic reactions of quantitative Zn metabolism on deficiency and subsequent repletion with Zn in ^{65}Zn labelled adult rats. Trace Elements and Electrolytes. 18, 128-133.

Windisch, W., 2003. Development of zinc deficiency in ^{65}Zn labelled, fully grown rats as a model for adult individuals. J. Trace Elem. Med. Biol. 17, 91-96.

Windisch, W. and M. Kirchgessner, 1994. Calcium- und Zinkbilanz ^{65}Zn-markierter Ratten bei defizitärer und moderat hoher Ca-Versorgung. J. Anim. Physiol. a. Anim. Nutr. 72, 184-194.

Windisch, W. and M. Kirchgessner, 1995. Anpassung des Zinkstoffwechsels und des Zn-Austauschs im Ganzkörper ^{65}Zn-markierter Ratten an eine variierende Zinkaufnahme. J. Anim. Physiol. a. Anim. Nutr. 74, 101-112.

Windisch, W., L. Zhao and M. Kirchgessner, 1996. The effect of Masson pine pollen on the quantitative zinc metabolism in ^{65}Zn-labeled growing rats. Trace Elements and Electrolytes 13, 186-189.

Windisch, W., S. Gabler and M. Kirchgessner, 1997. Effect of selenite, seleno cysteine and seleno methionine on the selenium metabolism of ^{75}Se labelled rats. J. Anim. Physiol. A. Anim. Nutr. 78, 67-74.

Windisch, W. and M. Kirchgessner, 1999. Zn absorption and excretion in adult rats at Zn deficiency induced by dietary phytate additions. J. Anim. Physiol. a. Anim. Nutr. 82, 106-115.

Windisch, W., A. Vikari and C. Hilz, 2003. Homeostatic response of Zn metabolism to dietary Zn supplements from sulfat, gluconate, orotate, aspartate or histidine in 65Zn labeled non-growing rats as a model to adult individuals. Trace Elements and Electrolytes. 20, 125-133.

Yu, B., Huang, W.J., Chiou, P.W.S., 2000: Bioavailability of iron from amino acid complex in weanling pigs. Anim. Feed Sci. Technol. 86, 39-52.

Zuberbuehler, C.A., R.E. Messikommer, M.M. Arnold, R.S. Forrer and C. Wenk, 2006. Effects of selenium depletion and selenium repletion by choice feeding on selenium status of young and old laying hens. Physiol. Behav. 87, 430-440.

Trace mineral requirements of fish and crustaceans

S.P. Lall and J.E. Milley

Keywords: fish, shrimp, trace minerals, bioavailability

Introduction

Unlike terrestrial animals, fish and crustaceans have unique physiological mechanisms to absorb and retain minerals from their diets and water. Calcium, magnesium, sodium, potassium, iron, zinc, copper, iodine, and selenium are readily absorbed from the water to satisfy part of the nutritional requirements of fish. About 29 naturally occurring elements are known to be essential for animal life and these elements are also considered important to fish. Most of these trace elements have been detected in fish tissues. Among the 15 known essential trace elements for animals, the requirement for only six elements (copper, iron, zinc, manganese, selenium and iodine) and their deficiencies have been investigated in certain fish. Some aspects for vanadium, boron, silicon, cadmium, arsenic and fluorine have been considered in fish metabolism; however, a physiological function of these trace elements has not been demonstrated. The exchange of ions from the surrounding water across the gills and skin of fish complicates the determination of the quantitative dietary requirement.

The main role of essential elements in fish and crustaceans include cellular metabolism (e.g. oxygen transport, respiration and enzyme activities), formation of skeletal structure, maintenance of colloidal systems (e.g. osmotic pressure, viscosity, diffusion), regulation of acid-base equilibrium and other important physiological functions (e.g. growth, reproduction, vision, immunity). They are important components of hormones, enzymes, and activators of enzymes. Calcium and phosphorus are required for the formation of skeletal structures of the body. Sodium, potassium, and chloride, along with phosphates and bicarbonates, maintain homeostasis and acid-base balance. Several trace metals (Fe, Mn, Cu, Co, Zn, Mo, Se) are firmly associated with a specific protein in metalloenzymes and produce a unique catalytic function. Calcium, magnesium, and manganese, are of particular significance as enzyme activators. Iodine is necessary for the biosynthesis of thyroid hormones, which in turn greatly affect development and metabolism in fish and other vertebrates. Some biologically important compounds contain mineral as an inherent part of their structure, e.g. haemoglobin and vitamin B_{12}.

Aquatic animals must maintain their osmotic balance between body fluids and in the aquatic environment which can vary from 0 to 35 $^o/_{oo}$ salinity between in fresh, brackish, and seawater. In the past two decades, significant progress has been made to develop a better understanding of mineral uptake by the fish gill from water; however, there have been limited studies on the absorption of trace elements from the gut. Most studies on trace element uptake from the gut and gills have been confined to freshwater and to a limited extent on marine fish.

In freshwater, the active uptake of salts through the gills enables them to maintain blood ion levels that are more concentrated than those of the external aquatic medium. Kidneys help in osmoregulation by maintaining a high glomerular filtration rate along with tubular and bladder reabsorption of filtered ions, producing dilute urine. Marine fish lose water through any permeable surface and thereby increase their level of salt. They replace water loss osmotically by drinking seawater. The gut actively absorbs monovalent ions and water into the blood. Excess monovalent ions derived from swallowed seawater and the passive uptake across the body surface is excreted, mainly through the gills. Lower trace element uptake from the gill in the marine environment suggests that intake of waterborne metals through drinking and food consumption is the main source of trace elements for marine fish (Rainbow, 1995). Preliminary studies on the transport mechanisms of trace elements such as iron, zinc and copper indicate that metal transport proteins play an important role in the acquisition of metals in fish through gills in freshwater and seawater and contribute significantly to their homeostasis (Bury *et al.*, 2003).

An excessive intake of minerals from either diet or gill uptake can cause toxicity and therefore a fine balance exists between mineral deficiency and surplus. It is vital for aquatic organisms to maintain their homeostasis either through increased absorption or by excretion. Studies on toxic elements have revealed that many aquatic organisms have the ability to regulate abnormally high concentrations by excretion of excessive metal intake and thereby maintaining relatively normal levels in the body (Handy, 1996). The soluble trace elements in water are more toxic than high dietary intake for minerals such as copper, cadmium, iron, and zinc. The toxicity mechanisms of metal ions include blocking of essential biological functional groups of enzymes, displacing the essential metal ion in the biomolecule (enzyme or protein), and modifying the active conformation of the biomolecule. This paper briefly reviews the current state of the knowledge on trace element nutrition of fish and crustaceans with an emphasis on iron, zinc, copper, manganese and selenium.

Essential trace elements for fish

Iron

Iron is an essential element in the cellular respiratory process through its oxidation-reduction activity and electron transfer. In the complex form, iron bound to proteins such as heme compounds (haemoglobin and myoglobin), heme enzymes (mitochondrial and microsomal cytochromes, catalase, peroxidase), and non-heme compounds (transferrin, ferritin, and iron-containing flavoproteins), it increases respiration, and oxygen binding and carrying capacity to transport oxygen to all tissues in aquatic organisms. Iron may also produce oxygen free radicals in cells due to its negative redox potential and that may be detrimental to cells and tissues of fish which contain high concentrations of polyunsaturated fatty acids. Excessive iron in water may flock and clog up the gills thus reducing critical cellular respiration in all aquatic organisms (Peuranen *et al.*, 1994).

Although there is relatively little information on absorption and metabolism of iron in fish, studies suggest that mechanisms of iron absorption from the digestive tract, and storage and excretion may be similar to those in other vertebrates (Lall, 2002; Bury *et al.*, 2003). Some absorption of Fe takes place across the gill membranes; however, the intestinal mucosa is considered the major site of Fe absorption. Food is regarded as the main source of Fe for metabolic purposes. In the acidic environment of the stomach, ferric ion is released from the ingested food materials and bind to mucin which may facilitate metal solubility in the small intestine of fish (Whitehead *et al.*, 1996). It is proposed that epithelial mucus secretion may play an important role in maintaining metal solubility in fish (Glover and Hogstrand, 2002).

Iron deficiency may be readily produced experimentally in certain fish fed low-Fe diets (reviewed by Lall, 2002); however, its deficiency is not commonly observed in fish cultured under hatchery or farm conditions. The most common deficiency signs of iron include microcytic anemia in brook trout, rainbow trout, Atlantic salmon, red sea bream, yellowtail, eels and carp. In most cases, the growth of fish was not influenced by Fe deficiency. In catfish, Fe deficiency suppressed haemotocrit, haemoglobin, and plasma iron levels and caused low transferrin saturation (Gatlin and Wilson, 1986). Dietary Fe toxicity has been induced in rainbow trout fed higher than 1380 mg Fe kg^{-1} (Desjardins, 1985). The major effects of Fe toxicity include reduced growth, poor feed utilization, feed refusal, increased mortality, diarrhoea and histopathological damage to liver cells.

The Fe requirement of certain fish but not crustaceans have been quantitatively established (Table 1 and Table 2). Some reports suggest that iron may not be essential for shrimp, *Paneus japonicus* (Kanazawa *et al.*, 1984). The Fe requirement for catfish and eel are 30 and 170 mg kg^{-1} of diet, respectively. A wide range of Fe requirement (33 to 100 mg kg^{-1}) values have been reported for Atlantic salmon reared in freshwater. The Fe supplied from a purified diet based on casein and gelatin (39 mg iron kg^{-1}) may not be sufficient to prevent iron deficiency in rainbow trout, indicating a need for dietary Fe supplement (Desjardins, 1985).

Table 1. Trace element requirements of certain finfish (amount per kg diet).

Species	Iron (mg)	Copper (mg)	Manganese (mg)	Zinc (mg)	Iodine (µg)	Selenium (mg)
Rainbow trout	R[a]	3	13	15-30	1.1	0.15-0.3
Atlantic salmon	30-60	5	10	37-67	R	R
Chinook salmon	R	R	R	R	0.6-1.1	R
Chum salmon	R	R	R	R	R	R
Channel catfish	30	3	2.4	20	1.1	0.25
Carp	150	3	13	15-30	R	R
Eel	170	R	R	R	R	R
Red sea bream	R	R	R	R	R	R

[a]Required, but quantitative requirement not determined.

Table 2. Mineral requirements of certain shrimps.

Minerals	*Paneus japonicus*	*Paneus monodon*	*Paneus merguiensis*
Calcium (%)	1.0[a]	1.0[c]	0.6[d]
Phosphorus (%)	1.0[a], 2.0[b]	1.0[c]	0.51[d]
Magnesium (%)	0.1 - 0.5[a]	R[f]	1-3[e]
Potassium (%)	0.9[a], 1.0[b]	R	R
Copper (%)	0.6[a]	R	R

[a]Kanazawa *et al.*, 1984.
[b]Deshimaru and Yone, 1978.
[c]Bautista and Baticados, 1989.
[d]Sick *et al.*, 1972.
[e]Aquacop, 1978.
[f] Required, but quantitative requirement not determined.

Copper

Copper (Cu) metabolism of fish is not clearly defined despite numerous investigations concerned with toxic effects. Cu is a vital component of several enzymes that are involved in oxidation-reduction reactions and occurs tightly bound to proteins in the cell rather than as free ions. Cu is associated with cytochrome oxidase of the electron transport chain in the cell and Cu metalloenzymes are involved in protection of cells from free radical damage (superoxide dismutase), collagen synthesis (lysyl oxidase) and melanin production (tyrosinase). Copper is also bound to the protein, ceruloplasmin, which occurs in the cell and plasma and is involved in iron utilization. Marine invertebrates, especially molluscs (mussel, oyster, clams) accumulate large quantities of Cu. Haemocyanin, a blue-colored Cu-containing complex, is distributed in the haemolymph of molluscs and crustaceans, where it serves as an oxygen carrier in the haemolymph of these organisms.

Diet is considered as a major source of Cu for optimum growth of fish; however, waterborne toxicity studies have demonstrated that the gill also contributes to a significant amount of Cu uptake (Taylor *et al.*, 2007). Copper enters the intestinal epithelium through a passive process with the rate limiting step in this process being the basolateral membrane extrusion (Clearwater *et al.*, 2000). Unlike terrestrial vertebrates, intestinal Cu uptake in fish primarily occurs in the mid-posterior region of the intestine. Excessive amounts of Cu supplied in the diet do not enter the body; however, it is retained in gut tissue by metallothionein (Olsen *et al.*, 1996). This metallothionein bound Cu may then be excreted into the faeces through sloughing off of the epithelial membrane (Clearwater *et al.*, 2000).

The clinical signs of Cu deficiency have not yet been reported for fish. A decrease in heart cytochrome *c* oxidase and liver copper-zinc superoxide dismutase activities have been observed in Cu-deficient catfish (Gatlin and Wilson, 1986). Fin erosion in rainbow trout has been linked to poor copper bioavailability from fish meal. Copper is widely distributed in feeds and water and therefore its deficiency would only occur in fish under extreme conditions. Copper toxicity arising from pollutants in fresh and sea water or produced experimentally may cause severe damage to the gills and necrotic changes in the liver and kidneys. The oral toxicity of this element was induced in rainbow trout fed 730 mg Cu kg^{-1} of diet (Lanno *et al.*, 1985). The toxicity signs of Cu include reduced growth and feed efficiency and elevated liver Cu levels. However, a diet of up to 665 mg of Cu kg^{-1} did not produce any toxicity signs in rainbow trout (Knox *et al.*, 1982; Lanno *et al.*, 1985).

The dietary Cu requirements of selected fish species have been reported in the following fish (mg per kg diet) and crustacean (% diet): rainbow trout and carp, 3; catfish, 5; and Atlantic salmon, shrimp, 0.6 (Table 2). Dietary Cu requirement may also depend on the physiological state of the animal, Cu concentration in the water, and also on certain elements considered metabolic antagonists of copper, such as iron, zinc, cadmium, and molybdenum. The antagonistic effects of zinc and copper were not observed in rainbow trout (Knox *et al.*, 1982).

Manganese

Manganese (Mn) is widely distributed in aquatic animal tissues. The highest concentration is found in fish bones; however, significant amounts are also present in liver, muscle, kidney, gonadal tissues and skin. Manganese, as a cofactor activating a large number of enzymes or as an integral part of certain metalloenzymes, is involved in carbohydrate, lipid, and protein metabolism. Although the uptake of Mn from water has been demonstrated, fish absorb Mn more efficiently from dietary sources. Mn absorption may be reduced by high levels of dietary calcium and phosphorus. The Mn requirements (mg per kg diet) of the following fish have been determined: channel catfish, 2.5; Atlantic salmon 7.5 to 10.5; carp and rainbow trout, 13 (Ogino and Yang, 1980). The Mn requirement of shellfish has not been investigated. Broodstock salmon and trout require considerably higher (> 30 mg kg^{-1} diet) amount of this element for their reproduction (Lall, unpublished).

Manganese deficiency causes reduced growth and skeletal abnormalities in rainbow trout, carp, and tilapia. In rainbow trout, low Mn intake decreases the activities of Cu-Zn-superoxide dismutase and Mn-superoxide dismutase in cardiac muscle and liver; it also suppresses Mn and calcium concentration of vertebrae (Knox *et al.*, 1981). Fish fed fish meal-based diets without Mn supplement had poor hatchability and low Mn levels of eggs of rainbow trout and Atlantic salmon (Takeuchi *et al.*, 1981; Lall, unpublished). Generally, Mn supplementation of practical broodstock diets is necessary because most feed ingredients used in the diets of salmonid fish do not supply sufficient Mn.

Zinc

The essentiality of zinc (Zn) for fish and other aquatic organisms is related to the important role of this element as an integral part of a number of metalloenzymes as well as a catalyst for regulating the activity of specific Zn-dependent enzymes in lipid, carbohydrate, and protein metabolism. The major routes of zinc absorption in fish are through the gill and intestinal track both in freshwater (Spry *et al.*, 1988) and seawater (reviewed by Lall, 2002). It is generally agreed, however, that Zn is more efficiently absorbed from the diet. In freshwater, even when dietary Zn levels are adequate, there is an active uptake of waterborne Zn (Spry *et al.*, 1988). The Zn status of fish is tightly controlled and surplus Zn is excreted via bile, the sloughing of intestinal mucosa in faeces and through gills (Handy, 1996). The accumulation of Zn in gills is also regulated through alteration in Zn uptake mechanisms limiting its excessive uptake (reviewed by Bury *et al.*, 2003). Zinc chelation with certain amino acids, such as histidine and cysteine that have high affinity for this element, may enhance Zn distribution in various tissues of fish (Glover and Hogstrand, 2002).

In rainbow trout, Zn deficiency causes growth depression, high mortality, lens cataracts, erosion of fins and skin and short body dwarfism. Excess minerals (total ash) present in white fish meal may affect Zn absorption and metabolism, resulting in lens cataract (Ketola, 1979). Zinc supplements (40 mg kg^{-1} or higher) in diets containing white fish meal alleviated dwarfism and cataract problems (Satoh *et al.*, 1987). Caudal fin Zn concentration is a good indicator of Zn status in rainbow trout. In catfish, diets low in Zn caused reduced growth, appetite, bone zinc and calcium levels, and serum Zn concentrations (Gatlin and Wilson, 1983). Zinc is considered as an essential trace element for reproduction of male and female fish. Broodstock diets low in Zn showed reduced egg production and hatchability of eggs (Takeuchi *et al.*, 1981). Elevated levels of dietary Zn (500 to 1,000 mg Zn kg^{-1}) caused reduced haemoglobin, hematocrit, and hepatic copper concentrations in rainbow trout (Knox *et al.*, 1984). Common carp accumulate much higher concentrations of Zn in their tissues, particularly in the viscera, than other fish studied, without any apparent toxicity signs (Jeng and Sun, 1981).

A dietary requirement for Zn (mg kg^{-1}) has been reported for several freshwater fishes (Table 1): rainbow trout and common carp, 15-30; Atlantic salmon, 37-67; channel catfish, 20; blue tilapia, 20; red drum, 20-25 ; guppy, 80. The minimum Zn requirement may vary with age, sexual maturity, composition of diet, water temperature and water quality. The Zn requirement of shrimp and other crustaceans has not been investigated.

Selenium

Although, the essentiality of selenium (Se) in farmed animals is well established, the importance of this element is also being recognized for fish. Se is an integral part of the enzyme glutathione peroxidase, which is known to destroy hydrogen peroxide and hydroperoxides to alcohol by reducing equivalents from glutathione, thereby protecting cells and membranes against peroxide damage. A low concentration of Se is found extensively in freshwater. The absorption, distribution and excretion of dietary Se have been reviewed by Hilton (1989). The uptake of Se across gills is very efficient at low waterborne concentrations. Liver and kidney play an important role in the excretory process of Se in trout, however, the major excretory routes appear to be the gills and urine.

Selenium deficiency causes growth depression in rainbow trout (Hilton *et al.*, 1980) and catfish (Gatlin and Wilson, 1984), but Se deprivation alone does not produce any pathological signs in these fish. Both Se and vitamin E are required to prevent muscular dystrophy in Atlantic salmon and exudative diathesis in rainbow trout. Glutathione peroxidase activity in plasma and liver is a useful index of Se status in fish.

The interaction of Se and vitamin E, polyunsaturated fatty acids, other dietary factors, and the concentration of waterborne Se significantly influences the requirement for Se. The Se requirement determined on the basis of optimum growth and maximum plasma glutathione peroxidase activity is estimated to be 0.15 to 0.38 mg Se kg^{-1} diet for rainbow trout (Hilton *et al.*, 1980) and 0.25 mg Se kg^{-1} for channel catfish (Gatlin and Wilson, 1984). Selenium toxicity occurs in rainbow trout and catfish when dietary selenium exceeds 13 and 15 mg kg^{-1} dry feed, respectively (Hilton *et al.*, 1980; Gatlin and Wilson, 1984). Reduced growth, poor feed efficiency, and high mortality are the major effects. Rainbow trout fed diets containing high levels of Se (10 mg kg^{-1}) show renal calcinosis (Hilton and Hodson, 1983).

Other trace elements

Information on the requirements of other trace elements is limited. Iodine is essential for the biosynthesis of thyroid hormones, thyroxine and iodothyronine in fish. Most fish and other aquatic organisms obtain iodine primarily from water through the gills and to a lesser extent from food sources. Iodine deficiency causes goiter in fish. Experimentally induced thyroid hormone deficiency induced by feeding thioglusinolates or rapeseed/canola meal caused growth depression in salmonid fishes (Higgs *et al.*, 1995). The minimum iodine requirement of most fish and crustacean species has not been studied. Marine fish have a much higher concentration of iodine in their body than freshwater fish. The level of iodine in common feedstuffs and fishery products is highly variable and a significant amount of iodine is lost during fish meal processing.

The importance of chromium, cobalt, fluorine and other elements is also recognized in fish nutrition. Many crustaceans including krill contain high concentrations of fluorine and arsenic. Increased dietary fluoride or krill supplementation enhances fluoride accumulation in the vertebrae of rainbow trout (Julshamn *et al.*, 2004). A role of boron in embryonic development of rainbow trout eggs has been demonstrated (Eckhert, 1998). Deficiency signs and requirement of these elements remains to be established.

Bioavailability

In order to determine how efficiently the body of fish utilizes dietary trace elements, one must know the availability of that element from a feed ingredient or complete diet. The bioavailability of an element can differ markedly when supplied from different feedstuffs and with the same element from feed in different diets. Generally, the bioavailability of a trace element depends on the following three major factors, a) physiological regulation by the animal, b) antagonist present in feed or feed ingredient, and c) chemical form in which it is supplied to the animal. In addition to the above factors, digestibility of the diet that supplies the element, particle size, physiological and pathological states of the animal, water chemistry, type of feed processing, and the species of animal being tested may also affect bioavailability of trace elements in fish. Limited studies have been undertaken on mineral bioavailability of fish and these studies have been mainly related to iron, zinc and selenium.

Aquafeeds of carnivorous fish contain a high proportion of fish meal and other marine by-products. Some minerals in these ingredients are found in concentrations above the recommended requirement levels. Excessive amounts of minerals, particularly calcium and phosphorus, reduce zinc bioavailability and have also been linked to cataract formation in juvenile salmonid fishes (reviewed by Lall, 2002). The increasing demands of the world's aquaculture production upon the finite quantity of high-quality protein provided from fish meals requires that fish feeds become increasingly comprised of alternate sources of highly digestible protein from plant and/or animal origin. There is a major emphasis to formulate fish feed based on alternate plant proteins, particularly soybean meal, canola meal and their concentrates, pulse meals, corn and wheat gluten meal and other oilseed proteins. In diets containing high levels of plant protein, mineral supplementation is necessary to improve growth and bone mineralization of carnivorous salmonid and marine fishes. Zinc is better absorbed from animal protein supplements than from plant protein sources. Although fish meal, fish solubles and other marine by-products represent the best natural source of selenium, certain fish meals (e.g. tuna), may have a poor biological availability of selenium due to complexing of this elements with heavy metals such as mercury.

Inorganic and organic minerals

Inorganic compounds of zinc, iron, copper, manganese and selenium are commonly used as supplements in feeds. These compounds are hydrolyzed in the digestive track. The formation of insoluble and non-absorbable substances in the gut may either hinder or facilitate mucosal uptake, transport and metabolism of an element. Thus the absorption of an element from a diet depends upon the molecular form in which the element is present, their valence states, and the ligands present when ingested from different diets and the pH of the gut. Once solubilised, the metals are sensitive to changes in oxidation state and they form complexes with the ligands of the chime. Inorganic supplements are therefore considered more susceptible to undesirable speciation into the forms that are less susceptible to digestive regulation. Certain inorganic elements may compete with the test element in the diet for important binding sites during these processes.

Generally oxide forms of Zn, Fe, Cu and Mn are poorly absorbed by fish. The concentration of Fe in common feedstuff is highly variable and influenced by the degree of contamination from ferrous metal during processing. Fish meal and marine by-products, the major sources of protein in aquatic animal feeds, contain approximately 150 to 800 mg Fe kg^{-1}. The bioavailability of iron may be also affected by age and species of test animals, intake of iron relative to the need and the amount and proportion of other dietary components with which Fe interacts metabolically. The utilization of ferrous sulfate and ferric chloride by fish are considered essentially the same. In red sea bream, ferrous and ferric chloride are more efficiently utilized than ferric citrate (Sakamoto and Yone, 1979). Biological availability of Fe measured by a haemoglobin regeneration assay in Atlantic salmon showed that the relative availability of Fe from ferric chloride, ferric oxide, blood meal and herring meal was 98.8, 17.8, 52.3 and 47.1%, respectively (Naser, 1997). However, haemoglobin slope ratio method showed that Fe in blood meal was more efficiently utilized than ferrous sulfate by Atlantic salmon (Anderson *et al.*, 1997). Ascorbic acid enhances iron absorption whereas phytate and tannic acid may decrease iron absorption. High concentrations of Cu (> 85 mg kg^{-1}) have been detected in condensed fish solubles and whey products but their availability to fish and crustaceans have not been studied. Processed feed ingredients of plant and animal origin show a variable Cu content resulting from contaminants. Moisture during processing of materials rich in protein provides a favorable condition for the uptake of Cu by the protein fraction.

The availability of Mn differs in various inorganic salt supplements. Manganese found in manganous oxide is poorly utilized by rainbow trout and Atlantic salmon (Watanabe *et al.*, 1997; Lall, unpublished). The availability of Mn is low in manganese carbonate for carp (Watanabe *et al.*, 1997). Zinc sulfate (ZnSO$_4$) and ZnNO$_3$ are effectively utilized by rainbow trout as Zn supplements (40 mg kg^{-1}) in diets containing white fish meal, (Satoh *et al.*, 1987). Some differences exist in Zn bioavailability from feedstuffs of plant and animal origin. Zinc is better absorbed from animal protein supplements than from plant protein sources. Cereals and other plant feedstuffs contain a number of substances, particularly phytate, which can bind zinc, and make it unavailable for absorption. Soluble phytates added to animal protein decrease Zn bioavailability and account for a large part of the low availability in oil seed protein. The bioavailability of Zn in fish meal is greatly affected by the tricalcium phosphate content (Satoh *et al.*, 1987). Higher levels of supplemental Zn is included in practical feeds to compensate for reduced Zn bioavailability caused by dietary phytate, calcium, and phosphorus.

Chelated minerals have been successfully used in farmed animal feeds. Studies conducted on some fish species show that chelated minerals may have higher bioavailability than inorganic forms. In organic trace element supplements, the metal ions are chelated with ligands of different nature e.g. amino acids, which limit losses due to precipitation in the gut and facilitates trace element uptake. Amino acid chelates of zinc and copper appear to be more readily available than inorganic sources of these minerals in rainbow trout (Apines *et al.*, 2003). Chelated forms of copper, iron, manganese, selenium, and zinc (proteinates) were shown to have higher bioavailability to catfish in purified and practical diets (Paripatananont and Lovell, 1997). An improvement in the availability of minerals in chelated forms rather than inorganic form in catfish diets has been attributed to inhibitory compounds such as phytate and fiber in feedstuffs for which chelated minerals are less affected. No apparent differences in the bioavailability of zinc sulfate and zinc methionine were noted in catfish (Li and Robinson, 1996). Use of citric acid and amino acid-chelated minerals in red sea bream diets improved growth, feed utilization, nutrient retention, and lowered nitrogen and P excretion (Sarkar *et al.*, 2005). Organic forms of selenium including selenomethionine and selenoyeast have been shown to have higher bioavailability than inorganic sodium selenite for channel catfish (Wang *et al.*, 1997). Selenomethionine supplementation in Atlantic salmon diets increased the muscle Se concentration but growth was not affected (Lorentszen *et al.*,

1994). Potentially the use of chelated minerals with higher bioavailability may allow for lower supplementation and reduced waste from unassimilated minerals. It is apparent from the above discussion that there is a need to use more than one criteria (growth, enzyme activity, mineral absorption and retention, blood or tissue concentration and prevention of certain deficiency sign) to properly establish the bioavailability of inorganic and organic iron, zinc, copper, manganese and selenium and there is a need to standardize experimental design and proper test diets.

References

Andersen, F., M. Lorentszen, R. Waagbø and A. Maage, 1997. Bioavailability and interactions with other micronutrients of three dietary iron sources in Atlantic salmon, *Salmo salar* L. smolts. Aquac. Nutr. 3, 239-246.

Apines, M.J., S. Satoh, V. Kiron, T. Watanabe and S. Fujita, 2003. Bioavailability and tissue distribution of amino acid-chelated trace elements in rainbow trout *Oncorhynchus mykiss*. Fish. Sci. 69, 722-730.

Aquacop, 1978. Study of nutrition requirements and growth of *Penaeus merguiensis* in tanks by means of purified and artificial diets. Proc. World Maricul. Soc. 9, 225-234.

Bautista, M.N. and M.C.L. Baticados, 1989. Dietary manipulation to control the chronic soft-shell syndrome in tiger prawn *Penaeus monodon* Fabricus. In: Hiranko, R. and I. Hanyu (Eds.) Proc. Second Asian Fisher. Forum, Tokyo, Japan 17-22 April, pp. 341-344.

Bury, N.R., P.A. Walker and C.N. Glover, 2003. Nutritive metal uptake in teleost fish. J. Exp. Biol. 206,11-23.

Clearwater, S.J., S.J. Baskin, C.M. Wood and D.G. McDonald, 2000. Gastrointestinal uptake and distribution of copper in rainbow trout. J. Exp. Biol. 203, 2455-2466.

Deshimaru, O. and Y. Yone, 1978. Requirements of prawn for dietary minerals. Bull. Jpn. Soc. Sci. Fish. 44, 907-910.

Desjardins, L.M., 1985. The effect of iron supplementation on diet rancidity and on growth and physiological response of rainbow trout. M. Sc. Thesis, University of Guelph, Canada.

Eckhert, C.D., 1998. Boron stimulates embryonic trout growth. J. Nutr. 128, 2488-2493.

Gatlin, D.M.III and R.P. Wilson, 1983. Dietary zinc requirement of fingerling channel catfish. J. Nutr. 113, 630-635.

Gatlin, D.M. III and R.P. Wilson, 1984. Dietary selenium requirement of fingerling channel catfish. J. Nutr. 114, 627-633.

Gatlin, D.M. III and R.P. Wilson, 1986. Characterization of iron deficiency and the dietary iron requirement of fingerling channel catfish. Aquaculture 52, 191-198.

Glover, C.N. and C. Hogstrand, 2002. *In vivo* characterisation of intestinal zinc uptake in freshwater rainbow trout. J. Exp. Biol. 205, 141-150.

Handy, R.D., 1996. Dietary exposure to trace metals in fish. In: E.W. Taylor (Ed.), Toxicology of Aquatic Pollution, Cambridge University Press, Cambridge, United Kingdom, pp. 29-60.

Higgs, D.A., B.S. Dosanjh, A.F. Prendergast, R.M. Beames, R.W. Hardy, W. Riley and G. Deacon, 1995. Use of rapeseed/canola protein products in finfish diets. In: Lim, C. and D.J. Sessa (Eds.), Nutrition and Utilization Technology in Aquaculture, AOCS Press, IL, pp.130-160.

Hilton, J.W., 1989. The interaction of vitamins, minerals and diet composition in the diet of fish. Aquaculture 79, 223-244.

Hilton, J.W. and F.V. Hodson, 1983. Effect of increased dietary carbohydrate on selenium metabolism and toxicity in rainbow trout (*Salmo gairdneri*). J. Nutr. 113, 1241-1248.

Hilton, J.W., J.V. Hodson and S.J. Slinger, 1980. The requirement and toxicity of selenium in rainbow trout, *Salmo gairdneri*. J. Nutr. 110, 2527-2535.

Jeng, S.S and L.T. Sun, 1981. Effects of dietary zinc levels on zinc concentrations in tissues of common carp. J. Nutr. 111, 134-140.

Julshamn, K., M.K. Malde, K. Bjornvatn and P. Krogedal, 2004. Fluoride retention of Atlantic salmon (*Salmo salar*) fed krill meal. Aquac. Nutr. 10, 9-13.

Kanazawa, A., S. Teshima and M. Sasaki, 1984. Requirements of potassium, copper, magnesium and iron. Mem. Fac. Fish. Kagoshima Univ. 33, 63-71.

Ketola, H.G., 1979. Influence of dietary zinc on cataracts in rainbow trout. J. Nutr. 109, 965-969.

Knox, D., C.B. Cowey and J.W. Adron, 1981. The effect of low dietary manganese intake on rainbow trout (*Salmo gairdneri*). Br. J. Nutr. 46, 495-501.

Knox, D., C.B. Cowey and J.W. Adron, 1982. Effects of dietary copper and copper:zinc ratio on rainbow trout *Salmo gairdneri*. Aquaculture 27, 111-119.

Knox, D., C.B. Cowey and J.W. Adron, 1984. Effects of dietary zinc intake upon copper metabolism in rainbow trout (*Salmo gairdneri*). Aquaculture 40, 199-207.

Lall, S.P., 2002. Minerals. In: J.E. Halver and R.W. Hardy (Eds.), Fish Nutrition, Academic Press, San Diego, CA, pp. 260-308.

Lanno, R.P., S.J. Slinger and J.W. Hilton, 1985. Maximum tolerable and toxicity levels of dietary copper in rainbow trout (*Salmo gairdneri*, Richardson). Aquaculture 49, 257-268.

Li, M.H. and E H. Robinson, 1996. Comparison of chelated zinc and zinc sulphate as zinc sources for growth and bone mineralization of channel catfish (*Ictalurus punctatus*) fed practical diets. Aquaculture 46, 237-243.

Lorentszen, M., A. Maage and K. Julshamn, 1994. Effects of dietary selenite or selenomethionine on tissue selenium levels of Atlantic salmon (*Salmo salar*). Aquaculture 121, 359-367.

Naser, N., 1997. Role of iron in Atlantic salmon (*Salmo salar*) nutrition: requirement, bioavailability, disease resistance and immune response. Ph. D. Thesis, Dalhousie University, Halifax, Canada.

Ogino, C. and G. Yang, 1980. Requirement for carp and rainbow trout for dietary manganese and copper. Bull. Jpn. Soc. Sci. Fish. 46, 455-468.

Olsen, P.-E., 1996. Metallothioneins in fish: induction and use in environmental monitoring. In: E.W. Taylor (Ed.), Toxicology of aquatic pollution, Cambridge University Press, Cambridge, United Kingdom, pp. 187-204.

Paripatananont, T. and R.T. Lovell, 1997. Comparative net absorption of chelated inorganic and inorganic trace elements in channel catfish, *Ictalurus punctatus* diets. J. World Aquacul. Soc. 28, 62-67.

Peuranen, S., P.J. Vuorinen, M. Vuorinen and A. Hollender, 1994. The effect of iron, humic acids and low pH on the gills and physiology of brown trout, *Salmo trutta*. Annales Zoologica Fennici 31, 389-396.

Rainbow, P.S., 1995. Physiology, physiochemistry and metal uptake – a crustacean perspective. Mar. Pollut. Bull. 31, 55-59.

Sakamoto, S. and Y. Yone, 1979. Availability of three iron compounds as dietary iron sources for red sea bream. Bull. Jpn. Soc. Sci. Fish. 45, 231-235.

Sarkar, S.A., S. Satoh and V. Kiron, 2005. Supplementation of citric acid and amino acid-chelated trace element to develop environment-friendly feed for red sea bream, *Pagus major*. Aquaculture 248, 3-11

Satoh, S., T. Takeuchi and T. Watanabe, 1987. Availability to rainbow trout of zinc in white fish meal and of various zinc compounds. Bull. Jpn. Soc. Sci. Fish. 53, 595-599.

Sick, L.V., J.W. Andrews and J.B. White, 1972. Preliminary studies of selected environmental and nutritional requirements for the culture of penaeid shrimp. Fish. Bull. Natl. Oceanic Atmos. Adm., Seattle 70, 101-109.

Spry, D.J., P.V. Hodson and C.M. Wood, 1988. Relative contributions of dietary and waterborne zinc in rainbow trout, *Salmo gairdneri*. Can. J. Fish. Aquat. Sci. 45, 32-41.

Takeuchi, T., T. Watanabe, C. Ogino, M. Saito, K. Nishimura and T. Nose, 1981. Effects of low protein-high calorie diets and detection of trace elements from fish meal diet on reproduction of rainbow trout. Bull. Jpn. Soc. Sci. Fish. 47, 645-654.

Taylor, L.N., J.C. McGreer, C.M. Wood and D.G. McDonald, 2007. Physiological effects of chronic copper exposure to rainbow trout (*Oncorhynchus mykiss*) in hard and soft water: Evaluation of chronic indicators. Environ. Toxicol. Chem. 19, 2298-2308.

Wang, C., R.T. Lovell and P.H. Klesius, 1997. Response to *Edwardsiella ictaluri* challenge by channel catfish fed organic and inorganic sources of selenium. J. Aquat. Anim. Health. 9, 172-179.

Watanabe, T., V. Kiron and S. Satoh, 1997. Trace minerals in fish nutrition. Aquaculture 151, 185-207.

Whitehead, M.W., R.P.H. Thomson and J.J Powell, 1996. Regulation of metal absorption in the gastrointestinal tract. Gut 39, 625-628.

Trace element status and immunity

V. Girard

Keywords: immunity, dairy, piglet, zinc, copper, selenium

Introduction

In North America, the National Research Council's Committee on Animal Nutrition has provided essential information to people raising animal for food, but generally have ignored requirements for immune functions. The immune system is not a particularly large organ and, quantitatively, its needs are small relative to processes such as muscle accretion in a growing animal or reproduction in an adult (Klasing and Calvert, 2000). As a result, trace elements (TE) requirements for immune functions are considered as part of the overall maintenance requirements. However, conditions challenging the immune system would theoretically change the TE 'state of balance', known as homeostasis, since the dietary level will be insufficient to cover both the metabolic and immunological needs of the animal.

The relationship between nutritional status and the immune system has been a topic of study for much of the 20th century. Dramatic increases in our understanding of the organization of the immune system and the factors that regulate immune function have demonstrated a remarkable and close concordance between host nutritional status and immunity. An important question is if the immune system is more vulnerable to marginal nutrient deficiencies relative to physiological process normally used to set dietary requirements (i.e. growth and reproduction). For such nutrients, the current recommended dietary levels should be considered inadequate for optimal disease resistance (Table 1). However, the immune system is a 'high priority' organ and leukocytes are endowed with transporters that are capable of obtaining these nutrients in preference to other tissues (e.g. skeletal muscle) (Klasing, 1998) but little attention has been devoted to its actual needs. The data presented in this paper will also cover multiple branches of the immune system; therefore, it is necessary to briefly outline two of the major categories of immunity and the tissues, cells, and cell products associated with these categories.

Table 1. Mechanisms where nutrition affects immunocompetence of animals (Klasing, 1998).

Mechanism	Examples
Substrate for the immune system	Amino acids, trace minerals, water soluble vitamins
Modulating signal transduction in leukocytes	PUFAs, vitamins A,D,E
Influencing hormonal milieu	Protein:energy ratio
Influencing intestinal dynamics	Fiber
Protecting against immunopathology	Antioxydants
Supplying critical nutrients to pathogens	Iron (mammals); Biotin (birds)

Non-specific immune system

Nonspecific Immunity Components of the nonspecific immune system include physical barriers (e.g. skin), phagocytic cells (e.g. macrophages and neutrophils), and soluble proteins derived from phagocytic cells (e.g. cytokines). The proinflammatory cytokines include tumor necrosis factor-α (TNF-α), interleukin-1β (IL-1), and interleukin-6 (IL-6). However, numerous other cytokines exist to support the immune response as well. These cytokines are initially secreted by leukocytes at the site of infection where they act locally to activate other immune cells. Furthermore, these cytokines and another group of leukocytic molecules called chemokines diffuse into the blood where they can activate and direct other leukocytes to the site of infection. Thus, the site of infection becomes a hotbed of activity where multiple leukocyte classes migrate into the tissue to help kill and clear pathogens from the infected tissue.

Components of nonspecific immunity are present both prior to and following antigen exposure and they typically are indiscriminant against most foreign substances, in other words, not 'specific' for any given antigen.

Specific Immunity Components of the specific immune system include the category of leukocytes called lymphocytes. Lymphocytes orchestrate specific immune responses within two broad categories: (1) Humoral immunity – mediated by circulating antibodies produced from B-lymphocytes and in response to specific antigen recognition. (2) Cell-mediated immunity – mediated by T lymphocytes which both recruit and stimulate phagocytic (nonspecific) activity as well as participate in direct lysis of infected cells (e.g. viral infected cells).

Stress hormones (particularly catecholamines and glucocorticoids) are potent modulators of immune function. They affect the cell-mediated immune response by shifting the balance between two distinct subsets of T helper cells that are responsible for either an inflammatory or non-inflammatory response, both being distinguished by the spectrum of cytokines they secrete (Merlot, 2004; Salak-Johnson and McGlone, 2007). Generally, stress increases the duration of inflammation.

On the other hand, nutritional insufficiency affects not only the whole cell-mediated immune response (Cunningham-Rundles *et al.*, 2005) but also the humoral immunity (Martin *et al.*, 2007; Bursch, 2004). As a result, the combination of nutritional insufficiency and stress further weakens the immune response, leading to altered immune cell populations and a generalized increase in inflammatory mediation (McCracken *et al.*, 1999).

In this review, we consider health as an organism's ability to efficiently respond to challenges (stressors) and effectively restore and sustain homeostasis. Under these conditions, being healthy is fundamentally being not stressed and well nourished. That's why this review use piglet's anorexia and cow's periparturient stress to demonstrate conditions where TE supplementation might strengthen the immune response.

Health and the piglet's intestinal function

Health has major impacts on both the well-being and the productive performance of the animal. Mucosal surfaces, especially those intestinal, are the most common entry pathway for pathogens. The gastrointestinal (GI) tract can be envisioned as an invagination of the environment into the host's body. Since the GI tract has much more surface area than the skin, the gut mucosa is indeed the largest host interface with the outside world. It has two functions: (1) It is a barrier to noxious outside stimuli; and (2) it is a site of active exchange (Rueda and Gil, 2000). Since molecules have to pass through the mucosal epithelium, this area lacks the mechanical protection provided by the multilayered and keratinised epithelium of the skin. It is becoming increasingly obvious that the immune mechanisms of the normal healthy gut are predominantly those of microflora regulation. It appears that apart from the inductive sites of the organised lymphoid tissue such as Peyer's patches and mesenteric lymph nodes, the diffuse lymphoid tissue of the small intestine is a major element in this regulatory system (Bailey and Haverson, 2006). The resulting actions of those mechanisms establish eubiosis: a balance between beneficial bacteria and potentially harmful bacteria in the GI tract.

To maintain eubiosis The GI tract is one of the largest immunological organs of the body, and it serves as the first line of defense against orally administered antigens (e.g. food proteins) and intestinal pathogens (e.g. bacteria, parasites). Gut-associated lymphoid tissues (GALT) make up 25% by weight of the gut mucosa and submucosa and thus constitute the largest extrathymic site of lymphocytes (McBurney, 1994).

Cells in GALT respond to intestinal pathogens by processing antigens for recognition by lymphocytes, by initiating a cascade of specialized immune responses to the antigens, by regulating the migration of immune mediators from the periphery to the infected gut and by participating directly in cytotoxic activities that limit parasite establishment and survival.

As a consequence, the GI tract is a site of high oxygen uptake, the latter amounting to 20-25% of the whole-body oxygen consumption, even in the post-absorptive or fasting state (Vaugelade *et al.*, 1994). This high energy demand parallels a high fractional rate of protein synthesis linked to a rapid turnover rate of epithelial cells and a high rate of intracellular protein synthesis (Duee *et al.*, 1995)

Generally, the response of weanling pigs to growth promoter is more pronounced in a conventional on-farm nursery than in an experimental nursery (Coffey and Cromwell, 1995; Zhao *et al.*, 2007). This suggests that differences in sanitation and subclinical pathogen exposure between locations are involved in the variable response (Le Floch *et al.*, 2006). In especially reared animals with no microbial flora (gnotobiotic), the weight of the GI tract is notably reduced (Gordon and Pesti, 1971) and growth promoters have no effect on their feed efficiency (Figure 1).

Sanitation might also modulate the effect of high concentrations of zinc (Zn) and copper (Cu) which have been associated with bactericidal effects similar to those of antimicrobial

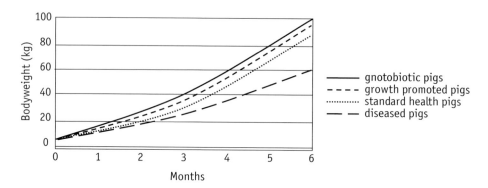

Figure 1. Growth curves of gnotobiotic pigs, growth promoted pigs, standard health pigs and diseased pigs (Bursch 2004).

compounds (Hojberg et al., 2005; Dunning et al., 1998; Fuller et al., 1960). Consequently without disease (under good hygiene and management), expression of disease resistance is questionable.

However when exploring a wide variety of protective functions, the complexity of immune systems becomes evident and the magnitude, breadth, and consistency of responses across assays vary among species (Albers et al., 2005; Matson et al., 2006). Ultimately, the real test of the efficacy of a food or food component that claims to improve immune function is a change in the incidence of infectious episodes or the severity or duration of symptoms of infection as this is the outcome of greatest clinical significance (Albers et al., 2005). Consequently growth response is often used for optimizing TE requirements. a doubtful choice if sanitation effect is not controlled.

Assays may be made for micronutrients status in body fluids, such as serum, plasma or urine. But, even in subclinical infection, immune response modify the blood concentration of some trace elements (Milanino et al., 1993; Powanda and Beisel, 2003; Le Floch et al., 2006;). Lower concentration is the result of sequestration within hepatocytes caused by the induced expression of metallothionein (MT) genes within these cells (Taylor, 1996). If sufficient enteric Zn is available, sequestration will not change its blood concentration (Milanino et al., 1993). Therefore it might be necessary to obtain concurrent information on inflammatory proteins like C-reactive protein and haptoglobin or ceruloplasmin (Tomkins, 2003) to assess blood TE status.

These non-specific acute phase protein are synthesized and released by the liver in response to activation by the inflammatory cytokine IL-6 (Wassell, 2000). It has been proposed to use haptoglobin to assess the health of pigs (Petersen et al., 2002) and dairy cows(Thielen et al., 2007). Unfortunately, for no micronutrient there is a linear relationship between change in inflammatory protein and change in micronutrient status (Milanino et al., 1993; Tomkins, 2003). Around 15 years ago Brown et al., 1993) concluded that the effect of concurrent infections may differ by nutrient, nutritional status of the population and prevalence and severity of the infection. Although this is specially true for Fe, Cu and Zn (Tomkins, 2003), there have been some recent developments in measurement technology and analysis.

Ambiguity might considerably be clarified by work in progress on the nutritional needs of the gastrointestinal mucosa. It is a rather novel finding, that the lower part of the digestive tract is most dependent on luminal nutrition for maintaining its integrity, structure and function. Anorexia, common during the weaning transition of pigs, might damage the gastrointestinal architecture (Spreeuwenberg *et al.*, 2001) and reduce absorptive capacity (McCracken *et al.*, 1999). Even the most complete parenteral nutrition regimen cannot, in the absence of adequate enteral nutrition, fully prevent the development of mucosal atrophy in the lower part of the digestive tract (Bengmark, 1996). The interaction between the gut-mucosa and the TE supply will result in a much more viable description of TE status. As such, the positive effect of high dietary Zn concentration on post-weaning piglets may be a secondary effect of improved mucosal cell function, differentiation, growth and repair, because of increased Zn held in the intestinal cells by the protein metallothionein (MT) (Carlson *et al.*, 2007).

Cytokines responsible for induction of acute-phase response, such as IL-1, regulate MT mRNA expression (Cousins and Leinart, 1988) and drastically alter zinc distribution kinetics among tissues. The injection of IL-1 produced diarrhea in 66% of Zn-deficient rats but in none of the well-fed animals (Cui *et al.*, 1997). This may explain the susceptibility to infectious diarrhea of Zn-deficient animal.

Mucosal MT mRNA but not leucocyte MT mRNA concentrations were significantly increased when piglets were fed 2500 ppm of Zn after 5-7 days (Carlson *et al.*, 2007). Over a longer period however, Allan *et al.* (2000) measured MT mRNA in T-lymphocytes from humans on a marginal Zn-deficient diet and found that it reflected differences in Zn intake, which was not detected by plasma Zn measurements.

In rodents, the effect of high dietary Zn concentration (2500 ppm) still increases in vitro lymphocyte proliferation during antigen stimulation (Denduluri *et al.*, 1997). Curiously, relatively few studies have examined the effects of Zn on the murine immune response in the parasite-infected host, and even fewer have considered the events occurring at the intestinal level. Since absorption of nutrients occurs where intestinal parasites reside, the gastrointestinal-associated lymphoid tissues play a role in directing both the local and the more systemic immune responses. Consequently, dietary Zn concentration is directly related with T_h1 and T_h2 cell populations proliferation in response to pathogen or parasite antigen (Scott and Koski, 2000; Shi *et al.*, 1998).

Bacteria-translocation facilitated by the loss of mucosal barrier integrity will, in turn, elicits the leukocyte release of toxic products. Amongst these, superoxid production may contribute to further damage of the cells membrane. Endogenous anti-oxidants superoxide dismutase (SOD), glutathione, and catalase are normally able to counteract oxidative stress in the intestinal mucosa. However, inflammation increases the demand for these important antioxidants and results in an imbalance between pro-oxidants and anti-oxidants, with subsequent mucosal damage. Researchers have studied the connection between oxidative stress and immune-regulated inflammatory factors. Selenium (Se) and glutathione-peroxidase (GSH-Px) activity were both decreased in patients with a chronic inflammatory bowel disease. These findings occurred in subjects who did not have evidence of malabsorption, indicating malabsorption is not the sole factor contributing to Se deficiency. The researchers concluded that «Se supplementation in deficient patient groups [should be] regarded as a potential protecting factor»(Reimund *et al.*, 2000).

In conclusion, most experiments focusing on piglets performances have made recommendations as to which of the indicators of micronutrient status is the most valuable. It is possible that increasing enteric Cu, Se or Zn concentrations during weaning anorexia could increase metallo-enzyme protection of piglet intestinal epithelial cell. Zn status, measured as MT mRNA is a useful marker of cell integrity. Assessment of micronutrient status is however, dependent on sanitation and requires novel research approaches if micronutrient assessment during inflammation is to be improved.

Health and cow's periparturient stress

Without question dairy cattle are 'stressed' during the transition period and this is at a maximum near time of parturition. However, it remains controversial whether lactating cows are physiologically stressed or not. One school of thought contends that high-producing dairy cows have been selected to manufacture high levels of milk and, as a result, are always close to metabolic disease. Hence, this school of thought believes that lactating cows are 'stressed'. Others contend that high production, per se, is indicative of a lack of stress and of a healthy cow. In favor of the former belief, several studies have reported that a variety of normal production practices cause 'stress' (Forsberg, 2004).

Mallard *et al.* (1998) reported that stress-associated immunosuppression is common in dairy cows and accounts for the high incidence of disease. A number of physiologic and metabolic responses to stress results in a decline in dry matter intake further compromising nutrient availability to support production.

The combination of stress and nutritional insufficiency is best illustrated by cows with negative energy balance and fatty infiltration of the liver. The results suggested that cows with more than 24% fat in the liver are slower in clearing *E. coli* from their mammary gland than cows with less fatty liver. (Hill *et al.*, 1985). Insufficient contents of TE in ruminant diets have been related with low disease resistance (Spears *et al.*, 1991). It is believed that the effects of stress are additive, thereby as stress situations accumulate, greater physiologic and metabolic changes occur ultimately resulting in abnormality seen as metabolic dysfunction or infectious disease. These stress responses will be more exaggerated in animals consuming an imbalanced diet, but may also overwhelm an animal consuming an adequate diet.

Several species of bacteria are found colonizing the teat canal and rarely cause clinical mastitis. Minor pathogens have even been credited with maintaining a higher than normal somatic cell count (SCC) and with increasing the resistance of the colonized quarter to invasion by a major pathogen (Radostits *et al.*, 2000; Jones, 1998). The prevalence of these bacteria is higher immediately after calving than in the remainder of the lactation (Radostits *et al.*, 2000). Consequently in the first weeks post-partum, the immune response in the mammary gland should imply similarities with those dictated for maintaining eubiosis in the piglet intestine.

An interesting challenge an animal has in fighting a mammary infection is related to the large dilution of immune cells which occurs when these cells enter the infected alveolus from the vasculature. As a result, the amounts of neutrophils which are secreted into infected mammary tissue are huge. The neutrophil is the primary immune cell entering alveolus to fight infections in ruminant animals. Concentrations of neutrophils in milk can reach levels as high as 4 million neutrophils/ml. In fact, neutrophils constitute the bulk of the somatic cell count (SCC). Rapid ability of a cow to 'pump' neutrophils into the alveolar space at the initiation of an infection correlates strongly with disease resistance (Burton and Erskine, 2003).

Physical barriers of the udder are anatomic features of the teat and associated structures that pose a physical blockade to invading bacteria at the teat sphincter, the point of entry. These anatomic features include the keratinised teat skin cells, teat sphincter muscle and keratin plug.

Up to 40% of the keratinised cells present before milking are removed by the shear-forces associated with milk flow (Capuco *et al.*,1997). This means that a proportion of the keratin is removed at every milking by desquamation and has to be replaced by new growth during the inter-milking period. The keratin plug is produced by skin lining the teat duct. Keratin is gummy, has bacteriostatic activity and completely occludes the teat canal.

Based on known effects on epithelial differentiation and growth, protein, vitamin A and Zn all influence epithelial health and can impact physical defense barriers of the udder. Protein status will also influence the integrity of the smooth muscle teat sphincter. Quality and quantity of the keratin plug may be influenced by protein, Zn and vitamin A.

Erskine and Bartlett (1993) showed that subclinical inflammation resulting from an intramammary infusion (IMI) of a minimum infective dose of *E. coli* resulted in a decline in mean serum concentrations of Zn, Fe, and Cu to 28, 35, and 52% of prechallenge concentrations, respectively. However in this case hypozincemia should not be attributed to Zn deficiency since it was not affected by supplemental Zn even at a high level (Lamand and Levieux, 1981).

Normally, a rise in serum Cu concentration is observed following parenteral administration of endotoxin. The increase in serum Cu concentration is attributed to increased synthesis of ceruloplasmin (CP) by the liver. In contrast, Erskine and Bartlett, (1993) documented a decrease in plasma Cu concentrations following experimental IMI with *E. coli*. They attributed this difference to the fact that, at the beginning of their study, their subjects were marginally Cu deficient and therefore unable to synthesize CP (Mills, 1987).

Cu has been shown to affect phagocytic function, but its impact on cell mediated and humoral immunity has been variable in cattle. It would be best to perform some testing to determine actual status in the animals prior to increasing dietary Cu as it potentially can be detrimental.

No single measurement provides a definitive simple diagnosis of Cu deficiency in cattle (Underwood and Suttle, 1999), and some authors have suggested the use of the ratio between CP and plasma Cu (CP/Cu) instead. However, different method to estimate CP activity exist and enzymatic methods are particularly dependent on laboratory conditions and so require coordinated testing before results from different laboratories can be considered equivalent (Laven and Livesey, 2006). Furthermore, a new interpretation of Scaletti et al. (2003) data shows that the ratio CP/Cu was not markedly affected by either calving or E. coli challenge and was not responsive to Cu supplementation (Laven et al., 2006). It is however possible that the ratio between plasma Cu and haptoglobin might be of interest since haptoglobin is a good marker of mastitis inflammation (Thielen et al., 2007).

Outcome of both cow's mastitis and piglet's diarrhea, bacteria and endotoxins might translocate into the circulation of the animal to cause sepsis. Integrity of the epithelial barrier during inflammation requires antioxidant, like Cu/Zn superoxide dismutase (SOD) to protect cell cytosol and membrane systems from free radicals (Boulanger et al., 2002). Cu-deficient beef cattle fed a basal diet plus 10 ppm ammonium molybdate had reduced SOD activities in red blood cells, neutrophils and whole blood (Xin et al., 1991). Around 15 y ago, DiSilvestro and Marten (1990) concluded that stress increase Cu demand for maintaining extra-cellular SOD at level of non-stressed animals, while erythrocyte SOD remained stable. In an experiment where weanling, male rats were fed diets for 6 weeks containing 3, 5 and 15 mg Zn/kg with dietary Cu set at 0.3, 1.5 and 5 mg Cu/kg at each level of dietary Zn. Serum extra-cellular SOD responded to changes in dietary Cu but not to changes in dietary Zn (Johnson et al., 2005). In time of periparturient stress, serum SOD is probably a good marker for Cu status.

One of the most studied nutrients relative to mastitis and immune function is Zn. A summary of 12 lactation trials addressing Zn supplementation showed a 33% reduction in milk SCC (Kellogg, 1990). Many of these studies were comparing a chelated form of Zn to inorganic Zn supplementation. However, not all organic forms of Zn showed positive effects on mastitis (Whitaker et al., 1997). There are differences in bioavailability among even the organic forms of minerals.

As a normal physiologic process, the blood Zn level declines around calving (Xin et al., 1993) due to reduced DMI, transfer of Zn to colostrum and increased stress at this time. Stressors such as parturition and microbial infections decrease the blood Zn concentrations due to redistribution of Zn from blood to tissues, especially the liver (Spears et al., 1991; Xin et al., 1993). Stress induces liver sequestration of Zn in MT, a protein associated with Zn metabolism (Xin et al., 1993).

Marginal intake of Zn and vitamin A is common during lactation and a deficiency of one micronutrient can result in a secondary deficiency of the other. However, the resistance of milk Zn concentration to changes in dietary Zn or vitamin A indicates tight regulation of mammary gland Zn transport. In an experiment with lactating rats (Kelleher and Lonnerdal, 2002) MT varied with vitamin A intake, although total mammary gland Zn was not affected. No references could be found on the potential use of MT expression in mammary cells or SCC for evaluating the cow's Zn status.

Beneficial health effects have also been consistently associated with Se supplementation in dairy cows (Pehrson *et al.*, 1999; Rayman, 2004; Malbe *et al.*, 2006). Glutathione peroxidase (GSH-Px) and thioredoxin reductase are selenoproteins whose activities were demonstrated in bovine mammary tissue, and which were influenced by Se status (Bruzelius *et al.*, 2007). These enzymes catalyse the reduction of hydrogen peroxides. For this reason, Se has a unique role as a cytosolic antioxidant preventing oxidative damages by free radicals (Reddy and Frey, 1990). Se is also considered to protect phagocytic cells from autoxidative damage when the respiratory burst is activated. Leakage of free radicals from the phagolysosomes, or failure to detoxify these products, could affect the microbiocidal and metabolic functions of phagocytic cells (Larsen, 1993; Rotruck *et al.*, 1973). Since the activity of these enzymes is related to the diet concentration of Se, Se status of the mammary gland might be an important regulator of selenoprotein expression. Higher activity of GSH-Px has also been reported in the erythrocytes of cows receiving supplemental Se compared with non-supplemented individuals (Ortman and Pehrson, 1999).

Previous reports have indicated that more GSH-Px can be obtained from feeding inorganic Se than organic Se (Butler *et al.*, 1990). One can understand that selenocysteine and selenomethionine, like their counterparts cysteine and methionine, are rapidely incorporated into body protein. Only toxic excesses of intra-cellular Se-cysteine (Stipanuk *et al.*, 2006) are catabolized and mineralized to selenide (Se^{--}). But they are also stored as glutathione, an antioxidant not affected by its Se form (Beld *et al.*, 2007).

On the other hand, most parenteral selenite (Se^{+IV}) is rapidly taken up by red blood cells (Suzuki and Ogra, 2002) and reduced to selenide. This selenide is activated as Se phosphate before being synthesized in seleno-cysteinyl, which is metabolically active for Se-GSH-Px (Suzuki and Ogra, 2002; Johansson *et al.*, 2005).

Contradictory evidence is based on the higher bioavailability of organic Se compared to inorganic Se (Smith and Picciano, 1987) particularly in ruminants (Spears, 2003).

A possible explanation resides in the up-regulation (Figure 2) of methionine transsulfuration to homocysteine, cystathionine and finally inorganic Se by oxidative (stress) conditions (Mosharov *et al.*, 2000). Consequently, significant methionine mineralization was demonstrated in piglets GIT tract (Riedijk *et al.*, 2007). Not only does this complicated metabolic process explain a delayed response of Se-GSH-Px to organic selenium supplementation, but it also provides information on how the response of Se-GSH-Px to organic selenium could be dependent upon the dietary provision of vitamin B6 (Yasumoto *et al.*, 1979; Matte, 2007).

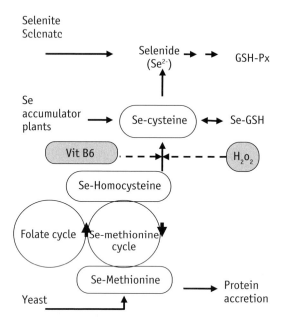

Figure 2. Regulation of homocysteine metabolic disposal by peroxides (H_2O_2) and other reactive oxygen species (Mosharov et al. (2000) adapted from Matte (2007)). Up-regulation: towards cysteine glutathione (GSH) and glutathione peroxidase (GSH-Px) system (transulfuration). Down-regulation towards methionine (transmethylation). For purposes of clarity, multistep pathways for transsulfuration, transmethylation, cysteine oxidation, GSH and GSH-Px synthesis have been condensed to single arrows.

With ruminants, limitation by vitamin B6 is less likely and in transported (stressed) young ruminant, the efficiency of organic Se for erythrocyte GSH-Px activity in the first week after arrival is similar to inorganic one (Figure 3). The equivalent efficiency at week 1 suggests that erythrocyte might be able to rapidly mineralyze selenoaminoacids (Malinow *et al.*, 1994). However, compared to inorganic Se, organic Se supplementation increases another plasma-selenoprotein than GSH-Px. Its function is still not completely understood but acting as an extracellular antioxidant seems most probable (Mostert, 2000). Validity of the ratio between GSH-Px activity and Se concentration in the mammary gland as indicator of mammary cells selenoprotein expression should be further investigated.

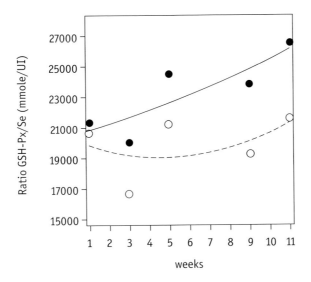

Figure 3. Ratio between GSH-Px (IU/L) and Se (mmole/L) in whole blood of growing beef cattle after feeding organic (O) or inorganic (●) supplemental Se (0.3 ppm/kg DM) for 11 weeks. Each point is the mean of 24 cattle. Except for week 1, efficiency of inorganic Se for GSH-Px activity was always superior (P<0.01).

Conclusion

Emerging knowledge the role of Zn, Cu and Se metalloprotein in cytokine signaling, apoptosis and differentiation should help to clarify methods for TE status under commercial sanitation procedures. Most experiments focusing on piglets performances have make recommendations as to which of the indicators of micronutrient status is the most valuable. Serum extra-cellular SOD responded to changes in dietary Cu but not to changes in dietary Zn. However Zn status, measured as lymphocyte MT mRNA was a useful marker of cell integrity. But no references could be found on the potential use of MT expression in mammary cells or SCC for evaluating the cow's Zn status. In cattle, ratios between erythrocyte GSH-Px and Se are useful indicators of Se bioavailability for the immune system. Furthermore, assessment of micronutrient status is dependent on sanitation and requires novel research approaches if micronutrient assessment during inflammation is to be improved. This review agrees with Tomkins, 2003 that « using agreed-upon assay methods and standards with a range of cutoff points for inflammatory proteins » is an obligatory prerequisite to assess the TE requirements needed to counteract the stress imposed by the commercial production environments.

References

Albers, R., J.M. Antoine, R. Bourdet-Sicard, P.C. Calder, M. Gleeson, B. Lesourd, S. Samartin, I.R. Sanderson, J. Van Loo, F.W. Vas Dias and B. Watzl, 2005. Markers to measure immunomodulation in human nutrition intervention studies. Br. J. Nutr. 94, 452-81.

Bailey, M. and K. Haverson, 2006. The postnatal development of the mucosal immune system and mucosal tolerance in domestic animals. Vet. Res. 37, 443-53.

Beld, J., K.J. Woycechowsky and D. Hilvert, 2007. Selenoglutathione: efficient oxidative protein folding by a diselenide. Biochemistry 46, 5382-90.

Bengmark, S., 1996. Econutrition and health maintenance – A new concept to prevent GI inflammation, ulceration and sepsis. Clin. Nutr. 15, 1-10.

Boulanger, V., X. Zhao and P. Lacasse, 2002 . Protective effect of melatonin and catalase in bovine neutrophil-induced model of mammary cell damage. J. Dairy Sci. 85, 562-9.

Brown, K.H., C.F. Lanata, M.L. Yuen, J.M. Peerson, B. Butron and B. Lonnerdal, 1993. Potential magnitude of the misclassification of a population's trace element status due to infection: example from a survey of young Peruvian children. Am. J. Clin. Nutr. 58, 549-54.

Bruzelius, K., T. Hoac, R. Sundler, G. Onning and B. Akesson, 2007. Occurrence of selenoprotein enzyme activities and mRNA in bovine mammary tissue. J. Dairy Sci. 90, 918-27.

Bursch, D., What can we expect when the growth promoters are finally withdrawn on January 1st 2006? The Pig Journal 54, 34-41. 2004 .

Burton, J.L. and R.J. Erskine, 2003. Immunity and mastitis. Some new ideas for an old disease. Vet. Clin. North Am. Food Anim. Pract. 19, 1-45, v.

Butler, J.A., P.D. Whanger, A.J. Kaneps and N.M. Patton, 1990. Metabolism of selenite and selenomethionine in the rhesus monkey. J. Nutr. 120, 751-9.

Carlson, D., J.H. Beattie and H.D. Poulsen, 2007. Assessment of zinc and copper status in weaned piglets in relation to dietary zinc and copper supply. J. Anim. Physiol. Anim. Nutr. (Berl.) 91, 19-28.

Coffey, R.D. and G.L. Cromwell, 1995. The impact of environment and antimicrobial agents on the growth response of early-weaned pigs to spray-dried porcine plasma. J. Anim. Sci. 73, 2532-9.

Cousins, R.J. and A.S. Leinart, 1988. Tissue-specific regulation of zinc metabolism and metallothionein genes by interleukin 1. FASEB J. 2, 2884-90.

Cui, L., Y. Takagi, M. Wasa, Y. Iiboshi, J. Khan, R. Nezu and A. Okada, 1997. Induction of nitric oxide synthase in rat intestine by interleukin-1alpha may explain diarrhea associated with zinc deficiency. J. Nutr. 127, 1729-36.

Cunningham-Rundles, S., D.F. McNeeley and A. Moon, 2005. Mechanisms of nutrient modulation of the immune response. J. Allergy Clin. Immunol. 115, 1119-28; quiz 1129.

Denduluri, S., M. Langdon and RK. Chandra, 1997. Effect of Zinc Administration on Immune Responses in Mice. J. Trace Elem. Exp. Med. 10, 155-162.

DiSilvestro, R.A. and J.T. Marten, 1990. Effects of inflammation and copper intake on rat liver and erythrocyte Cu-Zn superoxide dismutase activity levels. J. Nutr. 120, 1223-7.

Duee, P.H., B. Darcy-Vrillon, F. Blachier and M.T. Morel, 1995. Fuel selection in intestinal cells. Proc. Nutr. Soc. 54, 83-94.

Dunning, J.C., Y. Ma and R.E. Marquis, 1998. Anaerobic killing of oral streptococci by reduced, transition metal cations. Appl. Environ. Microbiol. 64, 27-33.

Erskine, R.J. and P.C. Bartlett, 1993. Serum concentrations of copper, iron, and zinc during Escherichia coli-induced mastitis. J. Dairy Sci. 76, 408-13.

Forsberg, M.E., 2004. Recent Insights Into Ruminant Immune Function: Effects of Stress and Immunostimulatory Nutritional Products. In: Proceedings of the 15th Florida Ruminant Nutrition Symposium. Gainesville, FL. pp. 81-91.

Fuller, R., L.G.M. Newland, C.A.E. Briggs, R. Braude and K.G. Mitchell, 1960. The normal intestinal flora of the pig. IV. The effect of dietary supplements of penicillin, chlortetracycline or copper sulphate on the faecal flora. J. Appl. Bacteriol. 23, 195-205.

Gordon, H.A. and L. Pesti, 1971. The Gnotobiotic Animal as a Tool in the Study of Host Microbial Relationships. Bacteriol. Rev. 35(4), 390-429.

Hill, A.W., I.M. Reid and R.A. Collins, 1985. Influence of liver fat on experimental Escherichia coli mastitis in periparturient cows. Vet. Rec. 117, 549-51.

Hojberg, O., N. Canibe, H.D. Poulsen, M.S. Hedemann and B.B. Jensen, 2005. Influence of dietary zinc oxide and copper sulfate on the gastrointestinal ecosystem in newly weaned piglets. Appl. Environ. Microbiol. 71, 2267-77.

Johansson, L., G. Gafvelin and E.S. Arner, 2005. Selenocysteine in proteins-properties and biotechnological use. Biochim. Biophys. Acta 1726, 1-13.

Johnson, W.T., L.A. Johnson and H.C. Lukaski, 2005. Serum superoxide dismutase 3 (extracellular superoxide dismutase) activity is a sensitive indicator of Cu status in rats. J. Nutr. Biochem. 16, 682-92.

Jones, M.J., 1998. Less recognized sources of mastitis infection. http://www.dasc.vt.edu/faculty/jones/ UncmmMas.htm Oct, 1998.

Kelleher, S.L. and B. Lonnerdal, 2002. Zinc transporters in the rat mammary gland respond to marginal zinc and vitamin A intakes during lactation. J. Nutr. 132, 3280-5.

Kellogg, D.W., 1990. Zinc methionine affects performance of lactating cows. Feedstuffs 62(35), 115.

Klasing, K.C., 1998. Nutritional modulation of resistance to infectious diseases. Poult. Sci. 77, 1119-25.

Klasing, K.C. and C.C. Calvert, 2000. The care and feeding of an immune system: an analysis of lysine needs. In: Proceedings of the VIIIth International Symposium on Protein Metabolism and Nutrition. Wageningen Press, Wageningen. pp. 253-264.

Lamand, M. and D. Levieux, 1981. Effects of infection on plasma levels of copper and zinc in ewes. Ann. Rech. Vet. 12, 133-6.

Larsen, H.J.S., 1993. Relations between selenium and immunity. Norwegian Journal of Agricultural Science 11, 105-119.

Laven, R.A. and C.T. Livesey, 2006. [Letter to the editor]Caeruloplasmin: plasma copper ratios in cows. Vet. Rec. 159, 608.

Laven, R.A., C.T. Livesey, R.J. Harmon and R. Scaletti, 2006. Factors affecting the relationship between caeruloplasmin activity and plasma copper concentration in cattle. Vet. Rec. 159, 250-1.

Le Floch, N., C. Jondreville, J.J. Matte and B. Seve, 2006. Importance of sanitary environment for growth performance and plasma nutrient homeostasis during the post-weaning period in piglets. Arch. Anim. Nutr. 60, 23-34.

Malbe, M., M. Attila and F. Atroshi, 2006. Possible involvement of selenium in Staphylococcus aureus inhibition in cow's whey. J. Anim. Physiol. Anim. Nutr. (Berl.) 90, 159-64.

Malinow, M.R., M.K. Axthelm, M.J. Meredith, N.A. MacDonald and B.M. Upson, 1994. Synthesis and transsulfuration of homocysteine in blood. J. Lab. Clin. Med. 123, 421-9.

Mallard, B.A., J.C. Dekkers, M.J. Ireland, K.E. Leslie, S. Sharif, C.L. Vankampen, L. Wagter and B.N. Wilkie, 1998. Alteration in immune responsiveness during the peripartum period and its ramification on dairy cow and calf health. J. Dairy Sci. 81, 585-95.

Martin, L.B. 2nd, K.J. Navara, Z.M. Weil, and R.J. Nelson, 2007. Immunological memory is compromised by food restriction in deer mice Peromyscus maniculatus. Am. J. Physiol. Regul. Integr. Comp. Physiol. 292, R316-20.

Matson, K.D., A.A. Cohen, K.C. Klasing, R.E. Ricklefs and A. Scheuerlein, 2006. No simple answers for ecological immunology: relationships among immune indices at the individual level break down at the species level in waterfowl. Proc. Biol. Sci. 273, 815-22.

Matte, J.J., 2007. Selenium metabolism, the glutathione peroxidase system and their interaction with some B vitamins in pigs. In: T.P. Lyons, K.A. Jacques and J.M. Hower (Eds.), Nutritional Biotechnology in the Feed and Food Industries Proceedings from Alltech's technical seminar series. Nottingham University Press, Nottingham, United Kingdom. pp. 87-96.

McBurney, M.I., 1994. The gut: central organ in nutrient requirements and metabolism. Can. J. Physiol. Pharmacol. 72, 260-5.

McCracken, B.A., M.E. Spurlock, M.A. Roos, F.A. Zuckermann and H.R. Gaskins, 1999. Weaning anorexia may contribute to local inflammation in the piglet small intestine. J. Nutr. 129, 613-9.

Merlot, E., 2004. Conséquences du stress sur la fonction immunitaire chez les animaux d'élevage. INRA Prod. Anim., 17, 255-264.

Milanino, R., M. Marrella, R. Gasperini, M. Pasqualicchio and G. Velo, 1993. Copper and zinc body levels in inflammation: an overview of the data obtained from animal and human studies. Agents Actions 39, 195-209.

Mills, C.F., 1987. Biochemical and physiological indicators of mineral status in animals: copper, cobalt and zinc. J. Anim. Sci. 65, 1702-11.

Mosharov, E., M.R. Cranford and R. Banerjee, 2000. The quantitatively important relationship between homocysteine metabolism and glutathione synthesis by the transsulfuration pathway and its regulation by redox changes. Biochemistry 39, 13005-11.

Mostert, V., 2000. Selenoprotein P: properties, functions, and regulation. Arch Biochem Biophys 376, 433-8.

Ortman, K. and B. Pehrson, 1999. Effect of selenate as a feed supplement to dairy cows in comparison to selenite and selenium yeast. J. Anim. Sci. 77, 3365-70.

Pehrson, B., K. Ortman, N. Madjid, and U. Trafikowska, 1999. The influence of dietary selenium as selenium yeast or sodium selenite on the concentration of selenium in the milk of Suckler cows and on the selenium status of their calves. J. Anim. Sci. 77, 3371-6.

Petersen, H. H., D. Dideriksen, B. M. Christiansen, and J. P. Nielsen. 2002. Serum haptoglobin concentration as a marker of clinical signs in finishing pigs. Vet. Rec. 151, 85-9.

Powanda, M.C. and W.R. Beisel, 2003. Metabolic effects of infection on protein and energy status. J. Nutr. 133, 322S-327S.

Rayman, M.P., 2004. The use of high-selenium yeast to raise selenium status: how does it measure up? Br. J. Nutr. 92, 557-73.

Reddy, P.G. and R.A. Frey, 1990. Nutritional modulation of immunity in domestic food animals. Adv. Vet. Sci. Comp. Med. 35, 255-81.

Reimund, J.M., C. Hirth, C. Koehl, R. Baumann, and B. Duclos, 2000. Antioxidant and immune status in active Crohn's disease. A possible relationship. Clin. Nutr. 19, 43-8.

Riedijk, M.A., B. Stoll, S. Chacko, H. Schierbeek, A.L. Sunehag, J.B. van Goudoever and D.G. Burrin, 2007. Methionine transmethylation and transsulfuration in the piglet gastrointestinal tract. Proc. Natl. Acad. Sci. USA 104, 3408-13.

Rotruck, J.T., A.L. Pope, H.E. Ganther, A.B. Swanson, D.G. Hafeman and W.G. Hoekstra, 1973. Selenium: biochemical role as a component of glutathione peroxidase . Science 179, 588-90.

Radostits O.M., C.C. Gay, D.C. Blood and K.W. Hinchcliff, 2000. Veterinary Medicine. Textbook of the Diseases of Cattle, Sheep, Pigs, Goats and Horses. 9the Edition. 2000. pp 603-700.

Rueda, R. and A. Gil, 2000. Influence of Dietary Compounds on Intestinal Immunity . Microbial Ecology in Health and Disease 1, 146-156.

Salak-Johnson, J.L. and J.J. McGlone, 2007 . Making sense of apparently conflicting data: stress and immunity in swine and cattle. J. Anim. Sci. 85, E81-8.

Scaletti, R.W., D.S. Trammell, B.A. Smith and R.J. Harmon, 2003. Role of dietary copper in enhancing resistance to Escherichia coli mastitis. J. Dairy Sci. 86, 1240-9.

Scott, M.E. and K.G. Koski, 2000. Zinc deficiency impairs immune responses against parasitic nematode infections at intestinal and systemic sites. J. Nutr. 130, 1412S-20S.

Shi, H.N., M.E. Scott, M.M. Stevenson and K.G. Koski, 1998. Energy restriction and zinc deficiency impair the functions of murine T cells and antigen-presenting cells during gastrointestinal nematode infection. J. Nutr. 128, 20-7.

Smith, A.M. and M.F. Picciano, 1987. Relative bioavailability of seleno-compounds in the lactating rat. J. Nutr. 117, 725-31.

Spears, J.W., 2003. Trace mineral bioavailability in ruminants. J. Nutr. 133, 1506S-9S.

Spears, J.W., R.W. Harvey and T.T.J. Brown, 1991. Effects of zinc methionine and zinc oxide on performance, blood characteristics, and antibody titer response to viral vaccination in stressed feeder calves. J. Am. Vet. Med. Assoc. 199, 1731-3.

Spreeuwenberg, M.A., J.M. Verdonk, H.R. Gaskins and M.W. Verstegen, 2001. Small intestine epithelial barrier function is compromised in pigs with low feed intake at weaning. J. Nutr. 131, 1520-7.

Stipanuk, M.H., J.E. Dominy Jr, J.I. Lee and R.M. Coloso, 2006. Mammalian cysteine metabolism: new insights into regulation of cysteine metabolism. J. Nutr. 136, 1652S-1659S.

Suzuki, K.T. and Y. Ogra, 2002. Metabolic pathway for selenium in the body: speciation by HPLC-ICP MS with enriched Se. Food Addit. Contam. 19, 974-83.

Taylor, A., 1996. Detection and monitoring of disorders of essential trace elements. Ann. Clin. Biochem. 33(Pt 6), 486-510.

Thielen, M.A., M. Mielenz, S. Hiss, H. Zerbe, W. Petzl, H.J. Schuberth, H.M. Seyfert and H. Sauerwein, 2007. Short communication: Cellular localization of haptoglobin mRNA in the experimentally infected bovine mammary gland. J. Dairy Sci. 90, 1215-9.

Tomkins, A., 2003. Assessing micronutrient status in the presence of inflammation. J. Nutr. 133, 1649S-1655S.

Underwood, E.J. and N.F. Suttle, 1999. Copper. In: The Mineral Nutrition of Livestock. 3rd edn.CAB International, Wallingford. pp. 251-282.

Vaugelade, P., L. Posho, B. Darcy-Vrillon, F. Bernard, M.T. Morel, and P.H. Duee, 1994. Intestinal oxygen uptake and glucose metabolism during nutrient absorption in the pig. Proc. Soc. Exp. Biol. Med. 207, 309-16.

Wassell, J., 2000. Haptoglobin: function and polymorphism. Clin. Lab. 46, 547-52.

Whitaker, D.A., H.F. Eayres, K. Aitchison, and J.M. Kelly, 1997. No effect of a dietary zinc proteinate on clinical mastitis, infection rate, recovery rate and somatic cell count in dairy cows. Vet. J. 153, 197-203.

Xin, Z., D.F. Waterman, R.W. Hemken and R.J. Harmon, 1991. Effects of copper status on neutrophil function, superoxide dismutase, and copper distribution in steers. J. Dairy Sci. 74, 3078-85.

Xin, Z., D.F. Waterman, R.W. Hemken and R.J. Harmon, 1993. Copper status and requirement during the dry period and early lactation in multiparous Holstein cows. J. Dairy Sci. 76, 2711-6.

Yasumoto, K., K. Iwami and M. Yoshida, 1979 . Vitamin B6 dependence of selenomethionine and selenite utilization for glutathione peroxidase in the rat. J. Nutr. 109, 760-6.

Zhao, J., A.F. Harper, M.J. Estienne, K.E. Webb Jr, A.P. McElroy and D.M. Denbow, 2007. Growth performance and intestinal morphology responses in early weaned pigs to supplementation of antibiotic-free diets with an organic copper complex and spray-dried plasma protein in sanitary and nonsanitary environments. J. Anim. Sci. 85, 1302-10.

Prions and trace elements

D.R. Brown

Keywords: prion, manganese, copper, scrapie, BSE

Introduction

Transmissible spongiform encephalopathies (TSEs) have cast a shadow of fear over the world especially for those who like a good piece of beef steak (Baker and Ridley, 1996). However, the reality of the variant CJD epidemic is that it has not happened (Will, 2004). Nevertheless, the possible transmission of a disease by a protein has sparked considerable interest in these diseases despite their very low occurrence. In addition the availability of animal models to study prion diseases has meant that they have become an exemplar for the study of neurodegenerative diseases in general (Prusiner, 1998). The family of TSEs or prion diseases include both animal and human forms. The first studied human disease became known as Creutzfeldt-Jakob disease after the German neuropathologists who first described the changes in human patients (Collinge, 1997). Kuru, a disease of natives of New Guinea was linked to the consumption of human brain that carried CJD (Gajdusek and Gibbs, 1971). Eventually, inherited human prion diseases such as Gerstmann-Sträussler-Scheinker syndrome (Ghetti *et al.*, 1995; Hsiao *et al.*, 1989) and Fatal Familiar Insomnia were also described (Lugaresi *et al.*, 1986). The inherited forms of disease are linked to point mutations or insertions in the gene that encodes the prion protein (*prnp*). The three most common animal forms of TSE are bovine spongiform encephalopathy (BSE) of cattle (Hope *et al.*, 1988), scrapie of sheep (Prusiner, 1982) and chronic wasting disease (CWD) of deer and elk (Browning *et al.*, 2004). BSE is still widespread in Europe despite attempts to eradicate it by preventing cattle from eating possibly infectious food sources. The origin of BSE remains unknown and attempts to link it to scrapie have proven to be an inadequate explanation (Smith and Bradley, 2003). Similary, scrapie and CWD have no clear cause and appear to develop sporadically, often associated with particular environmental locations (Brown and Sinclair, 2003). Experimental models of TSEs depend upon the inoculation of mice or hamsters with extracts from the brains of animals with one of the TSEs. This is therefore a form of experimental infection. Although such experimental transmission can occur in the laboratory, TSEs are not contagious diseases. Transmission between individuals largely does not occur, although it is possible that a small number of vCJD cases have been passed to other patients through blood transfusions (Ironside and Head, 2004). CJD is the most common form of human prion disease. It is different to variant CJD in that this latter form affects mostly young people with a specific genotype. To date, all patients with vCJD have two copies of the 129Met (coding methionine at codon129) allele of *prnp*.

All prion diseases are linked to the deposition of an abnormal isoform of the prion protein in the central nervous system of the affected individual (Prusiner, 1998). This abnormal isofrom (PrP^{Sc}) is accepted as the infectious agent of the diseases and is the probable cause of the pathology associated with the disease (Bolton *et al.*, 1982). PrP^{Sc} is highly rich in beta-sheet structure, is highly resistant to protease digestion and aggregates to form fibrils *in vitro*. Although central to the disease process, it is still not completely clear that this form of the protein is the one that can cause transmission of disease between individuals or that it is all that is essential for the disease. Without expression of the protein prion disease cannot developed because mice that have been transgenically altered to lack expression of PrP are resistant to TSEs (Büeler *et al.*, 1993). The 'protein-only' hypothesis remains largely accepted and has been supported by recent studies that show that recombinant PrP can be used to initiate a prion-like disease in mice and that this new disease can be transmitted to other mice. Yet, other researchers still argue that the protein itself is insufficient for a true TSE.

Although it is unclear what exactly causes prion disease and the mechanism of conversion of the normal cellular isoform (PrP^c) to PrP^{Sc} remains undescribed (Cohen and Prusiner, 1998), recent research has made strong inroads into understanding the nature of the substrate for prion diseases, the normal isoform of the prion protein. PrP^c is a glycoprotein (Rudd *et al.*, 1999) expressed by many cell types but especially by neurones (Kretzschmar *et al.*, 1986). The protein is concentrated at synapses (Salès *et al.*, 1998) and is therefore thought to have some special role in neuronal function. It is produced in cells a single monomeric polypeptide of around 250 amino acids in length. The regulation of its expression is still poorly characterised but it is highly regulated with not only a promoter but also by regulator regions in exons 1 and 2 encoded by the messenger RNA that are not transcribed into protein. The third exon of PrP encodes the whole open reading frame and contains highly conserved regions. Prion protein knockout mice develop normally and show little in the way of behavioural difference to wild-type (Büeler *et al.*, 1992). Nevertheless, there have been a large number of reported differences in the mice and especially there are distinct differences in the response of neurones cultured from the brains of the mice. These differences suggest that PrP^c aids to protect neurones from cellular responses to stress (Brown *et al.*, 2002).

The essential requirement of host prion protein expression for both transmission of the disease and the triggering of neuronal death implies that understanding the nature of the normal isoform of the prion protein is inseparable from discovering the mechanism of disease transmission and progression. The mechanism of conversion of the protein remains unknown but is thought to involve the formation of seeds or small aggregates of PrP which can then catalyse rapid conversion to further host protein to the abnormal isoform. This issue is further complicated by the existence of strains (Bruce, 2003). Strains of prion diseases (especially scrapie) are characterised by different characteristics of the resulting disease when applied to the same species of host. Thus the characteristics will remain the same when transmitted to different individuals of the same species of animal or even the same breed of mouse. These characteristics include location and severity of neuronal loss and gliosis, extent and location of PrP^{Sc} deposition, incubation time for the disease and duration of symptoms. These characteristics for BSE in cattle may change when the disease is transmitted to mice but within mice or cattle the characteristics are the same in different individuals after subsequent passages. This secondary information suggests that if the protein is the sole cause of the disease, then somehow this is encoded by the way the conformation is altered during conversion of PrP^c to PrP^{Sc}. There is currently no explanation for how this information could be encoded by the conformational change.

Prion diseases have only been identified in mammals. However, other vertebrates express homologues of PrP. These include birds, reptiles, amphibians and fish (Harris *et al.*, 1991; Wopfner *et al.*, 1999; Simonic *et al.*, 2000; Strumbo *et al.*, 2001; Oidtmann *et al.*, 2003). The sequences identified in reptiles and birds show very high homology to mammals except that in the N-terminus mammals have four or more repeats of an octomer while birds and reptiles have five or more repeats of a hexameric region. Amphibians and fish express one or more homologues of PrP but essential domains are quite different and it is unclear if these homologues would serve the same function in these lower vertebrates. The hexameric or octomeric repeat regions are the suggested binding place for copper to PrP^c. Investigation of the possible function of these regions lead to the suggestion that these sites could be copper bindings sites (Hornshaw *et al.*, 1995a,b). Early studies of peptides related to these regions were the first to demonstrate interaction of copper occurs at these sites. These findings triggered a wave of interest in the potential of PrP^c to be metalloprotein. This review will deal with current understanding of metal interactions with PrP^c and their possible consequence for the function of the PrP^c and disease progression in TSEs.

Prion protein and metals

The central event in TSEs pathogenesis is believed to be the post-translational conversion of a normal cellular prion protein (PrP^C) into an abnormal isoform called scrapie PrP (PrP^{Sc}). The disease form of PrP is partially resistant to proteases and can be passed between individuals, producing symptoms of TSE (Prusiner, 1998). An important and very current question in TSE research concerns the relationship between oxidative stress as a factor causing the symptoms and the role of the PrP protein as a possible cause of the disease. There has been considerable discussion about the normal function of the protein. It is likely that this function dependent on the metal binding capacity of the protein. Although the *in vivo* function of PrP^C remains to be confirmed, it has been demonstrated that both recombinant and brain-derived PrP^C have superoxide dismutase (SOD)-like activity when bound to Cu^{2+} (Brown *et al.*, 1999, 2001). The Cu^{2+}complexation in PrP is different, however, to that in cellular Cu/Zn-SOD (van Doorslaer *et al.*, 2001). It also appears that the protein transports Cu^{2+}, thereby increasing cellular resistance to Cu^{2+} toxicity (Brown *et al.*, 1998). Cu appears to be taken up in association with PrP^C into the cell (Pauly and Harris, 1998; Brown, 1999, 2004). However, PrP^C is not the main Cu-uptake protein expressed by cells. Binding of Cu^{2+} at the cell surface has been shown to be the main cause of PrP^C internalisation (Haigh *et al.*, 2006)

When PrP^C converts to PrP^{Sc}, this SOD-like function is lost (Thackray *et al.*, 2002). It is therefore of considerable interest to consider this property in relation to oxidative stress as a cause of neuronal damage in TSEs. Several lines of investigation have been taken to see (1) if the metal binding of PrP^C is altered in TSEs, (2) if metal imbalances also correlate with the loss of antioxidant function in PrP^C and (3) whether these alterations correlate with the disease phenotype, such as PrP^{Sc} and also the PrP genotype at codon 129, which influences the manifestation of the disease (Gambetti *et al.*, 1999).

NMR studies have shown that PrP^C consists of a structured C-terminal region, which is primarily α-helical, and an unstructured N-terminal region (Riek *et al.*, 1996; Zahn *et al.*, 2000). PrP^C exhibits high affinity, cooperative Cu^{2+} binding through a histidine-containing octapeptide repeat domain in the unstructured N-terminal region (Brown *et al.*, 1997). It has also been suggested to bind Cu^{2+} along the more structured C-terminal domain of the protein (Cereghetti *et al.*, 2001; van Doorslaer *et al.*, 2001). Continuous wave electron paramagnetic resonance studies demonstrated that Cu^{2+} first binds and fills the C-terminal binding sites before occupying the octarepeats at the N-terminus (Cereghetti *et al.*, 2001). Further study has not confirmed the existence of these C-terminal sites (Thompsett *et al.*, 2005) but a fifth site does exist close to the octameric repeat but more terminal (the so called 5th site). Recombinant PrP^C was also found to have the capacity to bind other metal ions such as manganese (Mn^{2+}) (Brown *et al.*, 2000) in both the octarepeats and the more C-terminal site (Jackson *et al.*, 2001). *In vitro* metal ion occupancy experiments showed that when Mn^{2+} replaced the Cu^{2+} ion in the prion protein, PrP^C altered its structure and took on a more PrP^{Sc}-like conformation (Brown *et al.*, 2000). The prion protein also lost its SOD-like function (Brown *et al.*, 2000).

Investigations on alterations in metal ion concentrations were carried out using mouse scrapie models (Thackray *et al.*, 2002) and in samples from sCJD cases (Wong *et al.*, 2001). Changes in the levels of Cu^{2+} and Mn^{2+} were detected in the brains of scrapie infected mice early in the disease, prior to the onset of clinical symptoms. In addition a major increase in blood Mn^{2+} was also noted in the early stages of disease. Analysis of purified PrP from the brains of scrapie infected mice also showed a reduction in Cu^{2+} binding to the protein and a proportional decrease in antioxidant activity between 30-60 days post infection.

A striking elevation of Mn^{2+} and to a lesser extent Zn^{2+} accompanied by a significant reduction in Cu^{2+} binding to purified PrP were found in subtypes of sporadic Creutzfeldt-Jakob disease (sCJD), the most common type of human prion disease. Studies were made using brain tissues and affinity purified PrP preparations (i.e. PrP^C, PrP^{Sc} and possibly other abnormal PrP species) obtained from four major subtypes of sporadic CJD. These were identified according to the genotype at codon 129 of the PrP gene and the PrP^{Sc} type as established by Parchi (Parchi *et al.*, 1999). Both Zn^{2+} and Mn^{2+} were undetectable in PrP^C preparations from control brain preparations. However, Cu^{2+} and Mn^{2+} changes were pronounced in sCJD subjects homozygous for methionine at codon-129 and carrying PrP^{Sc} type-1. It was also found that a decrease of up to 50% of Cu^{2+} and an approximately 10-fold increase in Mn^{2+} occurred in the brain tissues from sCJD subjects. Antioxidant activity of purified PrP was dramatically reduced by up to 85% in the sCJD variants, and correlated with an increase in oxidative stress markers in sCJD brains. These results clearly point to the fact that metal-ion occupancy alterations in PrP play a pivotal role in the pathogenesis of prion diseases. Since the metal changes differed in each sCJD variant, they may contribute to the diversity of PrP^{Sc} and disease phenotypes in sCJD (Wong *et al.*, 2001). These fascinating and significant results could also have bearing on potential approaches to the diagnosis of CJD. The increase in brain Mn^{2+} associated with prion infection is potentially detectable by MRI, and the binding of Mn^{2+} by PrP in sCJD might represent a novel diagnostic marker.

Manganese and animal TSEs

The two main animal prion diseases of interest are BSE and scrapie. We have studied the concentration of metals in cattle and sheep with either naturally occurring (field cases) or experimental TSEs (Hesketh *et al.*, 2007). The most consistent findings in this study were the elevation of manganese and the decrease in the levels of copper in both blood and regions of the central nervous system of animals with a TSE. Our results consistently and robustly demonstrate that elevated manganese is an indicator and thus a surrogate marker for the presence of a TSE infection.

Sheep show different levels of resistance or susceptibility to infection with scrapie and this has been linked to polymorphisms in the ovine *prnp* gene (Goldmann *et al.*, 1990; Belt *et al.*, 1995). These polymorphisms are referred to by the three amino acids that vary (residues 136, 154 and 171). The more resistant sheep genotype expresses a polymorphism commonly referred to as ARR. The existence of a resistant sheep genotype has lead to government policies promoting farmers to preferentially use sheep homozygous for the ARR polymorphism in PrP. The implication of this is that difference in the primary sequence of PrP alters the nature of the transmissible agent of scrapie. Understanding why different forms of PrP are more resistant to structural alteration by the scrapie agent remains unknown. However, understanding how polymorphisms can increase or decrease resistance to conversion is essential as it can inform on both the mechanism of infection and on prevention of the disease. There is some evidence that ARR sheep can have accumulation of PrP^{Sc} without scrapie symptoms (Buschmann *et al.*, 2004). Recent findings have also suggested that the ARR polymorphism may not necessarily be protective to other TSEs. There is evidence that such ARR homozygous sheep can be experimentally infected with BSE (Gonzalez *et al.*, 2005).

We analysed sheep that were homozygouse for the resistant (ARR/ARR) and susceptible polymorphisms (VRQ/VRQ) and a heterozygote form (VRQ/ARR). The sheep were experimentally challenged with scrapie. All the sheep showed a similar change in blood manganese levels that was apparent well before the onset of clinical signs (Hesketh *et al.*, 2007). ARR/ARR sheep that do not develop clinical scrapie showed the same change in their blood manganese levels. This implies that the change in blood manganese is a result of the scrapie challenge and is not a consequence of scrapie pathology. The immediate consequence of this finding is that although ARR/ARR sheep are considered to be 'resistant' to scrapie, they do show some indications that scrapie challenge results in similar metabolic changes as occur in non-resistant sheep. It should be noted that although, scrapie resistant sheep do not show symptoms of disease or pathology, in some instance challenge of resistant sheep does result in the presence of protease resistant PrP in various tissues including the brain (Buschmann *et al.*, 2004; Madec *et al.*, 2004; Gonzalez *et al.*, 2005; Le Dur *et al.*, 2005).

As well as experimental scrapie we also studied experimental BSE (Hesketh *et al.*, 2007). The analysis of bovine blood produced similar results to the blood from sheep. Once again manganese was altered. The change in manganese was dramatic and the change followed roughly a bell shape, occurring before onset of symptoms (at 27-29 months), peaking well before the terminal stage of the disease but decline again. These changes also suggest the elevation in manganese does not result from the major pathological changes but is a consequence of the infection. In experimental BSE there were no other significant changes in blood metals. In the field cases of both BSE and scrapie there were changes in selenium and molybdenum. Both metals were decreased in the TSE infected animals (Hesketh *et al.,* 2007). The difference between experimental and field cases of these diseases suggests a difference in the acquisition of the disease or possibly difference in susceptibility of animals to TSEs . The decreased levels of Mo and Se may not have been due to the TSE infection under farm conditions. Animals with decreased selenium and molybdenum levels could be inherently more susceptible to scrapie or BSE.

Manganese changes in the brains of sheep and cattle with scrapie or BSE have also been studied (Hesketh *et al.,* 2007). Field cases of TSEs all show changes in Cu throughout the regions analysed. Indicating the change is not specific of any particular region. The decrease in central nervous system (CNS) Cu levels was accompanied by a similar increase in the Cu concentrations of the liver. Changes in Cu levels in the brain are frequently accompanied by changes in liver Cu and such a change was noted for scrapie infected mice (Thackray *et al.,* 2002). This is suggestive of a decreased intake or retention of copper by the brain. Other changes in metals in the CNS are less consistent. Molybdenum and selenium are only altered in the cerebellum of scrapie field cases but not of BSE field cases. In experimental BSE only molybdenum was altered in the frontal cortex. The inconsistencies in these findings suggest that they might not be directly related to BSE or scrapie infection. In the case of BSE they might be related to the way in which the cattle were farmed. In the field cases of scrapie, as suggested for blood, these differences could be simply a bi-product of selection for scrapie sensitivity.

Origin of increased manganese

The origin of the increased Mn in the brains and blood of TSE infected animals remains unknown. The three possibilities are that there is decreased secretion of Mn from the body, release of Mn from other tissues or increased absorption of Mn from the environment. Currently, there is no evidence to clearly support any of these three possibilities. Several studies have linked elevated manganese in the soil with increased incidence of some TSEs (Ragnarsdottir and Hawkins, 2005). The two possible entry routes for increased Mn entry into the body from the soil are absorption through the diet or absorption through the respiratory and olfactory mucosae. In manganism, a disease common among manganese miners, Mn is absorbed through the lungs (Scholten, 1953; Roth, 2006). The amount of Mn in absorbed from the diet is around three percent of that present in the diet. Therefore it is very unlikely that Mn absorption through this route would have any significant effect on blood Mn levels. Absorption through the respiratory system, the known route of entry of excess Mn into the body, is the likely source of any possible increased Mn uptake/retention from the environment. It should be noted that in experimental cases of BSE and scrapie the elevation in manganese occurs independently of any increased manganese in the environment. In this case the change in Mn occurs as a result of the TSE infection. The elevated Mn is more likely to come from increased absorption through the lungs or retention of Mn rather than an increased availability from an environmental source. This does not imply that increased availability of manganese could not contribute but a rational approach to this possibility needs to be taken (see below).

Manganese binding to PrP is known to cause a conformational change in the protein (Brown *et al.*, 2000). Therefore increased brain manganese is likely to create an environment that would potentiate the rate of PrP conversion to the abnormal isoform. However, as it is well documented that the interaction between PrP^{Sc} and PrP^{c} induced the conversion to PrP^{Sc} in experimental TSE infections. It is nevertheless intriguing to speculate that prion infection could initially alter manganese metabolism in the brain which could in turn initiate conversion of host PrP^{c} to the abnormal isoform without the need for direct interaction between the introduced PrP^{Sc} and the host protein. These suggestions concerning experimental infection do not rule out the possibility that sporadic prion diseases could be triggered by a metal imbalance as previously suggested (Brown, 2001) although on the balance of probabilities, it is more likely that metal imbalances represent a risk factor than a causative agent. Indeed, despite the high correlation between increased Mn concentrations in CNS or blood, it is possible that elevated Mn is a result of TSE challenge and is not at all related to protein conversion or diseases progress.

Environmental issues

Considerable press coverage has been given to the issue of prion diseases and environmental factors. This has largely been initiated by publications suggesting that environmental levels of manganese could initiate TSEs. In particular a paper by Purdey (2000) has suggested that TSE hot-spots are associated with high environmental manganese and low copper. Sampling of soil and vegetation from areas of high and low risk appeared to show a correlation with manganese levels. Further study by geochemists using more appropriate sampling methodologies have largely confirmed these findings (Ragnarsdottir and Dawkins, 2006). My own interest in this theory led to my predicting that Japan would also have cases of BSE (Brown, 2001). As discussed above, it is quite possible that changes in manganese in TSEs are due to changes in the way manganese is handled by the body (e.g. excretion) rather than by an increased entry into the body. At the least, for increased uptake from the environment, it would have to involve a pathway that would lead to a substantial increase in manganese in the blood or brain or a pathological change that would increase retention or uptake of manganese. This may not require any further increase in manganese availability in the environment. With such a small proportion of ingested manganese entering the body, the amount of increased manganese in the diet would have to be many fold before this would increase the amount absorbed. This could potentially lead to manganese toxicity before it leads to prion disease.

A study by one group comparing the location of scrapie-infected farms and geochemical maps of manganese concentrations across the United Kindom concluded that there was no relationship between the two (Chihota *et al.*, 2004). This finding has largely been dismissed by geochemists because of the very poor resolution of the geochemical maps used. Also, analysis of manganese must include exclusion of that proportion that could not be absorbed and assess only the level of bioavailable manganese. An alternative possibility as to how soil manganese concentrations could result in increased incidence of BSE or scrapie could be through a role in increasing the amount of PrP^{Sc} trapped in soil. PrP^{Sc} is extremely resistant to degradation, especially in dry conditions and the presence of manganese could stabilise the protein even further. Ingestion of the protein from the soil could then potentially allow infection of further livestock.

It has also been suggested increased manganese in the diet of animals from mineral licks or other dietary supplements could be a potential source of manganese leading to prion disease. There is currently no evidence one way or another to conclude that this could happen. The limitations on absorption (described above) could mean that this increased diet would have no real effect on that absorbed.

The main route of entry into the body of excess manganese is through absorption by the lungs. Manganese borne by dust particles could enter the body through this route. Currently there has been no study of the relation between air-borne manganese and TSE incidence. This would be the most logical study for a link between TSEs and environmental manganese. Deer that develop the TSE chronic wasting disease are known to live in the vicinity of dry, eroded soils producing manganese rich dust (Brown and Sinclair, 2003).

Whether increased blood manganese and manganese induced misfolding of proteins results from increased absorption or causes that increase, a rational approach to the issue must be taken. Reports in the Media that manganese can cause prion disease are nonsensical and do not represent the facts as reported in the literature. Manganese is both ubiquitous and an essential trace element. Its potential role in prion disease must be in combination with other factors. Also any potential causal effect of manganese must be assessed in relation to what is known about its routes of entry into the body and its metabolism. Unfortunately manganese metabolism is poorly understood and a mechanistic explanation for the role of manganese in prion disease might have to wait until that metabolism is better understood.

References

Baker, H.F. and R.M. Ridley, 1996. What went wrong in BSE? From prion disease to public disaster. Brain Res. Bull. 40, 237-244.

Belt, P.B., I.H. Muileman, B.E. Schreuder, J. Bos-de Ruijter, A.L. Gielkens and M.A. Smits, 1995 Identification of five allelic variants of the sheep PrP gene and their association with natural scrapie. J. Gen. Virol. 76, 509-17.

Bolton D.C., M.P. McKinley and S.B. Prusiner, 1982. Identification of a protein that purifies with the scrapie prion. Science 218, 1309-1311.

Brown, D.R., K. Qin, J.W. Herms, A. Madlung, J. Manson, R. Strome, P.E. Fraser, T. Kruck, A. von Bohlen, W. Schulz-Schaeffer, A. Giese, D.Westaway, and H.A. Kretzschmar, 1997. The cellular prion protein binds copper *in vivo*. Nature 390, 684-687.

Brown, D.R., 1999. Prion protein expression aids cellular uptake and veratridine-induced release of copper. J. Neurosci. Res. 58, 717-725.

Brown, D.R., 2001. BSE did not cause variant CJD: An alternative cause related to post-Industrial environmental contamination. Med. Hypoth.57, 555-560.

Brown, D.R., 2004. Role of the prion protein in copper turnover in astrocytes. Neurobiol. Dis. 15, 534-543.

Brown, D.R., B. Schmidt and H.A. Kretzschmar, 1998. Effects of copper on survival of prion protein knockout neurones and glia. J. Neurochem. 70, 1686-1693.

Brown, D.R., C. Clive and S.J. Haswell, 2001. Anti-oxidant activity related to copper binding of native prion protein. J. Neurochem. 76, 69-76.

Brown, D.R., F. Hafiz, L.L. Glasssmith, B.-S. Wong, I.M. Jones, C. Clive and S.J. Haswell, 2000. Consequences of manganese replacement of copper for prion protein function and proteinase resistance. EMBO J. 19, 1180-1186.

Brown, D.R., R.St.J. Nicholas, and L. Canevari, 2002. Lack of prion protein expression results in a neuronal phenotype sensitive to stress. J. Neurosci. Res. 67, 211-224.

Brown, D.R. and K.Sinclair, 2003. Deer slaughter outrage. Vet. Times 33, 14-18.

Browning, S.R, G.L Mason, T. Seward, M. Green, G.A. Eliason, C. Mathiason M.W. Miller, E.S. Williams, E. Hoover and G.C. Telling, 2004. Transmission of prions from mule deer and elk with chronic wasting disease to transgenic mice expressing cervid PrP. J. Virol. 78, 13345-13350.

Bruce, M.E., 2003. TSE strain variation. Br. Med. Bull. 66, 99-108.

Büeler H., A. Aguzzi, A. Sailer, R.A. Greiner, P. Autenried, M. Aguet and C. Weissmann, 1993. Mice devoid of PrP are resistant to scrapie. Cell 73, 1339-47.

Büeler H., M. Fischer, Y. Lang, H. Bluethmann, H.P. Lipp, S.J. DeArmond, S.B. Prusiner, M.,Aguet and C. Weissmann, 1992. Normal development and behaviour of mice lacking the neuronal cellsurface PrP protein. Nature 356, 577-582.

Buschmann A., A.G. Biacabe, U. Ziegler, A. Bencsik, J.Y. Madec, G. Erhardt, G. Luhken, T. Baron and M.H. Groschup, 2004. Atypical scrapie cases in Germany and France are identified by discrepant reaction patterns in BSE rapid tests. J Virol. Methods 1171, 27-36.

Cereghetti, G.M., A. Schweiger, R. Glockshuber, and S. Van Doorslaer, 2001. Electron Paramagnetic Resonance Evidence for Binding of Cu^{2+} to the C-terminal Domain of the Murine Prion Protein. Biophys. J. 81, 516-525.

Chihota, C.M., M.B. Gravenor and M. Baylis, 2004. Investigation of trace elements in soil as risk factors in the epidemiology of scrapie. Vet Rec. 15426, 809-813.

Cohen, F.E. and Prusiner, S.B. 1998. Pathologic conformations of prion proteins. Annu. Rev. Biochem., 67, 793-819.

Collinge, J. 1997. Human prion diseases and bovine spongiform encephalopathy BSE. Hum. Mol. Genet.6, 1699-1705.

Gajdusek, D.C. and C.J. Gibbs, Jr., 1971. Transmission of two subacute spongiform encephalopathies of man Kuru and Creutzfeldt-Jakob disease.to new world monkeys. Nature 230, 588-591.

Gambetti P., R.B. Pettersen, P. Parchi, S.G. Chen, S. Capellari, L. Goldfarb, R. Gabizon, P. Montagna, E. Lugaresi, P. Piccardo and B. Ghetti, 1999.Inherited prion disease. Prion Biology and Disease. Edited by S.B. Prusiner, Cold Springs Harbour Laboratory Press, Cold Springs Harbour, USA.

Ghetti, B., S.R. Dlouhy, G. Giaccone, O. Bugiani, B. Frangione, M.R. Farlow, and F. Tagliavini, 1995. Gerstmann-Straussler-Scheinker disease and the Indiana kindred. Brain Pathol. 5, 61-95.

Goldmann, W., N. Hunter, J.D. Foster, J.M. Salbaum, K. Beyreuther and J. Hope, 1990. Two alleles of a neural protein gene linked to scrapie in sheep. Proc. Natl. Acad. Sci. USA. 877, 2476-80.

Gonzalez, L., S. Martin, F.E. Houston, N. Hunter, H.W. Reid, S.J. Bellworthy and M. Jeffrey, 2005. Phenotype of disease-associated PrP accumulation in the brain of bovine spongiform encephalopathy experimentally infected sheep. J. Gen.Virol. 86, 827-838.

Haigh, C.L., K. Edwards and D.R. Brown, 2005. Copper binding is the governing determinant of prion protein turnover. Mol. Cell Neurosci. 30, 186-196.

Harris, D.A., D.L. Falls, F.A. Johnson and G.D. Fischbach, 1991. A prion-like protein from chicken brain copurifies with an acetylcholine receptor-inducing activity. Proc. Natl. Acad. Sci. USA. 88, 7664-7668.

Hesketh, S., J. Sassoon, R. Knight, J. Hopkins and D.R. Brown, 2007.Elevated manganese levels in blood and CNS Occur prior to onset of clinical signs in scrapie and BSE. J. Anim. Sci. In press.

Hope, J., L.J. Reekie, N. Hunter, G. Multhaup, K. Beyreuther, H. White, A.C. Scott, M.J. Stack, M. Dawson and G.A. Wells, 1988. Fibrils from brains of cows with new cattle disease contain scrapie-associated protein. Nature 336, 390-392.

Hornshaw, M.P., J.R. McDermott and J.M. Candy, 1995a. Copper binding to the N-terminal tandem repeat regions of mammalian and avian prion protein. Biochem. Biophys. Res. Commun. 207, 621-629.

Hornshaw, M.P., J.R. McDermott, J.M. Candy and J.H. Lakey, 1995b. Copper binding to the N-terminal repeat region of mammalian and avian prion protein: structural studies using synthetic peptides. Biochem. Biophys. Res. Comm. 214, 993-999.

Hsiao, K., H.F. Baker, T.J. Crow, M. Poulter, F. Owen, J.D. Terwilliger, D. Westaway, J. Ott, and S.B. Prusiner, 1989. Linkage of a prion protein missense variant to Gerstmann-Straussler syndrome. Nature 338, 342-345.

Ironside, J.W. and M.W. Head, 2004. Variant Creutzfeldt-Jakob disease: risk of transmission by blood and blood products. Haemophilia 10, Suppl 4, 64-69.

Jackson, G.S., I. Murray, L.L. Hosszu, N. Gibbs, J.P. Waltho, A.R. Clarke, and J. Collinge, 2001. Location and properties of metal-binding sites on the human prion protein. Proc. Natl. Acad. Sci. USA, 98, 8531-8535.

Kretzschmar, H.A., S.B. Prusiner, L.E. Stowring, and S.J. DeArmond, 1986. Scrapie prion proteins are synthetized in neurons. Am. J. Pathol. 122, 1-5.

Le Dur, A., V. Beringue, O. Andreoletti, F. Reine, T.L. Lai, T. Baron, B. Bratberg, J.L. Vilotte, P. Sarradin, S.L. Benestad and H. Laude, 2005. A newly identified type of scrapie agent can naturally infect sheep with resistant PrP genotypes. Proc. Natl. Acad. Sci. USA. 102, 16031-16036.

Lugaresi, E., R. Medori, P. Montagna, A. Baruzzi, P. Cortelli, A. Lugaresi, P. Tinuper, M. Zucconi and P. Gambetti, 1986. Fatal familial insomnia and dysautonomia with selective degeneration of thalamic nuclei. New Engl. J. Med. 315, 997-1004.

Madec, J.Y., S. Simon, S. Lezmi, A. Bencsik, J. Grassi and T. Baron, 2004. Abnormal prion protein in genetically resistant sheep from a scrapie-infected flock. J. Gen. Virol. 85, 3483-3486.

Oidtmann, B., D. Simon, N. Holtkamp, R. Hoffmann and M. Baier, 2003.Identification of cDNAs from Japanese pufferfish Fugu rubripes.and Atlantic salmon Salmo salar.coding for homologues to tetrapod prion proteins. FEBS Lett. 538, 96-100.

Parchi, P., A. Giese, S. Capellari, P. Brown, W. Schulz-Schaeffer, O. Windl, I. Zerr, H. Budka, N. Kopp, P. Piccardo, S. Poser, A. Rojiani, N. Streichemberger, J. Julien, C. Vital, B. Ghetti, P. Gambetti, and H. Kretzschmar, 1999. Classification of sporadic Creutzfeldt-Jakob disease based on molecular and phenotypic analysis of 300 subjects. Ann. Neurol. 46, 224-233.

Pauly, P.C. and D.A. Harris, 1998. Copper stimulates endocytosis of the prion protein. J. Biol. Chem. 273, 33107-33110.

Prusiner, S.B. 1982. Novel proteinaceous infectious particles cause scrapie. Science 216, 136-144.

Prusiner, S.B. 1998. Prions. Proc. Natl. Acad. Sci. USA 95, 13363-13383.

Purdey, M. 2000. Ecosystems supporting clusters of sporadic TSEs demonstrate excesses of the radical generating divalent cation manganese and deficiencies of antioxidant co-factors; Cu, Se, Fe, Zn.Does a foreign cation substitution at PrP's Cu domain initiate TSE? Med. Hypoth. 54, 278-306.

Ragnarsdottir K.V. and D.P. Hawkins, 2006.Bioavailable copper and manganese in soils from Iceland and their relationship with scrapie occurrence in sheep. J. Geochem. Explor. 88, 228-234.

Riek, R., S. Hornemann, G. Wider, M. Billeter, R. Glockshuber and K.Wuthrich, 1996. NMR structure of the mouse prion protein domain PrP121-321. Nature 382, 180-182.

Roth, J.A. 2006. Homeostatic and toxic mechanisms regulating manganese uptake, retention, and elimination. Biol. Res. 391, 45-57.

Rudd, P.M., T. Endo, C. Colominas, D. Groth, S.F. Wheeler, D.J. Harvey, M.R. Wormald, H. Serban, S.B. Prusiner, A. Kobata and R.A. Dwek, 1999. Glycosylation differences between the normal and pathogenic prion protein isoforms. Proc. Natl. Acad. Sci. USA. 96, 13044-13049.

Salès, N., K. Rodolfo, R. Hässig, B. Faucheux, L. Di Giamberdino and K.L. Moya, 1998. Cellular prion protein localization in rodent and primate brain. Eur. J. Neurosci. 10, 2464-2471.

Scholten, J.M. 1953. On manganese encephalopathy; description of a case.Folia Psychiatr. Neurol. Neurochir. Neerl. 56, 878-884.

Simonic, T. S. Duga, B. Strumbo, R. Asselta, F. Ceciliani and S.Ronchi, 2000.cDNA cloning of turtle prion protein. FEBS Lett. 469, 33-38.

Smith, P.G and R. Bradley, 2003. Bovine spongiform encephalopathy BSE and its epidemiology. Br. Med. Bull. 66, 185-98.

Strumbo, B., S. Ronchi, L.C. Bolis and T. Simonic, 2001.Molecular cloning of the cDNA coding for Xenopus laevis prion protein. FEBS Lett. 508, 170-174.

Thackray, A.M., R. Knight, S.J. Haswell, R. Bujdoso and D.R. Brown, 2002. Metal imbalance and compromised antioxidant function are early changes in prion disease. Biochem. J. 362, 253-258.

Thompsett, A.R., S.R. Abdelraheim, M. Daniels and D.R. Brown, 2005. High affinity binding between copper and full-length prion protein identified by two different techniques. J. Biol. Chem. 280, 42750-42758.

Van Doorslaer, S., G. Cereghetti, R. Glockshuber and A. Schweiger, 2001.Unravelling the Cu2+ binding sites in the C-terminal domain of the murine prion protein : A pulse EPR and ENDOR study. J. Phys. Chem. B 105, 1631-1639.

Will, R. 2004. Variant Creutzfeldt-Jakob disease. Folia Neuropathol. 42 Suppl A, 77-83.

Wong, B.-S., S.G. Chen, M. Colucci, Z. Xie, T. Pan, T. Liu, R. Li, P. Gambetti, M.-S. Sy and D.R. Brown, 2001. Aberrant metal binding by prion protein in human prion disease. J. Neurochem. 78, 1400-1408.

Wopfner, F., G. Wiedenhöfer, R. Schneider, A. von Bunn, S. Gilch, T.F. Schwarz, T. Werner and H.M. Schätzl, 1999. Analysis of 27 mammalian and 9 avian PrPs reveals high conservation of flexible regions of the prion protein. J. Mol. Biol. 289, 1163-1178.

Zahn, R., A. Liu, T. Luhrs, R. Riek, C. von Schroetter, F. Lopez Garcia, M. Billeter, L. Calzolai, G.Wider, and K. Wuthrich, 2000.NMR solution structure of the human prion protein. Proc Natl Acad Sci USA. 97, 145-150.

Short communications

Chemical identity of crystalline trace mineral glycinates for animal nutrition

S. Oguey, A. Neels and H. Stoeckli-Evans

Keywords: trace mineral, chelate, complex, glycinate, glycine, X-ray diffraction

Introduction

Organic trace minerals are used for about 30 years in animal nutrition to replace traditional inorganic forms, like sulfates, oxides, carbonates or chlorides. Organic trace minerals are generally recognized to be more bioavailable (example: Ammerman *et al* 1995) than inorganic sources, but large variations may exist depending on the metal, the type of ligand (glycine, methionine, soy protein peptides, …) and the producer.

For 25 years, organic trace minerals have had a large drawback: their chemical identity was not precisely known and only theoretical formula could be used for a given product.

Since 2003, crystalline forms of organic trace minerals using glycine as ligand are produced on industrial scale (B-TRAXIM® 2C, Pancosma, Switzerland). In the E.U., these products are defined as 'chelate of glycine, hydrate' (European commission, 2006); outside the E.U., generally using the Association of American Feed Control Officials (AAFCO) definitions, these products are defined as 'specific amino acid complex'. Since crystalline forms open new analytical possibilities in identifying products, the aim of this study was to define the chemical structure of 4 crystalline trace mineral glycinates using X-ray diffraction analysis.

Material and methods

Four commercially available organic trace mineral sources (B-TRAXIM® 2C Fe, Zn, Cu and Mn, Pancosma, Switzerland) were analyzed by powder X-ray diffraction to measure their chemical structure and spatial organization.

Powder X-ray diffraction analysis: X-ray powder diffraction data were collected in transmission mode (0.5 mm rotating capillary) on a high resolution laboratory powder diffractometer (Stoe STADI P) using copper $K_{\alpha 1}$ radiation (1.5406 Å) and a curved germanium monochromator. Eight equivalent data sets were collected, each from 4 to 90° in 2θ with steps of 0.1° and a counting time of 30 sec. per step using a linear position sensitive detector (PSD). The latter is capable of measuring ca. 5° in 2θ for each step. No decomposition was observed during the measurements. The powder diffractograms were first indexed to reveal the crystal system, unit cell parameters and space group. The intensity data were then extracted from the profile and the structures solved by direct methods using the program EXPO (Altomare *et al*, 1997), and refined by Rietveld refinement using GSAS (Larson and Van Dreele, 1994). After the initial refinement of the scale, background and unit cell constants, the atomic positions were refined with soft constraints on bond distances (Int. Tables for Cryst. C, 1995) and angles for the lighter fragments of the structure (sulfate anion, glycine). The heavy atom was freely refined. In the final cycles of refinement the shifts in all the parameters were less than their estimated standard deviations.

Results and conclusion

The crystalline structure of trace mineral glycinates has permitted their identification by X-ray diffraction analysis (Figure 1). The four commercially available trace mineral glycinates (B-TRAXIM® 2C) have different hydration forms; The Mn product is anhydrous, the Fe product is trihydrated while Cu and Zn products are dihydrated and isomorphous. X-Ray diffraction gives an indisputable proof that each metal is chemically bound to its ligand glycine.

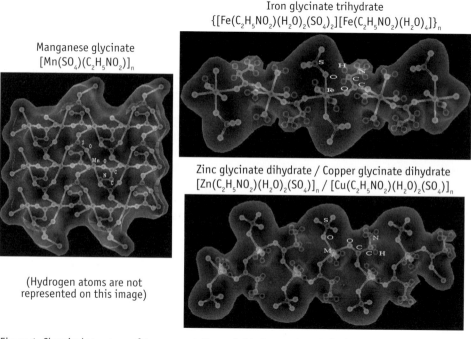

Iron glycinate trihydrate
$\{[Fe(C_2H_5NO_2)(H_2O)_2(SO_4)_2][Fe(C_2H_5NO_2)(H_2O)_4]\}_n$

Manganese glycinate
$[Mn(SO_4)(C_2H_5NO_2)]_n$

Zinc glycinate dihydrate / Copper glycinate dihydrate
$[Zn(C_2H_5NO_2)(H_2O)_2(SO_4)]_n / [Cu(C_2H_5NO_2)(H_2O)_2(SO_4)]_n$

(Hydrogen atoms are not represented on this image)

Figure 1. Chemical structure of 4 commercially available trace mineral glycinates.

References

Ammerman, C.B., D.H. Baker and A.J. Lewis, 1995. Bioavailability of nutrients for animals. Academic Press, Inc., San Diego, U.S.A.

European commission – Health and Consumer Protection, 2006, EC regulation n° 1831/2003, revision 4.

Stoe STADI P. Stoe & Cie GmbH, Darmstadt, Germany

Altomare, A., M.C. Burla, G. Cascarano,C. Giaccovazzo, A. Guagliardi, A.G.G. Moliterni and G. Polidori, 1997. EXPO. University of Bari, Italy.

Larson, A., R.B. Von Dreele, 1994. GSAS. Los Alamos National Laboratory, Los Alamos, NM.

International Tables for Crystallography Vol. C. ,1995. Kluwer Academic Publishers, Dordrecht, The Netherlands.

The use of the protein hydrolysation degree as analytical possibility to differentiate trace element chelates of amino acids

N. Helle and D. Kampf

Keywords: amino acid, trace element, chelate, differentiation, ultra filtration

Introduction

Under certain conditions, including organic trace elements (trace element chelates of amino acids) in animal feed leads to higher bio-availability for the animal than does providing inorganic bound trace elements in sulfate or oxide form (Kegley and Spears, 1994; Nockels et al., 1993; Wedekind et al., 1992). According to EU Commission Regulation Nr. 1334/2003 trace minerals chelated with soy protein are approved for use in feed for animal husbandry, but with some restrictions. The amino acids must be derived from hydrolyzed soy protein and the trace element (Zn, Cu, Mn, Fe) is allowed to be complexed to one to three amino acids. From this legislation, criteria regarding molecular weight and protein profile can be derived, to which commercially available chelates in the European market must conform. The present study describes the development of a simple methology for screening products presently available on the market.

Material and methods

Products for evaluation were obtained from commercial markets. Five-gram subsamples of each product were suspended in 75 ml of 1M sodium acetate/acetic acid buffer (pH 3.0), by stirring and ultrasound treatment, then filtered to remove carrier materials. The filtrates were then passed through an ultra filtration device (Amicon filter unit) with a 1,000 Dalton filter (Ghash and Cui, 2000; Millipore Corporation, 2006). Both in the original product as well as after ultra filtration the crude protein analysis took place according to Kjehldahl, trace mineral contents were determined with atom absorption spectrometry (Naumann and Bassler, 1993). The measured protein contents of the pure product and the smaller than 1,000 Dalton fraction were compared afterwards with the necessary protein portion needed in the trace element chelate, calculated on basis of the molecular weights of copper, zinc and soy protein amino acids (63.6g/Mol, 65.4g/Mol, 131.7g/Mol respectively). Therefore the content of hydrolyzed protein has to be at least two-times higher than the trace element content.

Results and conclusions

Figures 1 and 2 are showing the results of the screening procedure for respectively copper and zinc chelates. To get a better comparability all products were corrected to 10% trace element content. The examined commercial products varied clearly in crude protein (171 to 368 g/kg) and trace mineral contents (74 to 127 g/kg). In the smaller than 1,000 Dalton fractions of the examined products the protein contents showed an even bigger variation (11 to 250 g/kg) than the obtained trace mineral contents (58 to 131 g/kg). The comparison of the in the smaller than 1,000 Dalton fraction analyzed protein with the necessary portion of hydrolyzed protein for complete chelation of the available trace minerals (black dotted line in Figures 1 and 2) clarifies that the predominant part of the products contained an inadequate content of hydrolyzed protein. This correlates equivalently to an insufficient hydrolysis of the protein and thus with small portions of theoretically present chelate bonds (on basis of a 1:1 ratio of amino acid and trace element) in the product. In different products only 10% of the declared trace element can theoretically be chelated to soy protein amino acids. The main part of the analyzed products showed increasing proportions between protein and trace element contents after ultra filtration compared to the crude product. This can also be used as sign for an insufficient content of chelated trace element.

First investigations showed clear differences in crude protein content and hydrolyzed protein portion (<1000 Dalton) between individual trace element chelates of amino acids. These differences couldn't be obtained for the similar trace element contents. The necessary minimum content of hydrolyzed protein of at least 200 g/kg (based on a chelate product with 10% trace element content) was not reached at the majority of the examined products.

This analytical method enables a reliable classification of product quality of different commercial available trace element chelate products on soy protein basis.

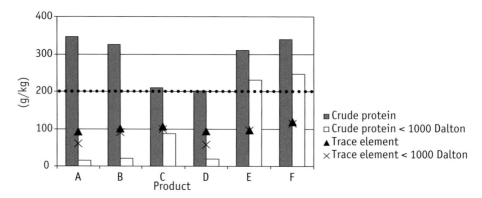

Figure 1. Crude protein content in the original product and after ultra filtration (1,000 Dalton) of copper chelates.

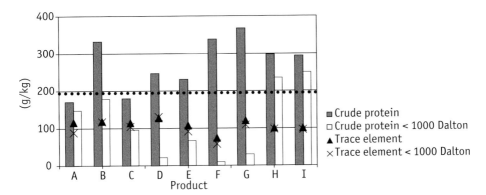

Figure 2. Crude protein content in the original product and after ultra filtration (1,000 Dalton) of zinc chelates.

References

Ghosh, R. and Z.F. Cui, 2000. Protein purification by ultra filtration with pre-treated membrane. J.Membrane Sci. 167, 47-53.

Kegley, E.B. and J.W. Spears, 1994. Bioavailability of Feed-Grade Copper Sources (Oxide, Sulfate or Lysine) in Growing Cattle. J.Anim.Sci. 72, 2728-2734.

Millipore Corporation, 2006. Ultrafiltration Handbook. http://www.millipore.com.

Naumann, C. and R. Bassler, 1993. Die chemische Untersuchung von Futtermitteln. Methodenbuch, Band III, VDLUFA-Verlag.

Nockels, C.F., J. DeBonis and J. Torrent, 1993. Stress Induction Affects Copper and Zinc Balance in Calves Fed Organic and Inorganic Copper and Zinc Sources. J.Anim.Sci. 71, 2539-2545.

Wedekind, K.J., A.E. Hortin and D.H. Baker, 1992. Methodology for Assessing Zinc Bioavailability: Efficacy Estimates for Zinc-Methionine, Zinc Sulfate, and Zinc Oxide. J.Anim.Sci. 70, 178-187.

First insight of copper and zinc speciation in pig slurry: sequential extraction and size fractionation studies

S. Legros, E. Doelsch, A. Masion, H. Saint-Macary and J.Y. Bottero

Keywords: pig slurry, trace elements, speciation, infrared spectroscopy

Introduction

Increased concentrations of Copper (Cu) and Zinc (Zn) have been measured in the surface horizon of soils on which pig slurry was applied (L'Herroux *et al*, 1997). Cu and Zn occur in pig slurry as a result of their use as growth promoters or biocides in animal feeds. A better prediction of the mobility and bioavailability of Cu and Zn from pig slurry spreading, requires the determination of the speciation of these elements. The objective of this study was to describe the speciation of Cu and Zn by using (1) a sequential extraction protocol and (2) a size fractionation.

Material and methods

Pig slurry (containing ca. 1800 mg/kg of Cu and 5000 mg/kg of Zn) was sampled on the Réunion Island. A sequential extraction procedure (Hall *et al*, 1996) was applied in triplicate to a dried pig slurry to determine operationally defined Cu and Zn speciation. Size fractionation of the pig slurry was also performed in triplicate using the following sieve sizes: 1000 µm, 630 µm, 355 µm, 200 µm, 50 µm, 20 µm, and 0.45µm. Chemical and infrared (FTIR) analyses of each fraction (e.g. PS0.45-20µm corresponding to the particle size between 0.45µm and 20 µm of pig slurry) were undertaken.

Results and discussion

Sequential extractions (Figure 1) performed on the pig slurry led to recoveries of 97% for Cu and 88% for Zn. There are marked differences between the two metals regarding the amount recovered after each individual extraction step.

High Cu concentrations were found in 2 fractions, viz. the oxide fraction (i.e. crystalline oxides and/or amorphous oxyhydroxides) at 1270 mg/kg (76%) and the organic matter fraction (325 mg/kg – 20%). Three fractions had high Zn concentrations: the adsorbed fraction (1640 mg/kg – 40%), the organic matter fraction (1420 mg/kg – 34%) and the oxide fraction (990 mg/kg – 24%).

The majority of the Cu and Zn was detected in the PS0.45-20µm fraction (Figure 2a), with 1540 mg/kg (78.2%) of Cu and 3120 mg/kg (75%) of Zn. The FTIR analysis of PS0.45-20µm fraction (Figure 2b) revealed the presence of aliphatic chains, organic functions (hydroxyls, acids and amines) and inorganic compounds (phosphates, carbonates and silicates).

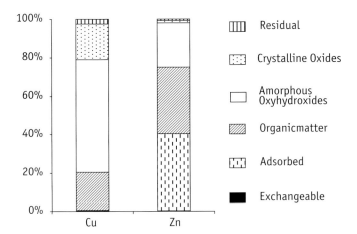

Figure 1. Cu and Zn distribution in pig slurry by sequential extractions.

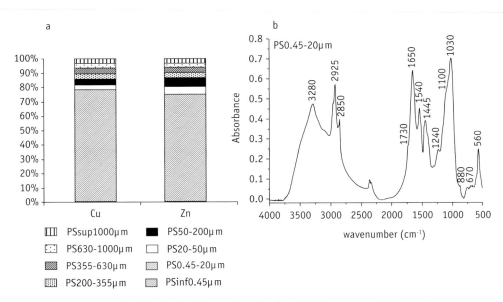

Figure 2. a. Cu and Zn distribution in different size fractions. b. PS0.45-20µm FTIR spectrum.

The mobility and bioavailability of Cu and Zn can be attributed mainly to the PS0.45-20µm fraction. In this fraction, several compounds play an important role in the speciation of Cu and Zn: Cu could be bound to oxides as strongly suggested by the high Fe and Mn concentrations measured in this fraction (10,000 mg/kg and 884 mg/kg, respectively). Most of the Zn was presumably bound to either carbonates (adsorbed fraction) or organic compounds. The FTIR data for the PS0.45-20µm fraction were consistent with this hypothesis. The main objective of these experiments was to determine the nature of the compounds associated with Cu and

Zn in pig slurry, therefore the mobility could not be quantified directly. Nevertheless the particle size and the nature of compounds associated with Cu and Zn suggest a relatively poor mobility.

References

Hall, G.E.M., G. Gauthier, J.C. Pelchat, P. Pelchat. and J.E. Vaive, 1996. Application of a sequential extraction scheme to ten geological certified reference materials for the determination of 20 elements. J. Anal. Atom. Spectro. 11, 787-796.

L'Herroux, L., S. Le Roux, P. Appriou and J. Martinez, 1997. Behavior of metals following intensive pig slurry applications to a natural field treatment process in Britany (France). Environ. Pol. 97, 119-130.

First insights of copper and zinc speciation within a pig slurry: an X-ray absorption spectroscopy study

E. Doelsch, S. Legros, J. Rose, A. Masion, O. Proux, J.-L. Hazemann, H. Saint Macary and J.Y. Bottero

Keywords: trace elements, speciation, X-ray fluorescence, X-ray absorption spectroscopy

Introduction

Studying the speciation of heavy metals instead of their total concentration in a complex matrix such as pig slurry is a scientific challenge that requires a combination of different analytical techniques. Previously, we compared the speciation of Cu and Zn within pig slurry (Legros *et al.*, 2007) by using sequential extraction and size fractionation. When sequential extraction procedures are used to study heavy metal speciation of wastes (Ure and Davidson, 2001), problems involving the non-selectivity of reagents or readsorption of elements following release are frequently reported (Martin *et al.*, 1987). Moreover, the extraction procedure allows determining only 'operationally' defined speciation which depends on the reagents and operating procedure. Recent advances in synchrotron-based X-ray absorption spectroscopy (XAS) have led to important innovations and increased our knowledge on trace element speciation. This analytical technique was used to describe the local environment of Cu and Zn within a pig slurry.

Materials and methods

An X-Ray absorption spectroscopy study was conducted with pig slurry described elsewhere (Legros *et al.*, 2008). Cu and Zn K-edge EXAFS measurements were performed in the transmission or fluorescence mode at the following synchrotron sources: (1) beamline FAME at the European Synchrotron Radiation Facility (Grenoble, France) (2) beamline X23A2, National Synchrotron Light Source (Upton, NY, USA).The x-ray absorption spectrum is typically divided into two regimes: x-ray absorption near-edge spectroscopy (XANES) and extended x-ray absorption fine-structure spectroscopy (EXAFS). XANES is strongly sensitive to formal oxidation state and coordination chemistry (e.g. octahedral, tetrahedral coordination) of the absorbing atom, while EXAFS is used to determine the distances, coordination number, and species of the neighbors of the absorbing atom. EXAFS data reduction was accomplished according to a procedure previously described (Doelsch *et al.*, 2006). Fourier transformation of the oscillatory fine structure (obtained after background subtraction) yields a radial function distribution (RDF) in real space with peaks revealing the local environment of the target atom.

Results and discussion

XANES spectra were used to infer a Cu oxidation state in the pig slurry sample. The comparison of pig slurry XANES spectra (Figure 1a) with three references clearly demonstrates that oxidation state of copper is Cu^{1+}. Moreover, the similarity between the pig slurry and the CuS reference XANES spectra indicates that the Cu neighbors are S atoms.

For Zn, the radial distribution function (RDF) consists of two main peaks resulting from Fe – backscatterer interactions (Figure 1b). The first peak at ≈1.5 Å (RDF distances uncorrected for phase shift function) corresponds to the contribution of O in the first ligand sphere surrounding Zn atoms. The second peak at 1.95 Å indicates the presence of a second contribution in the first ligand sphere of Zn. The comparison with ZnS reference indicates that this contribution results from S ligands.

It is worth noting that Cu and Zn speciation in the surface environment (soil or water) are mainly 2+ for oxidation state with O as first neighbors. Therefore, the speciation of Cu and Zn within pig slurry (Cu^{1+}-S; Zn^{2+}-O and Zn^{2+}-S) was unexpected. Metallothioneins (low molecular weight metalloprotein characterized by their high cysteine content), which have a high affinity for a wide range of transition-metal ions (especially Cu and Zn), could be the origin of Cu-S and Zn-S speciation within pig slurry. Indeed, the metal ions are bound through bridging and terminal thiolate groups of the cysteines (Chan *et al.*, 2002).

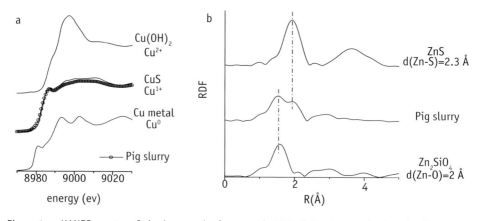

Figure 1. a. XANES spectra of pig slurry and references. b. RDF of pig slurry and mineral references.

References

Legros, S., E. Doelsch, A. Masion, H. Saint Macary and J.Y. Bottero, 2008, First insights of copper and zinc speciation within a pig slurry. In: P. Schlegel, S. Durosoy and A.W. Jongbloed (Eds.), Trace elements in animal production systems, Wageningen Academic Publishers, Wageningen, the Netherlands, pp.250-252.

Chan, J., Z. Huang, M.E. Merrifield, M.T. Salgado, M.J. Stillman, 2002. Studies of metal binding reactions in metallothioneins by spectroscopic, molecular biology, and molecular modeling techniques. Coordination Chemistry Reviews 233-234, 319-339.

Doelsch, E., I. Basile-Doelsch, J. Rose, A. Masion, D. Borschneck, J.L. Hazemann, H. SaintMacary and J.Y. Bottero, 2006. New Combination of EXAFS Spectroscopy and Density Fractionation for the Speciation of Chromium within an Andosol. Environ. Sci. Technol. 40, 7602-7608.

Macro and micronutrients in anaerobically digested pig slurry: recovery of Ca and Mg phosphates and nutrient fate in soil

C.E. Marcato, E. Pinelli, P. Pouech, A. Alric and M. Guiresse

Keywords: anaerobic process, copper, zinc, nutrients, phosphates, pig slurry

Introduction

Methane production from the biological conversion of organic waste is receiving increasing attention, both with a view to processing organic waste and as a source of energy (Pouech *et al.*, 2005). However, anaerobic digestion (AD) is a transformation process and not a destructive final treatment. The product obtained then has to be dealt with. In agricultural installations, the digestates are spread on the land. Applications are based on macro-nutrient loading, and trace metals may be overdosed with regards to crop requirements. In the current work, a livestock-manure AD plant was investigated to compare raw and anaerobically digested pig slurry. The fate of the macro and micro elements during the anaerobic treatment was followed. Mass balances were calculated for the AD plant and for the soil, after spreading of raw or digested slurry.

Material and methods

The present work was conducted on the ADÆSO experimental site which consists of five livestock buildings and a 150 m^3 AD plant (Pouech *et al.*, 2005). The anaerobic stirred tank reactor was run at 37 °C, with a hydraulic retention time of about 15 days. Macro and micro element composition of pig slurry was studied before and after AD. AD mass balance was calculated between the input and the output of the reactor.

Results

Mass balance showed that the AD process consumed 49% of the volatile solids (VS), which were most likely transformed into biogas. Hence the metal concentrations of the slurry nearly doubled after anaerobic digestion while nitrogen, the main nutrient, remained constant (Table 1). Most of the elements were conserved between the input and the output of the digester (N, K, Fe, S, Al, Cu and Zn). However, losses did occur for P, Ca, Mg and Mn due to crystallisation on the lining of the digester.

Scanning electron microscopy coupled with energy dispersive X-ray analysis (SEM-EDS) observations (Figure 1) clearly showed the crystals to be composed of Ca, Mg, Mn and P, and free of N. They were in fact amorphous magnesium phosphate (Figure 1, point 1) and well-crystallised calcium phosphate (Figure 1, point 2) and not the Struvite crystals often described (Suzuki *et al.*, 2002; Burns and Moody, 2002). They could be Whitlockite or Bobbierite nucleating at low temperatures on the stainless steel surface of the agitation system.

Table 1. Composition of pig slurry and mass balance over the whole anaerobic treatment (RS: raw slurry; DS: digested slurry).

	Flow	DM	Ntot	NH$_4^+$-N	VS	P	K	Ca	Fe	Mg	S	Al	Cu	Mn	Zn
		% fresh matter			g/kg DM							mg/kg DM			
RS		2.7	0.28	0.21	67.3	29.5	37.4	51.2	2.5	14.2	8.1	868	590	629	1507
DS		1.6	0.28	0.23	59.1	31.5	65.6	47.2	4.0	15.9	11.1	1641	1016	708	2628
	kg/day				kg/day							g/day			
RS	11297	301	31.6	23.7	202	8.9	11.2	15.4	0.5	4.3	2.4	261	177	189	453
DS	11121	182	28.9	24.5	107	5.7	11.9	8.6	0.5	2.9	2.0	298	185	129	478

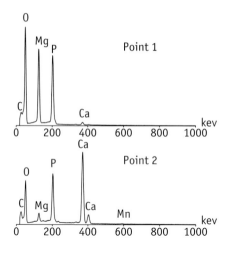

Figure 1. SEM picture and EDS spectra of Ca and Mg phosphates.

The quantities of slurry to be spread on the land are generally determined by fixing the levels of N. In French and European regulations, the recommended dosage of N is 170 kg N ha^{-1} yr^{-1}. From the composition of raw and digested slurry (Table 1) this will mean spreading 60 m^3 ha^{-1} yr^{-1}. Such applications will lead respectively to annual inputs of 47 and 22 kg P ha^{-1}, 64 and 60 kg K ha^{-1}, 13 and 9 kg S ha^{-1}, 82 and 34 kg Ca ha^{-1}, and 23 and 11 kg Mg ha^{-1}. K inputs more or less correspond to export by crops, but the P, S, Ca and Mg levels of input remain lower than the requirements of most crops. The slurry therefore contributes to the general nutrition of the plants. However, part of this amending value was lost during AD due to the crystallisation of the calcium and magnesium phosphates in the digester. Concerning the non-indispensable macro-elements, Al was the most abundant with inputs of 1.6 and

0.9 kg Al ha^{-1}, which is well below the quantities exported by the crops. To sum up, none of the macro-elements from the slurry will accumulate in arable land. The microelements presented a different picture and inputs were equivalent for both slurries. The levels of Fe and Mn were low, giving rise to soil inputs of 4 kg Fe ha^{-1} and about 1 and 0.5 kg Mn ha^{-1}, which is of the same order as the levels of export in crops. In contrast, Cu and Zn were very abundant and the input from slurry (about 0.9 kg Cu ha^{-1} and 2.5 kg Zn ha^{-1}) reached ten-fold typical export values.

References

Pouech, P., R. Coudure and C.E. Marcato, 2005. Intérêt de la co-digestion pour la valorisation des lisiers et le traitement de déchets fermentescibles à l'échelle d'un territoire. Journées Recherche Porcine 37, 39-44.

Suzuki, K., Y. Tanaka, T. Osada and M. Waki, 2002. Removal of phosphate, magnesium and calcium from swine wastewater through crystallization enhanced by aeration. Wat. Res. 36, 2991-2998.

Burns, R.T. and L.B. Moody, 2002. Phosphorus recovery from animal manures using optimized struvite precipitation. In Procedings of coagulants and flocculants: Global market and technical opportunities for water treatment chemicals, Chicago, Illinois.

Egg quality and layer performance as affected by different sources of zinc

H. Aliarabi, A. Ahmadi, S.A. Hosseini Siyar, M.M. Tabatabaie, A.A. Saki and A. Khatibjo

Keywords: zinc, source, layer hen, egg quality

Introduction

Zinc functions in the formation of eggs. Zinc deficiency affects the quality of the epithelium due to the role of zinc in protein synthesis. Activity of carbonic anhydrase, an essential enzyme for eggshell formation, has been directly related to zinc status of hens (Hudson *et al.*, 2004). Organic complexes of zinc have been proposed to be a more available source of zinc for layer hens (Cheng and Guo, 2004) and may be metabolized differently than inorganic forms (Spears, 1989). The main objective of the present study was to investigate the effect of concentrations and forms of dietary zinc on laying hen performance and egg quality.

Materials and methods

A total of 80 Leghorn layer hens at the age of 50 wk were assigned to 40 pens in a completely randomized design. The dietary treatments were: (1) basal diet (containing 29 mg kg^{-1} Zn), (2) basal diet supplemented with 50 mg kg^{-1} inorganic zinc as $ZnSO_4$, (3) basal diet supplemented with 50 mg kg^{-1} organic zinc (proteinated-Zn) and (4) basal diet supplemented with 25 mg kg^{-1} organic zinc (proteinated-Zn) and were fed for 6 weeks. The basal diet was formulated as per NRC (1994) recommendation to meet or exceed hen requirements except zinc. Feed consumption was recorded per pen. Daily egg production was recorded and expressed on a hen-day basis and weighed to calculate egg mass production. Feed conversion ratio (FCR) was calculated. Also all eggs produced on days 14, 28 and 42 were collected and measured for egg quality parameters. Data were analyzed by GLM Procedure of SAS software (SAS, 1997). Comparison between mean values was done using the Duncan Multiple Range Test.

Results and discussion

Zinc supplementation, regardless of its source had no effect on FCR, egg weight or production (*Table 1*). Cheng and Guo (2004) reported that different sources of zinc had no effect on egg production and egg weight. However, Khajarern *et al.* (2006) observed improvements in egg production, egg and egg shell quality for layers fed organic zinc. Carbonic anhydrase which is a zinc dependent enzyme plays a role in converting blood bicarbonate into calcium carbonate (Keshavarz, 2001). However, zinc supplementation had in the present study no effect on egg quality parameters except albumen height (P<0.006) as shown in *Table 2*. This may be due to the concentration of zinc in the basal diet (29 mg/kg) which possibly was sufficient for normal activity of carbonic anhydrase. Increased albumen quality in the present study, as measured by increased albumen height is in agreement with the report of Sahin and Kucuk (2003). Overall results showed that a zinc supplementation on the studied diet containing 29 mg/kg zinc did not affect layer performance, nor egg quality.

Table 1. Mean values of hen's performance as affected by zinc source and concentration.

Parameters	Treatment				P value	MSE
	Basal	Inorganic Zn 50 mg/kg	Organic Zn 50 mg/kg	Organic Zn 25 mg/kg		
Egg weight (g)	60.07	59.38	60.35	60.34	0.4511	19.85
Egg mass (g /hen/ day)	43.43	42.32	38.84	43.13	0.1778	12.02
Egg production (%)	70.60	73.75	66.46	71.40	0.1271	19.08
FCR	2.208	2.216	2.172	2.193	0.9848	0.038

Table 2. Effect of different sources and concentration of zinc on egg quality.

Parameters	Treatment				P value	MSE
	Basal	Inorganic Zn 50 mg/kg	Organic Zn 50 mg/kg	Organic Zn 25 mg/kg		
Albumen weight (g)	37.85	37.28	38.48	38.14	0.1658	12.29
Yolk weight (g)	16.86	16.73	16.63	16.97	0.4123	1.97
Shell weight (g)	5.32	5.22	5.29	5.28	0.5706	0.23
Shell thickness (mm)	0.396	0.394	0.387	0.388	0.1814	0.001
Albumen height (mm)	7.20 [c]	7.25 [bc]	7.49 [ab]	7.57 [a]	0.0058	0.67

Means bearing different superscripts in row differed significantly (P<0.05).

References

Cheng, T. and Y. Guo, 2004. Effect of Salmonella typhymurium lipopolysaccharide challenge on the performance, immune reponses and zinc metabolism of laying hens supplemented with two zinc sources. J. Anim. Sci. 17, 1717-1724.

Hudson, B.P., W.A. Dozier, J.L. Wilson, J.E. Sander and T.L. Ward. 2004. Reproductive Performance and Immune Status of Caged Broiler Breeder Hens Provided Diets Supplemented with Either Inorganic or Organic Sources of Zinc from Hatching to 65 wk of Age. J. Appl. Poult. Res. 13, 349-359.

Keshavarz, K, 2001. The influence of drinking water contaning sodium chloride on performance and egg sheel quality of modern colored laying strain. Corne. Poul. Point. 51, 12 13.

Khajaren, J., S. Khajaren, C.J. Rapp, T.A. Ward, J.A. Johnson and T.M. Falker, 2006. Effects of zinc and manganese amino acid complexes (Availa-Z/M) on layer production and egg quality. http://us.zinpro.com/research/ZPA/ZPA0048.htm.

National Research Council, 1994. Nutrient requirements of poultry 9th ed. Washington, DC. National Academy Press.

Sahin, K. and O. Kucuk, 2003. Zinc Supplementation Alleviates Heat Stress in Laying Japanese Quail. J. Nutr. 133, 2808-2811.

SAS Institute, 1997. SAS/STAT® User's Guide: Statistics, Version 6.12, SAS Institute Inc., Cary, NC.

Spears, J.W., 1989. Zinc methionine for ruminants: Relative bioavailability of zinc in lambs and effects on growth and performance of growing heifers. J. Anim. Sci. 67, 835-843.

Effects of the interaction between organic Zn and Mn on performance, mineral retention and immune response in young broiler chickens

G.S. Sunder, V.K. Chalasani, A.K. Panda, M.V.L.N. Raju, S.V. Rama Rao and A. Arun Kumar

Keywords: zinc, manganese, performance, mineral retention, immune response

Introduction

Zinc (Zn) and Manganese (Mn) are essential for growth, skeletal development and immune competence in broiler chickens. They are supplemented in diets to elicit positive response, particularly during early age. Recent research showed that organic Zn and Mn are relatively better available than inorganic sources. However, the availability of one mineral may possibly compliment or antagonize other minerals available in the diet (Collins and Moran, 1999). Therefore, the present study evaluated the interaction between organic Zn and Mn on the performance, tissue mineral concentration and immune response in broiler chickens up to 5 weeks of age.

Materials and methods

Day-old Cobb male broiler chicks (486) were distributed into 9 dietary groups with 9 replicates of 6 each and reared in stainless steel cages under standard managemental conditions. A basal diet of corn-soybean (32 and 30 ppm Zn and Mn, respectively) was prepared. Subsequently, nine test diets were formulated supplementing three levels each of organic Zn (40, 80 or 160 ppm) and organic Mn (60, 120 or 240 ppm), in a factorial pattern. Each diet (mash form) was fed *ad libitum* to one experimental group from 6-35 days of age. Data on individual body weights and leg abnormality scores (hock joint deformity) were recorded. Feed intake was measured replicate-wise at the end of each week, to calculate the feed conversion efficiency. Sheep red blood cells (SRBC), a non-pathogenic antigen was inoculated (12 birds/treatment) in the 5th week for evaluating the humoral immune response. The cell mediated immunity was assessed by measuring the hypersensitivity response of cutaneous basophils (CBH) to phytohaemagglutinin-P (PHA-P). At the end of experiment, 9 birds from each treatment were sacrificed by cervical dislocation for collection of tibia and liver for estimation of Zn, Mn and Cu concentrations (ppm/g ash or dried tissue). Data collected on different parameters were subjected to two-way factorial analysis (Snedecor and Cochran, 1989). The main factors, Zn and Mn levels and their interactions were statistically analyzed. Duncan's multiple range test was used to compare the means (P<0.05).

Results and discussion

Neither the interaction between Zn and Mn nor their main effects had any influence on body weight gain, feed conversion efficiency, leg abnormality score, tibia weight, tibia strength and bursa weight at 5 weeks of age. Similar findings were also reported in literature (Collins and

Moran, 1999). The interaction between Zn and Mn at 160 and 60 ppm respectively, improved (P< 0.05) Zn concentration in bone (271ppm) and liver (99.9 ppm) (Table 1). Similarly, Mn at 240 ppm with Zn at 40/80 ppm and 160 ppm levels improved respectively, the retention of Mn in bone (17.8 ppm) and liver (11.47 ppm). In contrast, Cu in bone was significantly (P< 0.05) reduced as the level of Mn increased in diets, while Zn was less reactive to Cu uptake.

Independently, Zn at 80 ppm level improved (P< 0.05) bone mineralization and antibody titers against SRBC over 40 ppm and was comparable to160 ppm (Table 2). A linear increase in the concentration of Zn and Mn in bone and liver was observed, with their levels of supplementation in diets. This observation was similar to our previous findings (Sunder *et al.*, 2006). Cu uptake by bone was antagonized by Mn and not Zn.

In conclusion, Zn and Mn at 160:60 and Mn and Zn at 240:40/80 ppm, respectively supported their individual retentions in bone and liver. However, Mn at higher levels of supplementation antagonized Cu uptake by bone, restricting its inclusion to 60 ppm in diets. Independently, Zn supplementation at 80 ppm improved bone mineralization and immune response in broilers at 5 weeks of age, suggesting optimum Zn and Mn levels in diets.

Table 1. Interaction effects of Zn and Mn on tibia mineral uptake and liver tissues.

Zn:Mn levels (ppm)	Mineral uptake by tibia (ppm/g ash)			Mineral uptake by liver (ppm/g DM)		
	Zn	Mn	Cu	Zn	Mn	Cu
40:60	214[cd]	9.45[e]	9.52[ab]	96.6[ab]	8.97[de]	16.21
40:120	211[d]	12.96[c]	9.54[ab]	91.3[c]	8.64[e]	16.99
40:240	203[d]	17.14[ab]	8.41[cd]	92.61[bc]	9.75[bcd]	16.32
80:60	240[bc]	10.22[e]	9.89[a]	95.98[abc]	10.15[bc]	17.84
80:120	213[cd]	15.92[b]	8.72[bc]	91.24[c]	9.32[cde]	16.48
80:240	253[ab]	17.80[a]	7.57[d]	95.95[abc]	10.49[b]	18.59
160:60	271[a]	11.62[d]	10.35[a]	99.86[a]	8.38[e]	19.67
160:120	172[e]	16.70[ab]	7.97[cd]	93.2[bc]	9.20[de]	18.35
160:240	260[ab]	14.05[c]	8.28[cd]	97.40[ab]	11.47[a]	19.70

Means in a column with different superscripts differ significantly (p≤ 0.05).

Table 2. Independent effects of Zn and Mn on mineral retention and antibody titers.

Zn and Mn levels (ppm)	Tibia ash (g/100g)	Tibia Zn (ppm/g ash)	Tibia Mn (ppm/ g ash)	Tibia Cu (ppm/ g ash)	Antibody titers (log$_2$)
Zn 40	47.7[b]	209[b]	13.33	9.14	5.08[b]
Zn 80	49.2[a]	235[a]	14.52	8.77	6.88[a]
Zn 160	49.6[a]	237[a]	14.03	8.93	6.40[ab]
Mn 60	48.6	242[x]	10.47[z]	9.94[x]	6.22
Mn 120	49.3	201[y]	15.14[y]	8.77[y]	6.11
Mn 240	48.5	237[x]	16.36[x]	8.10[z]	6.06

Means in a column with different superscripts differ significantly (p≤ 0.05).

References

Collins, N.E. and T. Moran Jr., 1999. Influence of supplemental manganese and zinc on live performance and carcass quality of broilers. J. Appl. Poult. Res. 8, 222-227.

Snedecor, G.W. and W.G. Cochran, 1989. Statistical Methods. Oxford and IBH Publishing Company, New Delhi.

Sunder, G.S., A.K. Panda, N.C.S. Gopinath, M.V.L.N. Raju, S.V. RamaRao and V.K. Chalasani, 2006. Effect of supplemental manganese on mineral uptake by tissues and immune response in broiler chickens. J. Poult. Sci. 43, 371-377.

Efficacy of three organic selenium sources for growing-finishing pigs

Y.D. Jang, H.B. Choi, S. Durosoy, P. Schlegel, B.R. Choi and Y.Y. Kim

Keywords: selenium, source, bioavailability , pig, pork

Introduction

Organic forms of Selenium (Se) have recently been introduced into animal nutrition for their higher bioavailability (absorbability and metabolic use of Se) or for their increased Se storage in muscle compared to inorganic Se sources. The objective of this study was to compare three dietary organic Se sources in growing-finishing pigs on their effect on growth performance, Se metabolism (serum Se) and Se retention (tissue Se).

Materials and methods

A total of 48 crossbred pigs, averaging 45.6 kg body weight (BW), were allotted to 12 pens on the basis of gender and BW. Pigs were housed in a conventional facility with half-slotted concrete floor. Pens were assigned to four dietary treatments on day 0 for 56 days. Treatments were: Negative Control (no Se supplementation), Se-yeast A, Se-yeast B, Se-proteinate (B-TRAXIM® Se, Pancosma S.A.). Part of the Se in Se-yeasts was in form of Selenomethionine. The three Se sources were supplied at 0.3 mg/kg Se into the basal corn/soy diet (0.06 mg/kg Se). Body weight, average daily gains (ADG), average daily feed intake (ADFI) were measured and feed efficiency (FCR) was calculated. Blood samples were collected from the anterior vena cava from two randomly selected pigs per pen. Collected blood samples were centrifuged at 3,000×g at 4 °C, sera were separated, frozen and analyzed for Se concentration. At the end of experiment, 3 pigs per treatment were selected and killed by exsanguination, and samples of loin, liver, kidney, pancreas and spleen were collected, frozen and analyzed for Se concentrations. Diet, serum, and the various tissues were analyzed for their Se content with the fluorometric method of AOAC (1995). The data obtained were subjected to a GLM analysis of SAS (1995). The pen was considered the experimental unit for growth performance data.

Results and discussion

Overall, ADFI and ADG were not affected by the addition of dietary Se (Table 1). Feed efficiency was improved (P<0.01) by 5.1% using either Se yeast A or B-TRAXIM®Se. The numerically reduced ADFI using Se yeast A was partly responsible for its improved feed efficiency.

Initial average Serum Se was 0.136 ppm. Serum Se was increased from day 0 to day 56 using B-TRAXIM®Se (P<0.01). Final serum Se concentrations (Table 2) were not different between Control, Se Yeast A and Se Yeast B. B-TRAXIM®Se had increased final serum Se compared to Control (+23.1%, P<0.01) and compared to Se-yeast B (+28.5%, P<0.01).

Table 1. Effect of dietary Se on growth performance of growing-finishing pigs.

	Control	Se-yeast A	Se-yeast B	B-TRAXIM	SEM
ADFI [kg]	2.88	2.82	2.87	2.87	0.38
ADG [g]	938	965	921	982	14.46
FCR	3.07[b]	2.92[a]	3.12[b]	2.92[a]	0.005

[ab]Different superscripts in the same row significantly differ (P<0.01).

Table 2. Effect of dietary Se on serum Se concentration of growing-finishing pigs [ppm].

	Control	Se-yeast A	Se-yeast B	B-TRAXIM	SEM
Final (day 56)	0.143[bc]	0.158[ab]	0.137[c]	0.176[a]	0.004

[abc]Different superscripts in the same row significantly differ (P<0.01).

Feeding organic Se sources resulted in reduced (P<0.05) kidney Se concentration with B-TRAXIM®Se (-7.0%) , increased (P<0.01) liver Se concentration with Se yeast A (+53.1%) and B-TRAXIM®Se (+57.1%) and in increased (P<0.01) loin Se concentration with Se yeast A (+56.8%) compared to the Control diet (Table 3).

The results of this study demonstrate that the efficiency of additional dietary organic Se sources were variable in growing-finishing pigs:
- Se yeast A improved Se status in plasma, liver and Se concentration in loin.
- Se yeast B did not differ from the Control and was therefore not an efficient Se source.
- B-TRAXIM®Se improved feed efficiency, reduced Se content in the kidney and improved Se status in plasma and liver.

Table 3. Effect of dietary Se on tissue Se concentration of growing-finishing pigs [mg/kg].

	Control	Se-yeast A	Se-yeast B	B-TRAXIM	SEM
Liver	0.37[b]	0.57[a]	0.34[b]	0.59[a]	0.03
Kidney	1.56[A]	1.62[A]	1.59[A]	1.48[B]	0.02
Pancreas	0.60	0.40	0.54	0.44	0.03
Spleen	0.27	0.30	0.18	0.24	0.02
Loin	0.18[b]	0.29[a]	0.20[b]	0.19[b]	0.01

[ab]Different superscripts in the same row significantly differ (P<0.01); [AB]significantly differ at P<0.05.

Knowing that reduced Se concentrations in kidney indicate lower Se excretion (Kim and Mahan, 1996), the liver is a Se-storage, but labile organ whereas loin Se is less readily available for the Se metabolism (Kim and Mahan, 2001), the following study indicates that Se-yeast A was an efficient source to store Se into meat and that B-TRAXIM®Se was a bioavailable source for the animal's Se metabolism.

References

AOAC, 1995. Official methods of analysis (16th ed). Association of official analytical chemists. Washington, DC.

Kim, Y.Y. and D.C. Mahan, 1996. Effect of inorganic and organic selenium at two dietary levels on reproductive performance and tissue selenium concentrations in first-parity gilts and their progeny. J. Anim. Sci. 74, 2711-2718.

Kim, Y.Y. and D.C. Mahan, 2001. Prolonged feeding of high dietary levels of organic and inorganic Se to gilts from 25 kg BW through one parity. J. Anim. Sci. 79, 956-966.

SAS, 1995. User's guide statistics, SAS Inst. Inc. Cary. NC. 27513.

Iron status of weaned piglets fed either dietary iron sulfate or iron glycinate

P. Schlegel, S. Durosoy and M. Dupas

Keywords: bioavailability, piglet, iron, source

Introduction

Anemia, related either to iron (Fe) or copper deficiency, is one of the major metabolic disorders in weaned piglets. Following the stress of weaning, piglet's iron status may be critical due to insufficient feed intake and digestive disorders. As an example, Gérard (2000) reported haemoglobin (Hb) levels of piglets on French farms averaging 10.6 g/dl, but the variation was between 3.0 and 14.6 g/dl, indicating high heterogeneity within herds. The aim of this study was to investigate the effect of two dietary iron sources (inorganic $FeSO_4$ and organic crystalline iron glycinate) on the iron status of weaned piglets. The organic crystalline iron glycinate (B-TRAXIM® 2C Fe) has recently been studied on Fe absorbability in weaned piglets and resulted in an increase of 33% compared to Fe from $FeSO_4$ (Ettle *et al.* 2007).

Materials and methods

A total of 384 twenty-one day old weaned piglets [(Large White×Landrace)×(Large White×Pietrain)] with average body weight (BW) of 6.0 kg, were assigned on BW, gender and litter origin to two blocs of 16 slotted pens each on weaning day (day 0). Piglets were fed, ad-libitum, a basal weaner diet (day 0-21) followed by a basal starter diet (day 21-42). Weaner and starter basal diets were formulated on wheat, barley, soy and contained 11.2 and 10.2 MJ/kg NE, 19.7 and 17.7 g/kg CP, respectively. Basal diets were not Fe supplemented. Dietary treatments were: (1) basal diet supplemented with 100 mg/kg Fe from $FeSO_4$; (2) basal diet supplemented with 100 mg/kg Fe from crystalline iron glycinate (FeGly, B-TRAXIM® 2C Fe, Pancosma S.A., Switzerland). Piglets had free access to drinking water (0.03 mg Fe/l). Prior study, piglets had free access to creep feed and were injected with 100 mg of iron dextran (day 3 to 5 after birth).

Blood samples were collected on two identified piglets (one female, one castrated male) per pen on days 0, 21 and 42 for Haemoglobin (Hb), hematocrit (Ht) and red blood cell count (RBC) analysis. In addition, Hb was measured on the same identified piglets with a kit (Hemocue®, Ängelholm, Sweden) on days 0, 2, 4, 7, 11, 14, 18 and 21.

Results were analyzed using GLM procedure of Statbox with treatment and bloc as factor. Hb values from Hemocue® were analyzed using repeated measure ANOVA procedure.

Results

Analyzed dietary Fe contents were 206 mg/kg Fe for the weaner and 233 mg/kg Fe for starter diets. Piglets fed FeGly had increased Hb levels (+15.5%, P<0.05); increased Ht (+12.3%,

P<0.001) and increased RBC (+5.1%, P<0.10) compared to FeSO$_4$ on day 21 (Table 1). On day 42, Hb and Ht were again increased (P<0.001) for FeGly compared to FeSO$_4$.

On weaning (day 0), the percent of anemic piglets (Hb<8.0 g/dl) was 53.1% and 40.6% for FeSO$_4$ and FeGly respectively. At the end of the study (day 42), 41.4% of the piglets fed FeSO$_4$ were still anemic, but only 10.7% were when FeGly was fed. The proportion of piglets with Hb concentrations above 10 g/dl was initially 12.5%. On day, 42, this proportion was reduced to 3.4% when FeSO$_4$ was fed and increased to 17.9% when FeGly was fed. The Hb evolution for the first 21 was hardly above 8 g/dl with FeSO$_4$ and hardly below 8 g/dl with FeGly (Figure 1).

Table 1. Hb, Ht and RBC on day 0, 21 and 42 from piglets fed two dietary iron sources.

Item	Day	Treatment		p - value	
		FeSO$_4$	FeGly	Treatment	Tr × bloc
Haemoglobin [g/dl]	0	8.08 ±1.40	8.59 ±1.59	n.s.	n.s.
	21	7.23 ±1.57	8.35 ±0.96	< 0.05	n.s.
	42	8.23 ±0.76	9.12 ±0.86	< 0.001	n.s.
Hematocrit [%]	0	27.8 ±4.2	29.7 ±5.2	n.s.	n.s.
	21	26.0 ±2.6	29.2 ±3.2	< 0.001	n.s.
	42	29.4 ±2.6	32.3 ±2.9	< 0.001	n.s.
Red blood cells [10^6/mm^3]	0	5.00 ±0.80	5.21 ±0.64	n.s.	n.s.
	21	5.68 ±0.51	5.97 ±0.59	< 0.10	< 0.10
	42	6.85 ±0.46	7.01 ±0.67	n.s.	n.s.

n.s. = not significant.

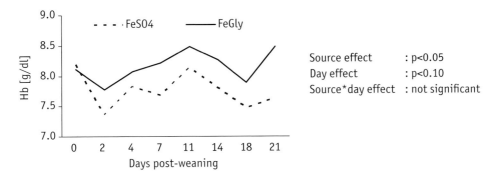

Figure 1. Haemoglobin evolution of piglets fed two dietary iron sources.

Conclusion

In summary, the capacity to maintain an adequate iron status in weaned piglets was improved when using 100 mg/kg of Fe from crystalline iron glycinate (B-TRAXIM® 2C Fe) compared to 100 mg/kg of Fe from $FeSO_4$.

References

Gérard, C., 2000. Les besoin de fer revus à la hausse. Réussir Porc, 58, 26-27.

Ettle, T, P. Schlegel and F.X. Roth, 2007. Investigations on iron bioavailability of different sources and supply levels in piglets. J. Anim. Phys. And Nutr. (in press).

Influence of source and level of supplemented copper and zinc on the trace element content of pig carcasses

A. Berk, M. Spolders, G. Flachowsky and J. Fleckenstein

Introduction

Zinc (Zn) and copper (Cu) are essential trace elements in the nutrition of plants and animals. In livestock production a deficiency of these elements can decrease animal performances and/or influence the health of animals (Mc Dowell, 2003). Normally these elements are supplemented to compound feeds of livestock. In areas with a high livestock density, the excreted amounts of Zn and Cu lead to discussions of environmental problems (KTBL, 2002). This was the reason for the derivation of new upper limits of some trace elements in the EU legislation valid from January 2004.

Especially when the supply of trace elements decreases because of environmental problems, the absorption rate (bioavailability) is of increasing interest. This bioavailability may be influenced by the source of trace elements (inorganic or organic binding) in the feed.

Material and methods

50 carcasses from all castrated males (initial LW 25.5 ± 0.95 kg) from a trial with 100 growing-fattening pigs (Berk *et al.*, 2003), were analysed to investigate the influence of source of trace elements (chemical binding form) on Cu- and Zn-concentration in the whole body at about 115 kg LW.

There were three levels of feed concentration of the trace elements: native feed content only (7 mg Cu/kg feed, 30 mg Zn/kg feed), and two levels of supplementation: (2 or 9 mg Cu/kg feed, Level 1, and 35 or 95 mg Zn/kg feed, Level 2) adequate to German recommendations (GfE, 1987) and to the possible EU - upper levels.

The supplementations of Cu and Zn were given as sulfate ($CuSO_4$ and $ZnSO_4$) or as an amino acid – trace element – complex (AATEC). Consequently, the design was created as shown in Table 1.

Table 1. Experimental design.

Group	1	2	3	4	5
n	10	10	10	10	10
Level	Native	1	1	2	2
Source	Native	Inorg.	Org.	Inorg.	Org.

When reaching the slaughter weight of approximately 115 kg live weight (LW), these 50 animals were slaughtered, divided in the three fractions blood and offal (b+o), soft tissue, and bones of the left carcass half (corrected by the whole carcass weight). These fractions and the organs liver, kidney and brain were analysed for Cu and Zn.

Results and discussion

The results of the performance are given by Berk *et al.* (2003). The mean live weight gain (LWG) was 895 ± 67 of the slaughtered 50 pigs. This performance could be considered as normal but on a high level.

The intake of Cu and Zn varied, depending on the 'native' level on the one hand and the 'highly supplemented' level on the other hand. But there was no significant influence of the trace element source on Cu- and Zn-intake (Table 2).

There were no significant differences dependent on trace element source in the whole body analysis (WBA) with regard to the contents of DM, CA, CP or EE. The higher level of trace elements led to significant differences in the amount of Zn in the fraction bones, but there was no significant influence of the source (Table 3).

Table 2. Mean intake of Cu and Zn (mg/pig).

Source	n	Cu	Zn
native	10	$1583^b \pm 130$	$7034^b \pm 578$
sulfate	20	$2575^a \pm 710$	$18857^a \pm 6795$
AATEC	20	$2256^a \pm 588$	$18327^a \pm 6125$

Tukey-Test, p < 0.05.

Table 3. Influence of source of trace elements on total amount of Cu/Zn (mg/animal).

Source	n	Soft tissue		Blood + offal		Bones	
		Cu	Zn	Cu	Zn	Cu	Zn
native	10	47±6	1127±125	26±4	197±29	19±29	378^b±117
sulfate	20	52±12	1156±108	28±6	213±36	13±10	497^a±94
AATEC	20	48±10	1191±133	26±3	217±32	12±6	501^a±89

Tukey-Test (P< 0.05).

The source of supplemented trace elements had no influence on the amount of Cu and Zn in the three fractions soft tissue, blood + offal and bones. That is also true if comparing the contents of Cu and Zn in the organs brain, kidney and liver. No significant differences in trace element contents of organs were measured due to trace element source (Table 4).

Table 4. Influence of source of supplementation on Cu/Zn content in organs (mg/kg DM).

Source	n	Brain		Kidney		Liver	
		Cu	Zn	Cu	Zn	Cu	Zn
native	10	12±2	48±5	17±4	83±6	23±5	150±56
sulfate	20	12±3	49±4	20±7	91±15	26±11	197±69
AATEC	20	11±2	49±4	19±4	91±9	22±5	205±73

Tukey-Test ($P< 0.05$).

Summary

The source and level of supplemented trace elements did not significantly influence pig performance of growing-finishing pigs from 25 to 115 kg LW ($p > 0.05$) (Berk et al., 2003). The addition of the studied form of organic binding trace elements had no advantages on pig performance and the trace element content in the organs brain, liver and kidney.

References

Berk, A., G. Flachowsky and J. Fleckenstein, 2003. Effect of supplemented phytase at different Zn- and Cu- feed contents in pig nutrition. Proc. 9. Symp. Vitamine und Zusatzstoffe in der Ernährung von Mensch und Tier, 23./24. 09.2003, Jena, 210-215.

GfE, 1987. Energie- und Nährstoffbedarf landwirtschaftlicher Nutztiere – Nr. 4 Schweine. DLG-Verlag Frankfurt/ Main.

KTBL, 2002. Fütterungsstrategien zur Verminderung von Spurenelementen/ Schwermetallen im Wirtschaftsdünger. Proc. KTBL-Workshop, 23./24. 04.2002, Göttingen, KTBL-Schrift 410, 162 S.

McDowell, L.R., 2003. Minerals in animal and human nutrition, 2nd ed., Elsevier, 644 p.

Effects of dietary inclusions of natural extracts on the mineral (Fe, Cu, Zn) status of weaned piglets

R.D. Criste, A. Untea, D. Torrallardona, N. Andres and I. Taranu

Keywords: piglets, weaning, carob, nucleotides, minerals

Introduction

In search of substances replacing antibiotics as growth promoters for weaned piglets, natural plants extracts have been proposed as possible alternatives (Van Nevel et. al, 2005). Possible interference with the absorption of essential nutrients is one of the areas of uncertainty (Harmuth -Hoene, 1980). Within this context, our study investigated the influence of two natural dietary extracts (NEs), pulp of carob and nucleotides from yeast extract, as replacement antibiotics on trace mineral status (copper-Cu, iron-Fe, and zinc-Zn) in plasma and organs of weaned piglets. Carob pulp is a by-product widely available in the Mediterranean area and it is a valuable source of fiber, tannins, poly phenols, minerals and vitamins (Lanza *et al.*, 2001). Nucleotides are ubiquitous molecules improving intestinal health and the development of the immune system. In our study, the mineral content was 5.1 ppm Cu; 38.7 ppm Fe; 10.1 ppm Zn in the carob pulp and 22.8 ppm Cu; 66.0 ppm Fe; 79.7 ppm Zn in the yeast extract nucleotides, respectively.

Material and methods

A total of 60 piglets weaned at 26 days were involved in a 14-d feeding trial. On d 0 of the trial 12 piglets were slaughtered and plasma and organ samples were taken in order to establish the initial trace mineral status (IS). The remaining 48 piglets were fed a basal barley-wheat meal diet providing vitamin and mineral supplements without Cu (NRC, 1998). The piglets were assigned to four groups as follows: *T1*-basal diet, negative control; *T2*-basal diet +nucleotides; *T3*-basal diet +carob pulp; *T4*-basal diet + nucleotides + carob. The mineral content of T1 was: 7.5 ppm Cu; 154.4 ppm Fe; 86.4 ppm Zn. The rate of NEs inclusion in the diet was 3% for carob and 1,500 ppm for yeast nucleotides. On d 6 and d 14 of the trial four piglets per group were slaughtered and plasma and organ samples were taken and assayed for Fe, Cu and Zn concentrations. Feed, liver and spleen samples were prepared for mineral analyses via dry digestion (by ashing at 550 °C) including the drying of samples at 65 °C. Plasma samples were deproteinized by addition of trichloracetic acid (40%) and hydrochloric acid (1M). Mineral analysis was done by flame atomic absorption spectrophotometry (Thermo Electron, Solaar M). All data are expressed as mean ± standard error of the mean (SEM). Statistical differences between groups for mineral concentrations were determined using an ANOVA test.

Results

Plasma response

Table 1 shows the effect of diets with NEs on plasma mineral concentration. After 14 d of feeding, plasma Cu concentration decreased below IS value while Zn increased slightly, but no difference between treatments was observed (p=NS). A time effect and a slight effect of diets (T2 and T3) at 14 d were observed in the case of plasma Fe, without significant differences, however (p=NS).

Table 1. Effect of dietary treatment on pig plasma mineral concentrations (µg/ml).

	IS	T 1		T 2		T 3		T4	
		d 6	d 14	d 6	d 14	d 6	d 14	d 6	d 14
Cu	0.74	0.71	0.65	0.74	0.63	0.71	0.64	0.84	0.62
	±0.05	±0.01	±0.05	±0.02	±0.01	±0.03	±0.03	±0.06	±0.05
Fe	1.19	1.23	1.28	1.21	1.54	1.28	1.52	1.45	1.33
	±0.10	±0.07	±0.03	±0.08	±0.03	±0.14	±0.13	±0.07	±0.07
Zn	2.07	1.28	2.26	1.31	2.10	1.4	2.17	1.43	1.99
	±0.15	±0.06	±0.14	±0.01	±0.21	±0.2	±0.07	±0.03	±0.04

Tissues response

In general, all mineral concentrations decreased compared to IS in measured organs. The analyses of liver Fe showed that IS concentration (d 0) was significantly higher (P≤ 0.05) than liver Fe concentration after 14 feeding days; these findings are in agreement with the observations of Rincker et al., 2004. No significant differences of liver Fe were observed between treatments (Figure 1). Under the feeding conditions of this trial we observed a slight increase in the spleen concentration of Zn at 14 d for T2 diet (Figure 2) and a significant increase for T4 (P<0.0006). Also a slight increase was observed for T3 and T4 in the case of Fe spleen concentration. No effect of the treatments was observed for spleen Cu (Figure2).

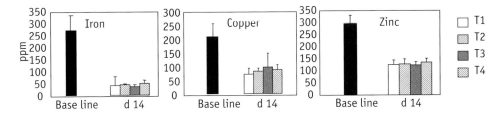

Figure 1. Effect of nucleotide and carob diet on liver mineral concentration (DM basis).

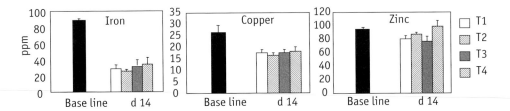

Figure 2. Effect of nucleotide and carob diet on spleen mineral concentration (DM basis).

Conclusions

Our results showed that the diets with NEs used in this trial had no significant effect on plasma and organs mineral concentration. Higher concentrations of dietary natural extracts and a longer duration of feeding should be investigated.

References

Harmuth -Hoene, A. and R. Schelenz, 1980. Effect of dietary fiber on mineral absorption in growing rats. J. Nutr. 110, 1774-1784.

Lanza, M., A. Priolo, L. Biondi, M. Bella and H. Ben Salem, 2001.Replacement of cereal grains by orange pulp and carob pulp in faba bean-based diets fed to lambs: effects on growth performance and meat quality. Anim. Res. 50, 21-30.

Rincker, M.J., 2004.Effects of dietary iron supplementation on growth performance, hematolo-gical status, and whole-body mineral concentration of nursery pigs. J. Anim. Sci. 82, 3189-97.

Van Nevel, C.J., J.A. Decuypere, N.A. Dierick and K. Molly, 2005. Incorporation of galactomannans in the diet of newly weaned piglets: effect on bacteriological and some morphological characteristics of the small intestine. Arch. Anim. Nutr. 59:123-38.

Bioavailability of copper from copper glycinate in steers fed high dietary sulfur and molybdenum

S.L. Hansen, P. Schlegel, K.E. Lloyd and J.W. Spears

Keywords: bioavailability, cattle, copper, growth

Introduction

Deficiencies of Cu are found around the world, and low Cu incorporation into enzymes necessary for many physiological processes may cause deficiency signs such as retarded growth, and hair depigmentation. Bioavailability of Cu from ruminant diets is depressed by the presence of antagonists such as molybdenum (Mo), sulfur (S), and iron (Spears, 2003). Certain organic Cu sources that are complexed or chelated to various ligands may be more bioavailable to cattle than traditionally fed inorganic Cu sulfate ($CuSO_4$). Results obtained with these organic sources have been variable with some studies indicating higher Cu bioavailability and others finding similar bioavailability, relative to $CuSO_4$. The present study was conducted to determine the bioavailability of copper glycinate (CuGly) relative to $CuSO_4$ in steers fed diets high in Mo and S. The Cu from CuGly used in the present study was bound to the amino acid glycine in a crystalline form (B-TRAXIM® 2C, Pancosma, Switzerland).

Materials and methods

Sixty Angus (n = 29) and Angus-Simmental cross (n = 31) steers, averaging 9 mo of age and 277 kg initial body weight, were used in this 148-d study. Steers were blocked by weight within breed and randomly assigned to one of five treatments: (1) control (no supplemental Cu), (2) 5 mg Cu/kg of DM from $CuSO_4$, (3) 10 mg Cu/kg of DM from $CuSO_4$, (4) 5 mg Cu/kg of DM from CuGly, and (5) 10 mg Cu/kg of DM from CuGly. Steers were individually fed a corn silage-based diet (analyzed 8.2 mg Cu/kg of DM), and supplemented with 2 mg Mo/kg of diet DM and 0.15% S for 120 d (Phase 1). Steers were then supplemented with 6 mg Mo/kg of diet DM and 0.15% S for an additional 28 d (Phase 2). Steers were fed once daily, with feed amounts based on what they would consume in a 24 h period. Jugular blood and liver biopsy samples were collected. Ceruloplasmin activity of fresh plasma was determined as described by Houchin (1958). Feed and liver samples were prepared for Cu analysis by wet ashing using microwave digestion (Mars 5™; CEM Corp., Matthews, NC). Copper content of plasma, feed and liver samples was determined by flame atomic absorption spectroscopy (Shimadzu Scientific Instruments, Kyoto, Japan). Statistical analysis of performance data was performed by ANOVA for a completely randomized design using the GLM procedure of SAS (SAS Inst. Inc, Cary, NC). Relative bioavailability of CuGly was determined, using $CuSO_4$ as the standard source, by means of multiple linear regression and the slope-ratio method (Littell *et al.*, 1997). Dependent variables (plasma Cu, plasma ceruloplasmin, and liver Cu) were regressed on supplemental Cu intake/d.

Results

Average daily gain and gain:feed were improved by Cu supplementation regardless of source (P = 0.01). Final ceruloplasmin, plasma Cu and liver Cu values were greater (P < 0.05) in steers fed supplemental Cu compared with controls. Plasma Cu, liver Cu and ceruloplasmin values were higher (P < 0.05) in steers supplemented with 10 mg Cu/kg of DM vs. those supplemented with 5 mg Cu/kg of DM. Based on multiple linear regression of final plasma Cu, liver Cu and ceruloplasmin values on dietary Cu intake in Phase 1 (2 mg Mo/kg of DM), bioavailability of Cu from CuGly relative to $CuSO_4$ (100%) was 140 (P = 0.10), 131 (P = 0.12) and 140% (P = 0.09), respectively. Relative bioavailability of Cu from CuGly was greater than from $CuSO_4$ following supplementation of 6 mg Mo/kg of DM for 28 d (Table 1).

Table 1. Estimated relative bioavailability of Cu sources in steers fed diets high in Mo and S (Phase 2)[1] based on multiple linear regression of Cu indices on supplemental Cu intake.

Cu indices	Cu source	Slope ± SE	P-value[2]	Relative bioavailability
Plasma Cu[3]	Sulfate	0.0000204 ± 0.0000632	0.01	100
	Glycinate	0.0000294 ± 0.0000075		144
Ceruloplasmin[4]	Sulfate	0.000646 ± 0.000135	0.01	100
	Glycinate	0.00101 ± 0.000158		156
Liver Cu[3,4]	Sulfate	0.0000332 ± 0.0000087	0.01	100
	Glycinate	0.0000499 ± 0.0000099		150

[1]Regression based on final measurements following 6 mg Mo/kg of DM supplementation.
[2]P-value for slope differences among Cu sources.
[3]Day 0 values used as a covariant.
[4]Data log transformed prior to regression analysis.

Conclusion

In summary, Cu from the CuGly complex tended to be more bioavailable than $CuSO_4$ when fed to steers receiving diets supplemented with 2 mg Mo/kg of DM. When supplemental Mo was increased from 2 to 6 mg Mo/kg of DM, bioavailability of Cu was greater from CuGly than from $CuSO_4$.

References

Houchin, O.B., 1958. A rapid colorimetric method for quantitative determination of copper oxidase activity (ceruloplasmin). Clin. Chem. 4, 519-523.

Littell, R.C., P.R. Henry, A.J. Lewis and C.B. Ammerman, 1997. Estimation of relative bioavailability of nutrients using SAS procedures. J .Anim. Sci. 75, 2672-2683.

Spears, J.W., 2003. Trace mineral bioavailability in ruminants. J. Nutr. 133, 1506S-1509S.

Influence of level and source of copper supplementation on immune response in growing Nellore lambs

P. Senthilkumar, D. Nagalakshmi, Y.R. Reddy, H.V.L.N. Swami and K. Sudhakar

Keywords: copper, copper proteinate, lambs, immunity

Introduction

Sheep are susceptible to various parasitic, bacterial and viral diseases. In order to obtain maximum profits it is essential to increase flock viability by improving disease resistance. Nutrition is the major decisive factor, which determines the expression of genetic potential of animals in terms of growth and immunity (Klasing and Barnes, 1988). Copper (Cu) deficiency has been associated with reduced immune function and increased susceptibility to disease (Stable and Spears, 1990). But the studies on amount of Cu supplementation required for optimum immune response is limited. An attempt has been made to determine the Cu requirement in growing lambs for optimum immunity from two different Cu sources.

Material and methods

Thirty male Nellore lambs (15.5 ± 0.06 kg) were randomly allotted to 5 dietary treatments viz., basal diet (BD) (7.4 ppm Cu) (no supplemental Cu), BD supplemented with 7 and 14 ppm Cu from copper sulfate ($CuSO_4.5H_2O$) and copper proteinate (10% Cu). Molybdenum and S content of the basal diet was 0.62 mg/kg and 1.5 g/kg, respectively. The lambs were fed the respective diet at 3.5 per cent of body weight to meet the nutrient requirements except Cu. The serum Cu in lambs at the start was 0.15+0.002 ppm. The humoral immune response against *Brucella abortus* and chicken RBC was assessed by standard tube agglutination test (Alton *et al.*, 1975), and direct HA test and 2-mercaptoethanol method (Wegmann and Smithies, 1966), respectively at 90 days of feeding. The cell mediated immune (CMI) response was determined by *in vitro* lymphocyte proliferation assay (Bounous *et al.*, 1992) after 180 days of the feeding. The data was subjected to ANOVA in a factorial design using SPSS student version 10.0.

Results and discussion

The humoral immune response against *B. abortus* was higher in Cu supplemented lambs (7, 14, 21 and 28d post-sensitization, PS), than with no effect of dose of supplementation. Lambs supplemented with Cu-proteinate had higher titers than $CuSO_4$ on 7 and 14 days of PS (Table 1). The total Ig and IgG levels against chicken RBC were higher in Cu supplemented lambs and the peak concentration of these immunoglobulins were observed on day 14 PS. On this day, the total Ig concentration was higher (P < 0.05) in lambs fed Cu-proteinate supplying 14 ppm Cu compared to the unsupplemented Cu diet. The optimum dietary Cu level for better humoral immune response was predicted as 16 ppm against *B. abortus* (y = 4.47+0.588x-0.0177x^2, R^2 0.54, SEM 0.55, P<0.001) and chicken RBC (y = 4.20+0.448x-0.0139x^2, R^2 0.25, SEM 0.75, P<0.02) by using regression equation. The CMI was maximal at

14 ppm Cu supplementation, with higher (P<0.01) response from Cu-proteinate. The lambs on 14 ppm Cu supplementation had higher immunity and the immune response against *B. abortus* and CMI was higher for Cu-proteinate than CuSO$_4$. It is concluded that the dietary Cu content of 16 ppm was predicted for optimal humoral immunity.

Table 1. Effect of dietary Cu concentration and source on immune response in lambs.

Days post inoculation	Basal diet (T$_1$)	Cu supplementation				SEM	P value			
		7 ppm		14 ppm			Basal vs. Cu	7ppm vs. 14ppm	T$_2$ vs. T$_3$	T$_4$ vs. T$_5$
		Inorg. (T$_2$)	Org. (T$_3$)	Inorg. (T$_4$)	Org. (T$_5$)					
Total Ig against chicken RBC (log$_2$ titres)										
7	6.83	7.67	7.83	7.17	7.23	0.154	0.10	0.10	0.73	0.89
14	6.66^b	7.50^{ab}	8.00^a	7.35^{ab}	7.61^a	0.149	0.01	0.32	0.25	0.55
21	6.23^b	7.33^a	7.33^a	6.74^{ab}	7.23^a	0.134	0.001	0.17	1.00	0.19
28	5.61^b	6.50^a	6.50^a	6.10^{ab}	6.23^{ab}	0.120	0.01	0.19	1.00	0.71
35[1]	4.23^b	5.17^{ab}	5.67^a	5.84^a	4.42^b	0.191	0.03	0.50	0.33	0.01
IgG against chicken RBC (log$_2$ titres)										
7	5.21^b	6.00^{ab}	6.33^a	5.64^{ab}	5.83^{ab}	0.145	0.04	0.20	0.45	0.66
14	5.46^b	6.17^{ab}	6.83^a	6.02^{ab}	6.43^{ab}	0.167	0.03	0.42	0.19	0.42
21	4.83^b	6.17^{ab}	6.17^a	5.21^{ab}	6.04^{ab}	0.161	0.01	0.09	1.00	0.06
28	4.81^b	5.50^{ab}	5.50^a	4.83^{ab}	5.23^{ab}	0.119	0.13	0.09	1.00	0.28
35	2.42^b	3.67^a	4.00^a	4.02^a	3.21^a	0.160	0.001	0.49	0.40	0.05
Immune response against *Brucella abortus* antigen (STAT titres, log$_2$)										
7	6.99^d	7.32^{cd}	8.52^a	7.72^{bc}	8.32^{ab}	0.149	0.01	0.77	0.001	0.10
14	7.49^b	7.99^b	8.74^a	7.92^b	8.72^a	0.139	0.01	0.75	0.05	0.04
21	7.99^b	9.15^a	9.35^a	8.92^a	9.12^a	0.140	0.001	0.88	0.60	0.59
28	8.32^b	8.99^{ab}	9.15^a	8.72^{ab}	8.92^a	0.122	0.04	0.33	0.66	0.60
35	7.49	7.15	7.54	7.32	7.52	0.135	0.77	0.34	0.40	0.66
Lymphocyte proliferation assay[1] (stimulation index)										
180d	0.16^e	0.29^d	0.58^c	0.84^b	0.95^a	0.056	0.001	0.001	0.001	0.001

[abcd]Means with different superscripts in a row differ significantly (P<0.05).
[1]Level × source (P< 0.05).

References

Alton, G.G, L.M. Jones and D.E. Pietz, 1975. Laboratory techniques in Brucellosis. 2nd edition, WHO Monograph Series, 55, WHO, Geneva.

Bounous, D.I, R.P. Campagnoli and J. Brown, 1992. Comparison of MTT colorimetric assay and tritiated thymidine uptake for lymphocyte proliferation assay using chicken splenocytes. Avian Diseases 36, 1022-1027.

Klasing, K.C and D.M. Barnes, 1988. Decreased amino acid requirements of growing chicks due to immunological stress. J. Nutr. 118, 1158-1164.

Stable, J.R and J.W. Spears, 1990. The effect of copper on immune function and metabolism. Plenum Publishing, New York pp 309-332.

Wegmann, N.T.G and O. Smithies, 1966. A simple haemagglutination system requiring small amounts of red cells and antibodies transfusion. Philadelphia 6, 67.

The effect of dietary molybdenum or iron on copper status and ceruloplasmin expression in sheep

A.M. Mackenzie, C.L. Williams, S.G. Edwards and R.G. Wilkinson

Keywords: copper status, molybdenum, ceruloplasmin expression

Introduction

Copper (Cu) deficiency represents an enormous cost to United Kingdom animal production with clinical manifestations including infertility, depressed growth, altered pigmentation and impaired immune function. It is reported to occur either as a primary or secondary deficiency, with the latter being due to antagonistic interactions between Cu and the minerals molybdenum-sulfur and iron-sulfur within the rumen. Williams *et al.* (2001) reported that these antagonists affect the parameters normally used to assess Cu status. In particular, a reduction in ceruloplasmin (CP) activity relative to plasma Cu and the increase in the non-ceruloplasmin copper pool within plasma. In order to determine the cause of these alterations in copper speciation, an experiment was conducted to investigate the effect of dietary antagonists on copper status and CP expression in the liver of Scottish Blackface sheep.

Materials and methods

Thirty Scottish Blackface wether lambs of approximately 14 months of age were blocked by liveweight and allocated to one of three dietary treatments. The lambs were individually housed in raised floor pens and offered a basal diet of straw pellets, rapeseed meal, barley and molasses formulated to supply; ME 10.7 MJ, CP 154 g, Cu 5.5 mg, Mo 0.67 mg per kg DM and support a live weight gain of 150 g/day. Lambs were offered either the basal diet (treatment C), the basal diet supplemented with 5 mg/kg DM Mo and 2 g/kg DM S (treatment Mo), or the basal diet supplemented with 500 mg/kg DM Fe (as $FeSO_4.7H_2O$) and 2 g/kg DM S (treatment Fe). Blood samples were obtained fortnightly for assessment of plasma copper (Pl-Cu) and serum ceruloplasmin activity (CP) for 12 weeks. After 12 weeks the sheep were electrically stunned and killed by exsanguination. Liver samples were immediately collected, frozen and stored at -80 °C. Liver trace element content was analysed by ICP-MS. Liver RNA was extracted using a SV Total RNA Isolation System kit (Promega, Maddison, WI, USA). CP gene expression was assessed by RT-PCR performed using the RT-PCR System kit (Promega, Maddison, WI, USA) with ß-actin as the internal standard. The PCR product was run on a 2% agarose gel and band intensity was measured as an unsaturated image using Molecular Analyst (Biorad) and the ratio of CP to β-actin (ba) was determined. The experiment was statistically analysed by ANOVA using Genstat.

Results

There was no significant difference between lambs on each treatment in health or performance. Over the 12 week period Pl-Cu was significantly lower (P<0.05) in lambs on treatment Fe compared to those on treatments C and Mo. However, CP activity was significantly higher (P<0.05) in lambs on treatment C compared with those on treatments Mo or Fe. The CP:Pl-Cu ratio are presented in Table 1. Lambs on treatment Mo had a significantly lower (P<0.001) ratio than those on treatment Fe, which had a significantly lower ratio than those on treatment C. At the end of the 12 week period liver Cu levels were significantly reduced in lambs on treatments Fe and Mo compared to treatment C.

There was no significant effect of dietary treatment on CP gene expression in the liver of the lambs (Table 2).

Table 1. Effect of molybdenum or iron supplementation on the ceruloplasmin to plasma copper ratio (CP: Pl-Cu) in Scottish Blackface wether lambs.

Week	Treatment			s.e.d	Significance
	Control	Fe	Mo		
0	1.23	1.23	1.23	0.065	NS
2	1.37^a	1.17^{ab}	1.03^b	0.106	*
4	1.26	1.18	0.97	0.065	***
6	1.10	0.89	0.67	0.050	***
8	1.29	0.98	0.69	0.073	***
10	1.20	0.85	0.59	0.058	***
12	1.13	0.80	0.59	0.062	***

NS P>0.05; * P<0.05; *** P<0.001; [a,b] Means with the same superscript are not significantly different (P<0.05).

Table 2. Effect of molybdenum or iron supplementation on mRNA CP/ba ratio.

Treatment			s.e.d	Significance
Control	Fe	Mo		
0.930	0.932	0.957	0.0854	NS

Conclusions

Both Mo and Fe had significant effects on the Cu status of lambs. CP activity was influenced by both antagonists. However, the effect of these antagonists is not at the gene expression level. CP activity may be inhibited either by inhibition of the cupro-enzyme in plasma, or inhibition of incorporation of Cu into the apo-enzyme. Further research is required to clarify the mechanisms of inhibition.

References

Williams, C.L., A.M. Mackenzie, D.V. Illingworth and R.G. Wilkinson, 2001, The effect of molybdenum, iron and sulphur supplementation on the growth rate and copper status of lambs. Proceedings of the British Society of Animal Science, 140.

Quantification of the effects of copper, molybdenum and sulfur on the copper status of cattle

A.W. Jongbloed and J. Kogut

Keywords: copper, molybdenum, sulfur, interaction, cattle

Introduction

In 2003, the European Union adopted new maximum authorized levels of trace minerals in animal rations (EC, 2003). It is questionable, whether the proposed lowered maximal levels of copper (Cu) in diets for cattle pose a risk that under certain conditions Cu will be marginal or even deficient. This may be the case at high dietary concentrations of molybdenum (Mo) and/or sulfur (S), which can be derived from the equations as proposed by Underwood and Suttle (1999). These equations are predominantly based on research with sheep. It is, therefore, questionable whether these equations hold true for cattle.

Material and methods

A study was carried out to quantify the effects on the Cu status of cattle at different dietary concentrations of Cu, Mo and S, and their interaction using a meta-analysis (Jongbloed *et al.*, 2005). As response variable the so-called response ratio was chosen, which is the quotient between liver Cu content at the end and liver Cu content at the start of the period. This ratio was subsequently ln-transformed. Furthermore, the data were restricted to an experimental period of 100 days, because in the long run of depletion hardly lower Cu liver concentrations are obtained. First a Fixed Effect Model was used to find out if there were outliers. Then the Random Effect Model was used to predict effects with random experiment effects to make model fit valid for all similar studies. It was impossible to include dietary Fe, Zn or protein content in the analysis, because in more than 50% of the experiments these data were not available. Finally, in total 22 experiments with cattle were used for the meta-analysis.

Furthermore, an inventory was made on concentrations of Cu, Mo, S in specific types of roughage, like grass silage, fresh grass and maize silage in Dutch practice (n > 24000). This was done to identify regions in The Netherlands with a high risk of Cu deficiency.

Results

By means of the meta-analysis on the data for beef cattle an accurate and reliable estimation could be obtained of change of liver Cu content as dependent on variation in dietary Cu, Mo and S concentrations. In contrast to the equations by Underwood and Suttle (1999), no significant interaction of dietary concentrations of Mo x S could be shown but an interaction between dietary concentrations of Cu and Mo together with main effects of dietary concentrations of Cu, Mo and S.

At a given Mo and S concentration in the diet it can be estimated at what dietary Cu content the liver Cu content is kept constant (Figure 1). In this way, one can better anticipate to local conditions of levels of Mo and S in roughages. Only in extreme situations (more than 5 g S/kg DM and 6 mg Mo/kg DM), is the required dietary Cu content higher than the maximal allowed concentration of 35 mg Cu/kg DM, as is given in the European Union for cattle. Due to lack of sufficient data on dairy cattle, no reliable estimation could be obtained of change of liver Cu content as dependent on variation in dietary Cu, Mo and S concentrations. The number of observations for dairy cattle was too low.

There were large differences in mineral concentrations in grass silages and fresh grass. This is due to season of the year and region. Based on the mineral concentrations in grass silages and fresh grass, and using the equation described above, regions were identified for a possible risk of suboptimal Cu supply to ruminants. The pattern of postal codes with a possible risk of suboptimal Cu supply, as obtained by the equation for beef cattle, was quite consistent with respect to postal codes in the same area. The most profound regions were found in areas with peat and young clay or sandy clay soils.

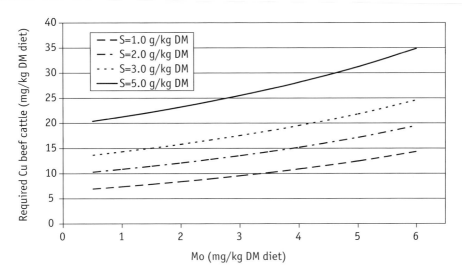

Figure 1. Estimated Cu requirement of beef cattle at several dietary Mo and S concentrations.

Conclusions

For beef cattle an accurate and reliable estimation could be obtained of change in liver Cu content as dependent on variation in dietary Cu, Mo and S concentrations. In contrast to equations by Underwood and Suttle (1999), no significant interaction of dietary concentrations of Mo x S could be shown but an interaction between dietary concentrations of Cu and Mo together with main effects of dietary concentrations of Cu, Mo and S were observed.

References

EC, 2003. Commission regulation (EC) No 1334/203 of 25 July 2003. Amending the conditions for authorisation of a number of additives belonging to the group of trace elements. L 187/11-L 187/15.

Jongbloed, A.W., P. Tsikakis and J. Kogut, 2004. Quantification of the effects of copper, molybdenum and sulphur on the copper status of cattle and sheep and inventory of these mineral contents in roughages. Report 04/0000637 ASG Nutrition and Food.

Underwood, E.J. and N.F. Suttle, 1999. The Mineral Nutrition of Livestock, 3rd edition. CABI Publishing, Wallingford, United Kingdom.

Effect of chelated vs inorganic zinc on vitamin A utilization in calves

H. Aliarabi, A. Chhabra and S.A. Hosseini Siyar

Keywords: chelated zinc, β-carotene, vitamin A, calves

Introduction

Zinc status influences several aspects of vitamin A metabolism, including its absorption, transport and utilization. Zinc deficiency can depress the synthesis of retinol-binding protein (RBP) in the liver and lead to lower transport of retinol from liver. Zinc also aids in intestinal vitamin A absorption (Rahman *et al.*, 2002). Zinc status affects conversion of β-carotene to vitamin A and zinc supplementation resulted in greater conversion of β-carotene to vitamin A (Chhabra, 1982). The objective of this investigation was to study the effect of chelated and inorganic zinc on vitamin A utilization in calves.

Materials and methods

Twenty cross bred one month old male calves were grouped randomly on the basis of their body weight as a completely randomized design. The calves were divided into four groups (T0, T1, T2 and T3). All the animals were fed on a diet containing 27 mg/kg zinc. Animals in groups T1, T2 and T3 were supplemented with 80 mg/kg of inorganic zinc ($ZnSO_4$), 80 mg/kg and 40 mg/kg of chelated zinc (chelating agent was partially hydrolyzed soy protein), respectively. Blood samples were collected on day zero and thereafter at fortnightly intervals. Plasma was separated and analysed for Zn, retinol and β-carotene. An HPLC method for simultaneous estimation of retinol and β-carotene in feeds, maize hay and plasma samples was adopted (Chawla and Kaur, 2001). Comparison between data was done using Duncan Multiple Range Test.

Results and discussion

Initial body weight of calves for treatment T0, T1, T2 and T3 was 32.08±1.90 32.40±1.50 32.26±1.70 32.42±1.14 kg, respectively and their final body weight was 88.75±3.82, 99.00±2.17, 106.60±4.37 and 101.40±3.96 kg, respectively. The animals were at a similar zinc status before the experimental feeding was started (Figure 1). After supplementing zinc, significant differences were observed among the treatments with a significantly lower value of plasma zinc in group T0 ($P<0.01$). Plasma zinc concentration of zinc supplemented calves increased sharply and it continued until middle of experimental period. Zinc depleted pigs showed decline in serum Zn and after Zn was supplemented, serum Zn increased but the response was similar for both Zn amino acid chelate and $ZnSO_4$ (Swinkels *et al.*, 1996).

Initial values of plasma β-carotene for groups T0, T1, T2 and T3 were similar (Table 1). Johnston and Chew (1984) reported that plasma β-carotene in cows varied from 2 to 4 µg/ml. Average plasma β-carotene during the whole period of growth remained similar in

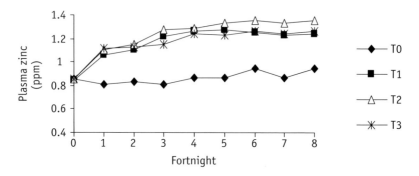

Figure 1. Periodical changes in plasma zinc concentrations.

Table 1. Plasma retinol and β-carotene concentrations of calves.

	T0	T1	T2	T3
Initial plasma retinol (μg/ml)	0.53[a] ±0.04	0.57[a] ±0.04	0.57[a]±0.05	0.56[a] ±0.03
Average plasma retinol (μg/ml)	0.53 [a] ±0.02	0.73 [b] ±0.03	0.75 [b]±0.03	0.72[b]±0.02
Initial plasma β-carotene (μg/ml)	1.73 [a] ±0.07	1.71 [a] ±0.11	1.74 [a] ±0.07	1.68 [a] ±0.06
Average plasma β-carotene (μg/ml)	1.68 [a] ±0.02	1.68 [a] ±0.01	1.72 [a] ±0.03	1.70 [a] ±0.02

*Means ± SEM bearing different superscripts in row differed significantly (P<0.01).

all animals and no significant differences were observed between treatments. Before the experimental feeding, no significant differences were observed among the treatments for plasma retinol level (Table 1). Plasma retinol concentration of cows was 0.54 μg/ml (Goff and Stable, 1990). Average plasma retinol concentration of treatment T0 was significantly lower than treatments T1, T2 and T3 (P<0.01). No significant differences were observed between the treatments T1, T2 and T3. It was evident that supplementation of zinc regardless of its source and level had a highly significant positive effect on plasma retinol concentration. The elevation of plasma retinol level in the zinc supplemented groups observed in the present study was in agreement with reports of earlier workers (Chhabra, 1982; Rahman *et al.*, 2002). In kids, dietary zinc helped in the mobilization of vitamin A from liver hepatocytes through greater synthesis of RBP (Chhabra and Arora, 1985).

The overall conclusion is that zinc supplementation regardless of its source increased significantly plasma retinol and zinc concentrations of calves.

References

Chawla, R. and H. Kaur, 2001. Isicaratic HPLC method for simultaneous determination of β-carotene, retinol and α-tocopherol in feeds and blood plasma. Ind. J. Dairy Sci. 54, 84-90.

Chhabra, A. 1982. Effect of dietary zinc on the conversion of β-carotene to vitamin A, on superoxide dismutase activity and keratinization in ovaries and testes. Ph.D. Thesis, National Dairy Research Institute (Deemed university), Karnal, Haryana.

Chhabra, A. and S.P. Arora, 1985. Effect of zinc deficiency on serum vitamin A level, tissue enzymes and histological alterations in goats. Livest. Prod. Sci. 12, 69-77.

Goff, J.P. and J.R. Stable, 1990. Decreased plasma retinol, α-tocopherol and zinc concentration during the preparturient period: Effect of milk fever. J. Dairy Sci. 73, 3195-3199.

Johnston, L.A. and B.P. Chew, 1984. Prepartum change of plasma and milk vitamin A and β-carotene among dairy cows with or without mastitis. J. Dairy Sci. 67, 1832-1840.

Rahman, M.M., M.A. Wahed, G.J. Fuchs, A.H. Baqui and J.O. Alvarez, 2002. Synergistic effect of zinc and vitamin A on the biochemical indexes of vitamin A nutrition in children. Am. J. Clin. Nutr. 1, 92-98.

Swinkels, J.W., E.T. Kornegay, W. Zhou, M.D. Lindemann, K.E. Webb and M.W.A. Verstegen, 1996. Effectiveness of a zinc amino acid chelate and zinc sulfate in restoring serum and soft tissue zinc concentrations when fed to zinc depleted pigs. J. Anim. Sci. 74, 2420-2430.

Effects of level and form of dietary zinc on dairy cow performance and health

C.M. Atkin, A.M. Mackenzie, D. Wilde and L.A. Sinclair

Keywords: zinc, somatic cell counts, milk amyloid A

Introduction

Mastitis is a complex and costly disease for the dairy industry. Zinc (Zn) is involved in maintaining the status of the immune system and it has been suggested that a deficiency in Zn could result in increased mastitis and somatic cell counts (SCC). Traditionally cattle diets are supplemented with inorganic minerals (e.g. zinc oxide), but these may be poorly absorbed, resulting in an economic and environmental cost due to excess of minerals being excreted. It is claimed that organically chelated minerals are able to resist interaction at the absorption site in the small intestine, which may result in a lower dietary inclusion required whilst maintaining the health and performance of the dairy cow. The objective of the current experiment was to investigate the effect of an organically-bound source of Zn as a replacement for inorganic Zn on dairy cow health and performance when supplemented at or below the recommended supplemental level.

Materials and method

A total mixed ration based on corn silage, grass silage, urea-treated wheat, soyabean meal and rapeseed meal, was offered at 1.05 of *ad libitum* intake. The basal diet was predicted to supply 811 mg of Zn per day and was supplemented with one of four concentrates differing in their level and form of dietary Zn. The concentrates provided an additional 600 mg Zn/day (to supply recommended levels (NRC, 2001); (H) or 120 mg Zn/d (to supply 0.2 of the recommended supplemental daily level; L), either supplemented as ZnO (I) or organically bound Zn (O) (Bioplex Zn^{TM}; Alltech Inc., Nicholasville, USA) and were fed at the rate of 2 kg/cow/day in one meal through parlour feeders. Forty-four Holstein-Friesian dairy cows, that were on average 31 days (s.d. ± 11.4) into lactation were allocated to one of the four treatments. Milk yield and feed intake were recorded daily and milk samples were taken at the beginning of the experiment and then fortnightly for subsequent analysis of SCC and milk amyloid A. All cows remained on treatment for 14 weeks. The data was analyzed by analysis of variance as a 2 × 2 factorial design.

Results

Cows supplemented with organically-bound zinc at the recommended level of inclusion (HO) had a higher (P<0.05) milk yield than those fed inorganic zinc at the recommended level (HI) or organically bound zinc at the low level (LO), but was not different from the low level of inorganic zinc (LI) (Table 1). No effects on DM intake (P>0.05). Animals that received the low level of Zn (LI and LO) had higher SCC (P<0.05) at weeks 12 and 14 of the study (Table 2).

Table 1. Effect of level and form of dietary zinc on dairy cow performance.

	Treatments				s.e.d	Significance		
	HI	HO	LI	LO		Level	Form	L×F
Total DM intake (kg/d)	22.8	23.7	23.1	24.0	0.82	0.590	0.139	0.862
Milk yield (kg/d)	35.2[a]	37.6[b]	36.0[ab]	35.2[a]	0.96	0.268	0.247	0.026

[ab]within a row, means without a common superscript letter differ (P<0.05).
Level = level of dietary zinc; form = form of dietary zinc and L×F = interaction between level and form.

Table 2. Effect of level and form of dietary zinc on somatic cell count (x1000/ml) (no s.e.d reported due to data being back transformed).

Week	Treatment				Significance		
	HI	HO	LI	LO	Level	Form	L×F
2	57	54	66	122	0.126	0.348	0.268
4	43	65	65	81	0.260	0.237	0.712
6	55	41	57	69	0.150	0.815	0.189
8	67	53	84	81	0.202	0.598	0.693
10	51	57	78	72	0.163	0.960	0.664
12	56[ab]	41[a]	92[b]	59[ab]	0.030	0.062	0.707
14	73[a]	32[a]	84[ab]	108[b]	0.029	0.336	0.075

[ab]within a row, means without a common superscript letter differ (P<0.05).
Level = level of dietary zinc; form = form of dietary zinc and L×F = interaction between level and form.

There was also a trend for cattle on the low Zn inclusion rate to have a higher milk amyloid A at week 12 (P=0.057) (Table 3).

Table 3. Effect of level and form of dietary zinc on milk amyloid A (μg/ml).

Week	Treatment				s.e.d	Significance		
	HI	HO	LI	LO		Level	Form	L×F
2	1.16	0.55	1.26	2.80	1.138	0.151	0.561	0.195
4	1.55	1.92	2.05	2.41	0.580	0.239	0.376	0.991
6	0.65	0.63	0.77	0.91	0.186	0.150	0.653	0.534
8	0.47	0.34	0.93	0.66	0.390	0.168	0.476	0.805
10	0.03	0.17	0.18	2.12	1.371	0.291	0.295	0.360
12	1.91	1.90	3.08	2.07	0.481	0.057	0.141	0.149
14	0.75	0.74	0.84	1.02	0.212	0.216	0.593	0.519

Conclusions

At the recommended level of inclusion supplementing Zn in an organically chelated form increased milk yield by 2.4 kg/d, but there was no effect of form of Zn on yield at the lower level of supplementation. Supplementing Zn at lower levels leads to an increase in immune response to infection in the mammary gland as indicated by the SCC and milk amyloid A.

Acknowledgements

Financial support of Alltech (UK) Ltd and DEFRA is gratefully acknowledged.

References

National Research Council. 2001. Nutrient Requirements of Dairy Cattle, Seventh Revised Edition, National Academy Press.

Effect of manganese on reproductive performance of beef cows and heifers

S.L. Hansen and J.W. Spears

Keywords: cattle, manganese, reproduction

Introduction

Manganese is an essential trace element that is required for numerous biological processes, including production of the cholesterol precursor squalene, and activity of enzymes such as glycosyltransferases which are involved in skeletal development. Some early studies suggested that diets low in Mn may result in impaired reproductive performance of beef cattle, including delayed onset of estrus, and an increase in the number of services to conception (Bentley and Phillips, 1951). However, several limitations to these studies exist, due to low animal numbers per treatment and the limited dietary Mn levels they examined. Our laboratory has conducted several experiments examining the effects of dietary Mn concentration on reproductive performance of beef cows and heifers, the results of which are summarized in this paper.

Materials and methods

Study 1 Eighty Angus (n = 40) and Simmental (n = 40) heifers, approximately 10 months old were individually fed a control diet (analyzed 15.8 mg Mn/kg DM) for 196 days (Hansen *et al.*, 2006a). Heifers were randomly assigned to treatments and supplemental Mn sulfate was provided at levels of 0, 10, 30 or 50 mg Mn/kg DM. Heifers were artificially inseminated at 13 months of age following synchronization of estrus by 2 doses of Lutalyse (Pfizer Animal Health, New York, NY), and rectal palpation was performed on day 196 for pregnancy determination. Blood samples for cholesterol were taken every 28 days.

Study 2 Twenty pregnant Angus (n = 9) and Simmental (n = 11) heifers (17 months of age) from Study 1 were continued on their previous treatments (0 or 50 mg supplemental Mn/kg DM) through pregnancy and early lactation (Hansen *et al.*, 2006b). Heifers were housed in pens of 2 and bunk fed by treatment (control diet analyzed 16.6 mg Mn/kg DM) during the 267 day study. Heifers calved from d 179 to 226 of the study. Within 24 hours of birth calves were weighed, bled, and observed visually for detection of deficiency signs.

Study 3 A 274 day study examining the effects of manganese on weaning weights and reproductive performance of beef cattle was conducted using 93 pregnant Angus and Simmental cows and heifers that were blocked by age and breed and randomly assigned to one of two treatments. Treatments consisted of a free choice mineral containing either no supplemental Mn or 3000 mg Mn/kg mineral from Mn sulfate. Cows were grazed on tall fescue pasture and supplemented with free choice hay, limited corn silage and corn gluten feed when forage was limiting. Cattle started on trial approximately one month prior to calving, and continued on treatment until calves were weaned. Cow weights were taken on days 0,

30, 86, 183, 203, 254, and 274 with calf weights taken on days 86, 183, 254, and 274 of the study. Cow weight changes between day 0 and 274 (overall) and day 86 and 274 (post partum) and calf average daily gain (weaning weight -birth weight/weaning age) were calculated as indicated. Cows were artificially inseminated following a synchronization protocol and cleanup bulls were used.

Results

Study 1 Addition of 0, 10, 30 or 50 mg Mn/kg DM to a diet containing 15.8 mg Mn/kg DM did not affect pregnancy rate, conception rate, or services to conception. Cholesterol concentrations were not affected by treatment (Hansen *et al.*, 2006a).

Study 2 Calves born to heifers receiving supplemental Mn had greater ($P < 0.05$) whole blood Mn levels and were heavier at birth (39 vs. 32 kg; $P < 0.05$). Some heifers not receiving supplemental Mn gave birth to calves that were disproportionately small compared to their supplemented counterparts ($P < 0.05$), or were unsteady or weak at birth ($P < 0.05$). Several calves born to unsupplemented heifers suffered from varying degrees of superior brachygnathism ($P < 0.01$), likely due to reduced activity of glycosyltransferases (Hansen *et al.*, 2006b).

Study 3 Pregnancy rate and services to conception were not affected by addition of 3000 mg Mn/kg DM to a free choice mineral. Performance of cows and calves are shown in Table 1.

Table 1. Weight change and calf performance of beef cows in Study 3.

	Supplemental Mn		SE	*P* value
	0	3000		
Overall weight change, kg	-56.2	-54.5	4.3	0.8
Post partum weight change, kg	-6.6	-1.2	3.7	0.3
Calf weaning weight, kg	232.2	234.7	4.4	0.7
Calf average daily gain, kg	0.82	0.85	0.02	0.3

Conclusion

A diet of at least 16 mg Mn/kg DM is sufficient for successful conception in heifers. However this level of dietary Mn is not sufficient for proper fetal development when fed to gestating beef heifers. Also, addition of supplemental Mn to a practical forage-based diet in North Carolina, USA did not improve pregnancy rates of cows or weaning weights of their calves.

References

Bentley, O G. and P.H. Phillips, 1951. The effect of low manganese rations upon dairy cattle. J. Dairy Sci. 34, 396-403.

Hansen, S.L., J.W. Spears, C.S. Whisnant and K.E. Lloyd, 2006a. Growth, reproductive performance, and manganese status of beef heifers fed varying concentrations of manganese. J. Anim. Sci. 84, 3375-3380.

Hansen, S.L., J.W. Spears, K.E. Lloyd and C.S. Whisnant, 2006b. Feeding a low manganese diet to heifers during gestation impairs fetal growth and development. J. Dairy Sci. 89, 4305-4311.

Injectable trace elements increase reproduction efficiency in dairy cows

K. Mitchell, W.A. Smith, A. Storch, N. Michael and J. Els

Keywords: injectable trace elements, reproduction, dairy cows

Introduction

The utilization of injectable trace elements before critical events such as drying off, calving and breeding could play a vital role to optimize trace element functions at these critical events to increase immunocompetence and reproduction efficiency, especially conception rates, in dairy cows. Injecting trace elements circumvents the effect of antagonisms between minerals in the gut, and the excretion of elements into the environment is reduced. It has been shown that injectable sodium EDTA chelates are utilized efficiently and are an effective manner of circumventing the naturally poor absorption mechanisms in the ruminant and meeting the demands of intensive animal husbandry practices. Research at Texas A&M showed that beef cows maintained a highly significant (P<0.01) advantage in liver copper concentration for 252 days after two treatments with copper in Multimin® (sodium-copper-EDTA) (Daugherty *et al.*, 2002). The objective of these two studies was to investigate trace mineral supplementation techniques, other than feeding, in an effort to enhance trace element status and dependent functions at critical events.

Materials and methods

In the first trial conducted at a top 10 dairy in the Tulare County of California, the herd averaged a heat detection rate (HDR) of 57% and a pregnancy rate of 18% and had excellent average milk production (10886 kg/year, 2x milking) and no serious cow health or production problems. Cows with even numbers were injected subcutaneously in the neck area with 5ml Multimin® (10mg Mn, 40mg Zn, 5mg Se and 15mg Cu/ml) at moving into the close-up group and again at six weeks fresh. In total, 615 cows were treated with Multimin® (even numbers), while 635 cows were in the control group (odd numbers).

In the second trial a well managed dairy herd in Waunakee, Wisconsin (750 cows, 12,610kg average/annum; 3x milking) was selected. The herd averaged a heat detection rate of 65% and an annual pregnancy rate of 16%. Cows with odd numbers (n=278) were injected subcutaneously with 5ml Multimin® 3 to 4 weeks prior to expected calving date and 3 to 4 weeks before the end of the voluntary waiting period (VWP=45DIM). Cows with even numbers (n=311) were taken as the control group and were injected subcutaneously with 8ml of MuSe® (5mg Se, 68 IU Vit E/ml) at 3 to 4 weeks before expected calving date. The feeding regimen, consisting of, a TMR and supplementary trace element/vitamin program was regarded as adequate for both herds. Chi-squared tests were performed to test the null hypothesis of independence between two factors (x-treatment and y-pregnancy result). When the chi-squared value is significant (P<0.05) the null hypothesis of independence is rejected. Regression analysis was used to determine the median days open for each treatment and calculate the respective 95% confidence limits.

Results and discussion

In trial 1, the conception rates (cows pregnant as a percentage of cows bred within a specific time frame) of all Multimin® treated cows was significantly (P<0.05) higher than the conception rate of control cows at days 70, 80 and 90 days in milk (Figure 1).

In trail 2, increased conception and pregnancy production resulted in a decrease in median days open from 119 days in the Mu-Se group to 99 days in the Multimin® group (P<0.05) (Figure 2).

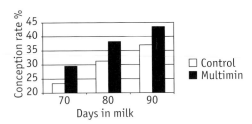

Figure 1. The effect of Multimin® on conception rate in all cows on days 70, 80 and 90 in milk (Trial 1).

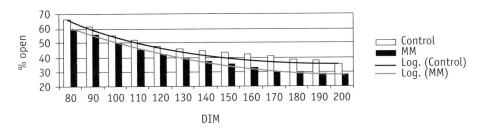

Figure 2. The effect of Multimin® on days open in all cows between 80 and 200 days in milk (Trial 2).

Conclusion

Results from these two trials clearly illustrate that utilization of injectable trace elements before calving and breeding could significantly increase conception rate and also significantly decrease median days open.

Injectable trace minerals could play a larger role in production systems, to optimize trace element functions, especially reproduction, in dairy cows, with limited oral intakes at critical times.

References

Boila, R.J., T.J. Devlin, R.A. Drysdale and L.E. Lillie, 1984. Injectable copper complexes as supplementary copper for grazing cattle. Can. J. Anim Sci. 64, 365-378.

Daugherty, S.R., G.E. Carstens, D.B. Herd, K.S. Barling and R.D. Randel, 2002. Effects of prenatal and prebreeding trace mineral/Vitamin E injections on calf health and reproductive performance of Beef Cows. Texas A&M 2002/2003 Beef Cattle Research in Texas, 39-43.

Effect of organic or inorganic selenium and zinc supplementation of ewes on ewe and lamb performance and mineral status

R.G. Wilkinson, A.M. Mackenzie, S.E. Pattinson and D. Wilde

Keywords: organic, selenium, zinc, ewes, lambs

Introduction

Selenium (Se) and zinc (Zn) are important trace elements that are metabolically active as prosthetic groups in a number of enzyme systems required for animal health and productivity. Selenium is a component of the enzyme glutathione peroxidase, which catalyses the removal of hydrogen peroxide arising from oxidative stress, whereas Zn has been shown to be involved in over 300 different enzyme systems. Traditionally, animal diets are supplemented with inorganic mineral compounds such as sodium selenite (Na_2SeO_3) and zinc oxide (ZnO). However, these are inefficiently absorbed, resulting in high levels of mineral excretion. Recent evidence suggests that organically-bound minerals such as mineral proteinates may be more readily absorbed, increase animal performance and reduce mineral excretion. The objective of the experiment was to investigate the effects of including organic or inorganic Se and Zn supplements in the diet of ewes during late pregnancy and early lactation on the performance and mineral status of ewes and lambs.

Materials and methods

Forty twin-bearing Suffolk x Mule ewes were blocked according to age, liveweight and condition score and allocated at 103 days of gestation to one of four dietary treatments in a 2 x 2 factorial experiment. All ewes were fed a basal concentrate (SC Feeds Ltd, Stone Staffordshire, United Kingdom) supplemented to provide Se and Zn at 0.3 and 50 mg/kg respectively, using either organic (SelPlex® and Bioplex Zinc®, Alltech Ltd) or inorganic sources (sodium selenite and zinc oxide). Concentrates were fed twice daily (08:00 and 16:30) with the feeding level increasing by 0.1 kg/day from 0.8 kg/day 6 weeks before lambing to 1.3 kg/day at lambing, with ewes continuing to receive 1.3 kg/day during lactation. Throughout the experiment ewes were housed individually and offered straw and water *ad-libitum*.

Ewe liveweight and condition score were recorded weekly and blood was sampled by jugular venepuncture into vacutainers containing lithium heparin and EDTA four weeks prior to lambing (week -4), and four weeks post lambing (week +4). Lamb liveweight was recorded at birth and then weekly until four weeks of age. Three weeks post lambing, the lambs were separated from the ewes; the ewes were injected with oxytocin and milked out completely. After four hours they were re-injected with oxytocin and milked out again. The four hour milk production was used to estimate daily milk yield. Blood was sampled from one lamb in each litter at seven and 21 days of age. Whole blood glutathione peroxidase activity was determined using a Ransel kit (Randox Laboratories) and plasma zinc and milk selenium concentration were determined by atomic absorption spectrophotometry. The results were statistically analysed by ANOVA using Genstat.

Results

There were no significant differences between treatments in ewe liveweight or condition score. Ewes offered zinc oxide (InZn) with sodium selenite (InSe) had a significantly higher milk yield than those offered InZn with Selplex (OSe) (P<0.05). Organic Se supplementation (OSe) significantly increased milk Se concentration (P<0.01). There were no other significant effects on ewe performance. However, plasma glutathione peroxidase activity and plasma Zn concentration was numerically higher in ewes offered OSe and Bioplex Zinc (OZn) respectively (Table 1).

There were no significant differences between treatments in lamb birth weight, plasma glutathione peroxidase activity or zinc concentration. However, lambs reared by ewes offered OZn had a higher growth rate than those reared by ewes offered InZn (Table 2).

Table 1. Effect of Se and Zn source on plasma glutathione peroxidase activity (GSHPx, u/ml PCV), Zn concentration (umol/l), milk yield (ml/day) and milk selenium concentration (mg/l).

	OSe		InSe		s.e.d	Significance		
	OZn	InZn	OZn	InZn		Se	Zn	Int
Plasma GSHPx								
Week -4	109.2	126.5	94.8	103.2	14.91	NS	NS	NS
Week +4	132.4	138.2	127.1	143.9	12.38	NS	NS	NS
Plasma Zn								
Week -4	10.95	9.16	8.19	9.05	1.431	NS	NS	NS
Week +4	10.24	8.81	9.25	8.26	0.876	NS	NS	NS
Milk yield	2118	1638	1974	2292	249.0	NS	NS	*
Milk Se content	0.047	0.049	0.032	0.040	0.0108	**	NS	NS

NS P>0.05; * P<0.05; ** P<0.01.

Table 2. Effect of Se and Zn source on lamb birth weight (kg), daily gain (kg/day), plasma glutathione peroxidase activily (GSHPx, u/ml PCV) and zinc concentration (umol/l).

	OSe		InSe		s.e.d	Significance		
	OZn	InZn	OZn	InZn		Se	Zn	Int
Plasma GSHPx								
Day 7	163	117	112	123	29.2	NS	NS	NS
Day 21	261	236	235	216	26.4	NS	NS	NS
Plasma Zn								
Day 7	15.2	14.3	16.7	11.5	2.77	NS	NS	NS
Day 21	19.1	18.0	19.4	18.7	2.36	NS	NS	NS
Birth weight	4.33	4.78	4.69	4.64	0.351	NS	NS	NS
Daily gain	0.262	0.218	0.265	0.247	0.0177	NS	*	NS

NS $P>0.05$; * $P<0.05$.

Conclusion

Organic Se and Zn supplementation had no effect on ewe performance, but enhanced ewe mineral status compared to inorganic supplementation. Organic Zn supplementation of ewes also increased lamb growth rate, however the mechanism requires further investigation.

Selenium metabolism in lambs supplied with different selenium sources and levels

F.A. Paiva, M.A. Zanetti, F.R. Martins, L.B. Correa, G.R. Del Claro and A. Saran-Netto

Keywords: metabolic balance, organic mineral sources, sheep

Introduction

Usually it is accepted that inorganic selenium is less absorbed by ruminants than by monogastrics because the rumen microorganisms transform inorganic Se into insoluble forms. Results from research, however, are quite variable. Koering *et al.* (1997) found a higher Se absorption for the inorganic than the organic source, using Se[82] in lambs. There are several factors that affect the Se absorption. Se from inorganic sources can be better absorbed than the organic sources in a high concentrate diet (Underwood and Suttle, 1999). Se from organic sources can be absorbed through amino acids transport and directly incorporated into body protein, improving its absorption and retention as related to inorganic sources (Peter *et al.*, 1982), but it depends on the kind of diet. The aim of this research was to compare the utilization of organic Se sources at high levels to sodium selenite in lambs receiving a high concentrate diet, through Se metabolic balance studies.

Material and methods

This research was carried out at the College of Animal Science and Food Engineering (University of São Paulo, Pirassununga, SP, Brazil). Forty Suffolk lambs (3 months old and 19.8 ± 2.1kg BW) were submitted to the following treatments: 1) – no Se supplement; 2, 3 and 4) – 0.2, 0.8 and 1.4 mg/kg of supplementary Se from sodium selenite, respectively; 5, 6 and 7) – 0.2, 0.8 and 1.4 mg/kg of supplementary Se from Seleno-Yeast, respectively; 8, 9 and 10) – 0.2, 0.8 and 1.4 mg/kg of supplementary Se from Selenomethionine, respectively. Lambs were fed a high concentrate diet (corn-54.7%, extruded soybean-18.0%, cottonseed hulls (roughage)-25.0%, limestone-1.3% and vitamin mineral supplement-1.0%), and finished during an 84-day period. During the last five days of experiment, a metabolic balance of Se was performed with daily total collection of faeces and urine, through metabolism cages and bags. The evaluated parameters were: selenium intake (mg/day), and percentage of absorbed and retained Se in relation to the ingested Se. The experiment was arranged in a randomized complete block design. Regression analysis was performed using the PROC REG of SAS (SAS, 2000).

Results and discussion

Se sources did not affect apparent Se absorption and retention (P>0.05). However, the supplementary Se levels affected Se absorption and retention, and the response for the levels was linear (P< 0.001) for absorption and retention (Figure 1). The high concentrate levels could have affected the Se utilization from the different sources. The increase in apparent Se absorption probably is due to the fact that faecal endogenous excretion is approximately the same in all treatments, and increasing the dietary Se level reduced the contribution of endogenous Se to total faecal Se excretion. Se absorption was high, probably due to the high dietary levels used in this study.

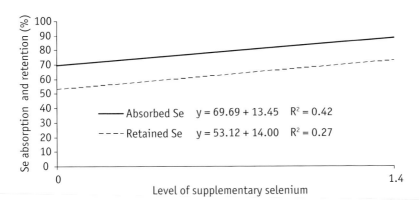

Figure 1. *Selenium absorption and retention (%) in lambs supplied with different levels and sources of selenium.*

Conclusion

Selenium apparent absorption and retention from organic and inorganic sources were similar in sheep fed high concentrate diets. Selenium absorption and retention increased with increasing dietary Se levels.

References

Koenig, K.M., L.M. Rode, R.D.H. Cohen and W.T. Buckley, 1997. Effects of diet and Chemical form of selenium on selenium metabolism in sheep. J. Anim. Sci. 75, 817-827.

Peter, D.W., P.D. Whanger, J.P. Lindsay and D.J. Buscall, 1982. Excretion of selenium, zinc and copper by sheep receiving continuous intraruminal infusions of selenite or selenomethionine. Proc. Nutr. Soc. Aust, 7, 178.

SAS Institute Inc., 2000. SAS Stat Guide, Version 8.2 Edition. Cary. NC: SAS Institute Inc.

Underwood, E.J. and N. Suttle, 1999. The mineral nutrition of livestock, 3 ed. Edinburg United Kingdom, 624 p.

Distribution of total selenium and selenized amino acids within the edible tissues of beef cattle fed graded additions of selenized yeast

D.T. Juniper, R.H. Phipps and G. Bertin

Keywords: cattle, selenium, selenomethionine, selenocysteine, seleno-yeast

Introduction

Diets for ruminant animals are often of plant origin, the selenium (Se) contents of which can be extremely variable. Consequently, dietary concentrations of Se can be deficient and the addition of supplementary Se may be required. Selenium supplements are in two principal forms, inorganic mineral salts such as sodium selenite (Na_2SeO_3) and sodium selenate (Na_2SeO_4), or organic forms such as Se enriched yeasts. Only in 2007 was the use of Se enriched yeasts permitted within the EU and supplementary Se had usually been derived from inorganic sources. Absorption and metabolism of these two Se sources can be different; sodium selenate transport may be active within the rat gastrointestinal tract (Turner *et al.* 1990) whereas the absorption of selenite is far less efficient, as simple diffusion is probably the main route of absorption (Wolfram *et al* 1985). Conversely selenomethionine (SeMet), the predominant form of Se within selenoyeasts (SY), is transported via methionine transporter mechanisms (Weiss, 2003). Furthermore, SeMet can be incorporated non-specifically into general body proteins forming a Se pool, whereas Se from inorganic sources is either incorporated into functional selenoproteins or excreted (Suzuki and Ogra, 2002). The aim of this study was to investigate the effects of Se source and dose on total Se and the proportion of total Se as either SeMet or Selenocysteine (SeCys) within the tissues of beef cattle fed graded additions of SY.

Materials and methods

Thirty-two castrated Limousine cross cattle were enrolled onto the study with mean weight 489 ± 7.6 kg, blocked by live weight and then randomly allocated to one of four dietary treatments. Treatments were an unsupplemented control (BM1; 0.15 mg total Se/kg DM), control diet supplemented with 0.15 (BM3; 0.30 mg total Se/kg DM) or 0.35 mg Se/kg DM (BM4; 0.50 mg total Se/kg DM) as SY produced by a specific strain of *Saccharomyces cerevisiae* CNCM I -3060 (Sel-Plex®, Alltech) or control diet supplemented with 0.15 mg/kg DM (BM2; 0.30 mg total Se/kg DM) of an inorganic selenium source (Na_2SeO_3). Dietary treatments were offered on an individual, *ad-libitum* basis comprising of a predominantly maize silage-based TMR that differed either in the level of organic Se supplementation (BM1, BM3 and BM4) or Se source (BM2 vs BM3). Following 112 d exposure cattle were euthanased and samples of heart, liver and skeletal muscle tissue retrieved for determination of total Se content and the proportion of total Se comprised as either SeMe or SeCys. Total Se was determined by ICP MS using the method of standard additions following mineralisation. Selenized amino acids were quantified by reversed phase HPLC-ICP MS using a collision cell detector following the proteolytic digestion and purification by size exclusion HPLC. Data was analysed by ANOVA using the GLM procedure

Results

There were significant treatment (P<0.01) and linear dose effects (P<0.005) in concentrations of total Se within heart, liver and skeletal muscle (*M. Psoas Major* [PM] and *M. Longissimus Thoracis* [LT]) with increasing dietary inclusion of SY. Total Se was numerically higher in liver (1.44 vs. 1.9 mg Se/kg DM) and PM (0.29 vs. 0.36 mg Se/kg DM) and significantly higher (P<0.05) in heart (0.72 vs. 0.86 mg Se/kg DM) and LT (0.31 vs. 0.46 mg Se/kg DM) in SY supplemented animals when compared to those receiving a comparable dose of selenite (Figure 1). Selenocysteine was the predominant selenized amino acid within the tissues of cattle receiving treatment BM1 (no additional Se). As supplementary Se was added to the diet there was little change in the proportion of total Se comprised as SeCys but there were notable increases in the proportion as SeMet. Furthermore, changes in the proportion of total Se as SeMet were more marked in those diets where SY formed the Se supplement when compared to a comparable dose of selenite. This would indicate that the increases seen in total tissue Se in animals offered SY are attributable to both the improved uptake of SeMet and the non-specific incorporation of SeMet into body tissue.

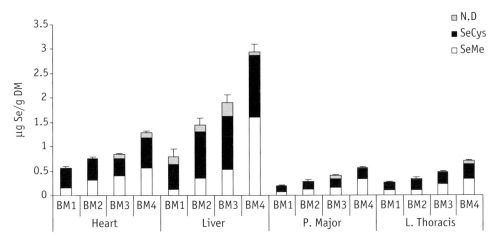

Figure 1. Total Se and the proportion of total Se comprised as SeMet or SeCys in the tissues of beef cattle fed graded additions of SY.

References

Suzuki, K.T and Y. Ogra, 2002. Metabolic pathway for selenium in the body: speciation by HPLC-ICP MS with enriched Se. Food Addit. Contam. 19, 974-983.

Turner, J.C., P.J. Osborn and S.M. Mc Veagh, 1990. Studies on selenate and selenite absorption by sheep ileum using an everted sac method and an isolated vascularly perfused system.Comp.Biochem.and Physiol.A,Comp. Physiol. 95, 297-301.

Weiss, W.P., 2003. Selenium nutrition of dairy cows: comparing responses to organic and inorganic selenium forms. In: Lyons, P. and K. Jaques (Eds.) Proc. 19th Alltech Annual Symposium Nutritional, Biotechnology in the Feed and Food Industries. Nottingham University Press, Nottingham, United Kingdom, pp.333-343.

Wolframm, S., E. Anliker and E. Scharrer, 1986. Uptake of selenate and selenite by isolated brush border membrane vesicles from pig, sheep and rat; Biol.Trace Elem.Res. 10, 293-306.

Iodine and selenium antagonism in ruminant nutrition

L. Pavlata, S. Slosarkova, P. Fleischer, A. Pechova and L. Misurova

Keywords: goat, glutathione peroxidase, metabolism, potassium iodide

Introduction

Apart from the primary deficiencies due to trace elements deficiencies in feed, secondary deficiencies can also occur. They are caused mostly by a negative action of an overload of other agents. Selenium (Se) interactions with other elements rate among the factors bearing on Se metabolism in an organism. There are descriptions of interactions of Se with sulfur, cadmium, arsenic, copper, cobalt, manganese, lead, and iron (Shamberger, 1983, Ivancic and Weiss, 2001). Because deficiencies of iodine, Se and other trace elements occur in many regions of the world, including the Czech Republic, frequent (Travnicek and Kursa, 2001, Pavlata et al., 2005) attention is paid to their sufficient supply in both veterinary and human medicines. From this perspective, there is still a lack of information on the potential interactions between highly supplemented iodine and other trace elements. The aim of this study was to explore the effect of varied iodine supply on Se metabolism in goat kids.

Material and methods

The study included 7 clinically healthy 14-day-old kids of the white short-haired goat breed from mothers with high iodine supplementation (group E) and 7 clinically healthy kids from mothers with mild hypoiodemia (group C). Kids in group E were administered potassium iodide orally from 14 to 90 days of age. During the experimental period, the group E kids had a total daily iodine intake (from the feeding ration and from the per os potassium iodide administration) of 440 - 590 μg per head/ day in comparison with 150 - 190 μg per head/ day in the group C kids (only from the feeding ration; no potassium iodide administration). In kids of both groups, Se concentration and glutathione peroxidase (GPx) activity were monitored in whole blood.

Results and discussion

Animals receiving iodine supplementation had statistically lower concentration of Se in blood at 75 and 105 days of age and lower activity of GPx at 105 days (Table 1). Results show the possible negative impact of higher dosage iodine supplementation on the Se metabolism, based on the Se concentration and the GPx activity in whole blood. Concentration of Se as well as the GPx activity increased with age in kids not supplemented with iodine but the increase in blood Se and GPx activity with increasing age was less in iodine supplemented kids. The concentration of Se as well as the GPx activity in the blood of the highly iodine supplemented kids reached almost 70% of the value in kids with no iodine supplementation at the end of the experiment. This significant lower GPx activity in kids highly supplemented with iodine thus indicates a possible negative impact of the increased iodine supply on

Table 1. The Se concentration (μg/l) and the GPx activity (μkat/l) in whole blood of the kids supplemented with iodine (E) and not supplemented with iodine (C).

Day		14		45		75		105	
		Mean	SD	Mean	SD	Mean	SD	Mean	SD
Se	E	65.7	11.2	66.1	14.1	76.0[ae]	11.7	88.1[ad]	10.9
	C	65.5	7.8	77.3[b]	11.1	101.4[bce]	11.9	131.8[cd]	23.2
GPx	E	247.6[h]	60.8	335.8[hi]	117.9	408.2[ij]	88.5	484.0[jl]	125.4
	C	235.2[g]	28.5	414.0[gk]	84.6	523.0[kf]	120.1	713.3[fl]	153.3

[a-f] within a row, means without a common superscrpt letter differ (P<0.01); [h-l] within a row, means without a common superscrpt letter differ (P<0.05).

important selenoproteins as well. They mediate biological functions of Se in the organism (Birringer *et al.*, 2002).

Results of this study suggest that high dietary iodine had a negative impact on absorption or further Se metabolism in kids. It has been noted that during passage through the intestines, neither Se nor iodine display any significant interactions with the other diet components and that regulation impacting their intestinal absorption is minimal. The main homeostatic regulation for Se and iodine in the organism is by their renal excretion (Windisch, 2002, Boldizarova *et al.*, 2003). In ruminants however, this process can to a certain degree be under the influence of the changes during the ruminal fermentation. One can therefore speculate that daily supply of potassium iodide could have an influence on the composition and function of the ruminal microflora in a manner that resulted in lower biological availability of Se in the subsequent sections of the gastrointestinal tract. The results indicate that increased iodine supplementation may have a negative effect on selenium metabolism and/ or selenium status in ruminants.

Acknowledgements

This work supported by the MSMT of the Czech Republic (Project No. MSM6215712403).

References

Birringer, M., S. Pilawa and L. Flohe, 2002. Trends in selenium biochemistry. Nat. Prod. Reports 19, 693-718.
Boldizarova, K., L. Gresakova, S. Faix, M. Levkut and L. Leng, 2003. Urinary selenium excretion in selenite-loaded sheep and subsequent Se dynamics in blood constituents. Reprod. Nutr. Dev. 43, 385-393.
Ivancic, J. jr.and W.P. Weiss, 2001. Effect of dietary sulfur and selenium concentrations on selenium balance of lactating Holstein cows. J. Dairy Sci. 84, 225-232.
Pavlata, L., A. Podhorsky, A. Pechova and P. Chomat, 2005. Differences in the occurrence of selenium, copper and zinc deficiencies in dairy cows, calves, heifers and bulls. Vet. Med.-Czech 50, 390-400.

Shamberger, R.J., 1983. Biochemistry of Selenium. Plenum Press, New York, 334 pp.

Travnicek, J. and J. Kursa, 2001. Iodine concentration in milk of sheep and goats from farms in South Bohemia. Acta Vet. Brno 70, 35-42.

Windisch, W., 2002. Interaction of chemical species with biological regulation of the metabolism of essential trace elements. Anal. Bioanal. Chem. 372, 421-425.

Fluorine availability from rock phosphate in sheep

M.A. Zanetti, F.O. Miller, J.A. Cunha, L.H.O. Silva and H.C. Humberto

Keywords: bone, alkaline phosphatase, sodium fluoride, rock phosphate, phosphorus

Introduction

Phosphorus is one of the most important minerals for grazing cattle and is deficient in large areas of the world, especially in acid iron-rich soils which are common in hot climate countries like Brazil. Natural phosphorus sources like rock phosphate are usually rich in fluorine, and may cause fluorine toxicity. Therefore it is necessary to chemically process it in order to use this kind of source. In Brazil, it is possible to use the natural rock phosphate for cattle up to 1/3 of the inorganic phosphorus supplemented. Average fluorine content in Brazilian rock phosphate which is not from sedimentary origin but from igneous origin is 1.5%, and the fluorine in this source of rock phosphate is probably less soluble than the sedimentary one. As most of available research on the fluorine effect on animals is from rock from sedimentary origin, it is important to study the igneous rock. The present study was conducted to determine the availability of fluorine from igneous rock phosphate relative to sodium fluoride.

Material and methods

Twenty-one lambs, three months old, were submitted to three treatments: (1) Negative control diet with 13 mg of F/kg DM (from dicalcium phosphate); (2) Diet supplemented with 98 mg of F/kg DM (from sodium fluoride) and (3) Diet supplemented with 98 mg of F/kg DM (from rock phosphate). The fluorine level in the rock phosphate was 16900 mg/kg DM. The phosphorus level in the total diet was 0.18%, for all treatments. The research lasted 150 days and at the end the animals were killed and a rib was sampled for fluorine analysis by a specific electrode (Dolan *et al.*, 1978). Blood was sampled at the beginning and at 28 days intervals for alkaline phosphatase analysis (Labtest®). In the last five days, a metabolic balance study was carried out with total faeces and urine collection for fluoride analysis. Data were statistically analyzed as a randomized complete design: for phosphatase the data were analyzed as a repeated measures design using PROC MIXED procedure of SAS (2000) and for fluorine balance it was used the PROC GLM.

Results and discussion

The fluorine from sodium fluoride and rock phosphate increased ($P < 0.01$) the fluorine bone content, while the animals supplemented with sodium fluoride had higher levels than those receiving the rock phosphate ($P < 0.01$). The period used (150 days) was long enough to increase significantly the fluorine bone levels in the rib of the lambs. The bone fluorine values were 188 mg/kg DM (dry fat-free basis) for the control group, 1760 mg/kg DM for the sodium fluoride supplementation and 1350 mg/kg DM for the rock phosphate supplementation. Based on bone fluorine concentrations, fluorine from rock phosphate was approximately 76% as available as sodium fluoride. During the 5-day balance period, fluorine retention for sodium fluoride was 74.8% ± 1.9, for the dicalcium phosphate 39.7% ± 4.1 and for the rock phosphate 28.7% ± 6.1. Fluorine absorption was also higher (P<0.001) for the sodium fluoride being 85.1% ± 1.1, for the dicalcium phosphate it was 51.5% ± 5.8 and for the rock phosphate 37.4% ± 7.0. According Ewing and Charlton (2005), up to 90% of the ingested fluorine can be absorbed, a value close to that observed in the sodium fluoride group. From the first to the last period of 28 days, the alkaline phosphatase levels (Figure 1) were higher in the rock phosphate than in the others treatments (P<0.01). On day 140, alkaline phosphatase level in the sodium fluoride supplemented group was higher than in the dicalcium phosphate supplemented group (P<0.10).

High alkaline phosphatase levels are usually related with low phosphorus levels in the diet (Underwood and Suttle, 1999) and the higher levels for the rock phosphate group could be due to phosphorus deficiency as the phosphorus in the diet was low and this source has a low bioavailability (Ammerman *et al.*, 1995). However, the fluorine supplementation also slightly increased the alkaline phosphatase (only in the last period) in animals receiving the dicalcium phosphate as phosphorus source.

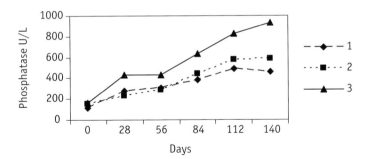

Figure 1. Serum alkaline phosphatase (U/L) for control (1), supplemented with sodium fluoride (2) and supplemented with rock phosphate (3).

Conclusions

Fluorine bone level of lambs receiving sodium fluoride was higher than from rock phosphate. Fluorine absorption and retention were higher for the sodium fluoride than for the other sources. Alkaline phosphatase was higher in animals that received rock phosphate.

References

Ammerman, C.B., D.H. Baker and A.J. Lewis, 1995. Bioavailability of nutrients for animals. Academic Press, New York, 441 p.

Dolan, T., L. Legette and J. McNeal, 1978. Determination of fluoride in deboned meat. J. Assoc. Off. Anal. Chem. 61, 982-985.

Ewing, W.N. and S. Charlton, 2005. The minerals directory. Contex Publications, Leicestershire, England, p. 10a-10e.

Underwood, E.J. and N.F. Suttle, 1999. The mineral nutrition of livestock, 3 ed. Cabi Publishing, 614 p.

Feed supplementation of selenium enhance growth and disease resistance of Indian ornamental fish 'dwarf gourami' *Colisa lalia*

N. Felix

Keywords: gourami, selenium, growth, disease resistance

Introduction

Selenium (Se) is a trace mineral, which has recently received a considerable amount of attention in animal nutrition. Selenium is a component of the enzyme glutathione peroxidase (Rotruck et al., 1973). This enzyme catalyzes reactions necessary for the conversion of hydrogen peroxide and fatty acid hydroperoxides into water and fatty acid alcohols by using reduced glutathione, thereby protecting cell membranes against oxidative damage. The requirement of Se has been quantified in only few fish species. Dwarf gourami, an indigenous ornamental fish of India has great marketability. The purpose of this study was to estimate the Se requirement for maximal growth and disease resistance of gourami using organic selenomethionine supplementation.

Material and methods

Casein, fish oil, corn oil and corn starch were used for the preparation of purified diet. Organic Se (selenomethionine, Sigma Chemical) was used as the Se source and supplemented at 0, 0.1, 0.2, 0.4, 0.6 and 1 mg Se/kg purified diet. The Se concentrations of the experimental diets were determined by hydride generation atomic spectrophotometer (Tinggi, 1999). The experiment was conducted in a closed recirculation system consisted of a series of 100 l glass aquaria. Each aquarium was stocked with 12 fish with initial mean weight of 3.26±0.12 g. Half of the water in the system was exchanged daily. A photoperiod of 12 h light (08:00-20:00 h), 12 h dark was used. The fish were fed test diets at 3% of their body weight per day for an 8-week period. At the end of the experiment, body weight gain in each aquarium, feed efficiency (FE) and survival were calculated. After the final weighing, the fish body Se content was determined by hydride generation atomic absorption spectrophotometer (Tinggi, 1999). The hepatic Se concentration (Tinggi, 1999) reduced glutathione (GSH) and oxidized glutathione (GSSG) content (Griffith, 1983) and hepatic glutathione peroxidase (GPx) (Bell *et al.*, 1985) and glutathione reductase (GR)(Goldberg and Spooner, 1983) activities were also determined. The remaining fish were used in a challenge against bacterial strain *Aeromonas sobria*. Results were analyzed by one-way analysis of variance (ANOVA) with Duncan multiple range test.

Results

Weight gain (256.12±1.87 g) and feed efficiency (0.95±0.03) were highest ($P<0.05$) in fish fed diet with 0.4mg Se/kg diet, followed by fish fed diets with 0.2, 0.1 and 0.6 mg Se/kg diet, and lowest in fish fed diet with 1 mg Se/kg diet (weight gain of 165.45±3.65 g and feed efficiency of 0.62±0.02). Survival was not affected by the dietary treatment. Hepatic

Se concentration was highest in fish fed diet with 1 mg Se/kg, followed by 0.6 mg Se/kg diet, then 0.4 mg Se/kg, and lowest in fish fed the basal diet (Table 1). Hepatic glutathione peroxidase (GPx) activity was highest in fish fed diets with 1 mg Se/kg, followed by 0.4 and 0.6 mg Se/ kg, and lowest in fish fed diet with 0.1 mg Se/kg. Fish fed the basal diet had a higher hepatic glutathione reductase (GR) activity than fish in all other dietary groups. The hepatic GSH/GSSG ratio in fish was not affected by the dietary treatment.

Survival of fish after the challenge was highest ($P<0.05$) in fish fed diet with 0.4mg Se/kg diet (95.83±7.34), followed by fish fed diets with 0.6 and 1 mg Se/kg diet, and lowest in fish fed basal diet (58.34±3.56).

The results of the present study clearly indicate that gourami have a requirement for Se that cannot be met without supplementation of selenium.

Table 1. Hepatic selenium concentration, glutathione peroxidase (GPx) and glutathione reductase (GR) activities, reduced glutathione (GSH)/oxidized glutathione (GSSG) ratio and survival (after challenge against Aeromonas sobria) of gourami fed different diets for 8 weeks.

Se supplementation (mg Se/kg)	Se concentration (µg/g tissue)	GPx activity (µmol NADPH/ min/mg protein)	GR activity (nmol NADPH/ min/mg protein)	GSH/GSSG ratio	Percentage survival of fish after challenge
0 (0.15)[1]	0.62±0.09[a]	0.45±0.09 [a]	42.37±5.43 [b]	1.55±0.20	58.34 [a]
0.1 (0.24)	0.89±0.10 [ab]	1.18±0.04 [b]	22.74±1.56 [a]	1.52±0.12	75.00 [b]
0.2 (0.38)	1.14±0.04 [b]	1.20±0.01 [b]	27.80±5.78 [a]	1.75±0.33	75.00 [b]
0.4 (0.59)	1.78±0.13 [c]	1.48±0.03 [c]	30.54±3.65 [a]	1.62±0.23	95.83 [c]
0.6 (0.82)	1.91±0.05 [cd]	0.42±0.05 [c]	27.45±4.32 [a]	1.50±0.17	83.33 [bc]
1.0 (1.29)	2.08±0.24 [d]	1.49±0.07 [c]	29.56±3.45 [a]	1.78±0.16	83.33 [bc]

[1]analytically determined concentration of selenium.
Values are means ± SD from three groups of fish (n=3).
Different superscripts indicate significant ($P<0.05$) difference between different dietary Se levels.

References

Bell, J.G., C.B. Cowey, J.W. Adron and A.M. Shanks, 1985. Some effects of vitamin e and selenium deprivation on tissue enzyme levels and indices of tissue peroxidation in rainbow trout (Salmo gairdneri). Br. J. Nutr. 53, 149 -157.

Goldberg, D.M. and R.J. Spooner, 1983. Glutathione reductase, 3rd edn. In: H.V. Bergmeyer, J. Bergmeyer and M. Grassl (Eds.), Method of Enzymatic Analysis, vol. 3. Weinhein, Deerfield Beach, pp. 258 -265.

Griffith, O.W., 1983. Glutathione and glutathione disulphide, 3rd edn. In: H.V. Bergmeyer, J. Bergmeyer and M. Grassl (Eds.), Method of Enzymatic Analysis, vol. 3. Weinhein, Deerfield Beach, pp. 521 -529.

Rotruck, J.T., A.L. Pope, H.E. Ganther, A.B. Swanson, D.G. Hafeman and W.G. Hoekstra, 1973. Selenium: biochemical role as a component of glutathione peroxidase. Science 179, 585-590.

Tinggi, U., 1999. Determination of selenium in meat products by hydride generation atomic absorption spectrophotometry. J. AOAC Int. 82, 364-367.

Concluding words

Summary and conclusions

A.W. Jongbloed

Keywords: environment, trace minerals, livestock, fish, nutrition

Introduction

The human population continues to grow worldwide at a high rate. Farm animal populations grow as a consequence to cover consumers' demand for high quality animal products. Therefore, the livestock and aquaculture sector emerge as one of the top two or three most significant contributors to the most serious environmental problems, at every scale from local to global. Studies by FAO suggest that there should be a major policy focus when dealing with problems of land degradation, climate change, air pollution, water shortage and pollution, and loss of biodiversity. Livestock's contribution to environmental problems is on a massive scale and its potential contribution to their solution is equally large. The impact is so significant that it needs to be addressed with urgency. Therefore, the first International Symposium on Trace Elements in Animal Production Systems was held in Geneva, Switzerland to focus specifically onto the trace mineral subject in which soil scientists, hydrologists, chemists, nutritionists, technologists, policy makers and environmentalists could exchange information and share their knowledge. The conference was named OTEANE 2007 for 'Organic Trace Elements in Animal Nutrition and the Environment' which was sponsored by the firm of Pancosma. The first day, a series of papers were presented dealing with livestock production and environment, while on the second day papers were focused on nutrition and environment. In addition, 25 posters were presented on trace minerals covering chemistry of mineral sources, manure composition, and effects of several trace mineral sources on performance, absorption and utilization in poultry, pigs, ruminants, fish and crustaceans.

Summary

In their paper, Gerber and Steinfeld of the FAO showed an overview of the worldwide growth of animal production and its environmental consequences. The livestock sector is by far the single largest user of land. The total area occupied by grazing is equivalent to 26 percent of the ice-free terrestrial surface of the planet. In addition, the total area dedicated to feed crop production amounts to 33% of total arable land. In all, livestock production accounts for 70% of all agricultural land, and 30% of the land surface of the planet.

Growing human population and income, along with changing food preferences, are rapidly increasing the demand for livestock products. Global production of meat is projected to more than double from 229 million tons in 1999/01 to 465 million tons in 2050, and that of milk to grow from 580 to 1043 million tons. The major environmental concerns comprise land degradation, atmosphere and climate, quality of water, and biodiversity. The environmental impact per unit of livestock production must be cut by one half, just to avoid the level of damage worsening beyond its present level. These developments urgently require improved efficiency, thus reducing the land area required for livestock production.

The paper by Verstraete who represented the EU legislative authorities, showed clear concerns about high concentrations of trace minerals in animal and fish feeds in relation to their environmental impact. In the first place, the maximum content of trace elements authorised in feedingstuffs in the EU takes into account the physiological requirements of animals. Following progress in scientific and technical knowledge, the maximum levels of trace elements in feedingstuffs are reviewed when necessary and were recently reduced in order to reduce the harmful effects caused by animal excretions and also to minimise the adverse effect on human health and environment. Maximum levels for heavy metals (and dioxins) in trace elements have also been established in the EU on undesirable substances in animal feed. The presence of high levels of heavy metals, in particular lead and cadmium, are regularly reported through the EU Rapid Alert System for Feed and Food (RASFF).

Several papers were presented concerning the accumulation of heavy metals on farm, regional and European country scale. Heavy metals which enter agro-ecosystems through various agricultural and industrial activities can accumulate in soils and may have long-term implications for soil health and function, the quality of agricultural produce and the wider environment. Eckel *et al.* showed the results of the Assessment and Reduction of Heavy Metal Input into Agro-ecosystems (AROMIS) project financed by the EU, on a national assessment of heavy metals in European agriculture. Several input sources relevant for agricultural such as mineral and organic fertilisers, feedstuffs, feed supplements, other input sources like atmospheric deposition, as well as background concentrations in soils were calculated. The evaluation of the farm gate balances showed that the input of heavy metals in most cases exceeds the outputs leading to a net surplus in the system. The balance results are dominated by the feeding regime in the case of livestock farming systems and by the choice of fertilisers, i.e. the use of sewage sludge (not allowed in various countries) and inorganic P fertilizers. A recent overview of heavy metal balances across EU member states (Eckel *et al.*, 2005) shows that net accumulation rates for Cu vary from -138 to +908 g ha^{-1} yr^{-1} (median value: +109 g ha^{-1} yr^{-1}) and from -131 to +3523 g ha^{-1} yr^{-1} (median: +389 g ha^{-1} yr^{-1}) for Zn. A negative net accumulation only occurred at sites without application of animal manure (or sewage sludge).

Also in various countries of South East Asia, a fast growing intensive pig and poultry production has developed. Therefore, to analyse the present situation, to monitor the future development, to support farmers in the correct management of manure and fertilisation and to assist policy makers, the Nutrient Flux model for Area Wide Integration (NuFlux-AWI) was developed to quantify nutrient fluxes, manure quality and value and environmental effects, especially for pig and poultry production. This model was presented by Menzi showing that it can be used for calculations such as nutrient and heavy metal balances when livestock manure is recycled on crops. The inclusion of Cu and Zn in NuFlux allows for the evaluation of the accumulation and will support the awareness of this problem.

The paper by Lall and Milley briefly reviewed the impact of trace elements from aquafeeds used for intensive aquaculture on freshwater and marine environments. Aquaculture has been the world's fastest growing food production system over the last two decades. It currently involves the production of over 195 species in freshwater, brackish water and seawater. Major environmental impacts of aquaculture have been associated mainly with high-input high-output intensive systems for carnivorous marine fish and salmonids cultured in raceways and cages. This includes the discharge of suspended solids (faeces and uneaten food), antibiotics, other drugs, disinfectants, algaecides and other chemical compounds leached in water. Minerals and organic enrichment of recipient waters causes the build-up of anoxic sediments, changes in benthic communities (alteration of seabed fauna and flora communities) and eutrophication of water. Carnivorous fish feeds contain a high proportion of fish meal and marine by-products supplemented with trace elements at a higher concentration than required due to limited information on their requirements and bioavailability from feed ingredients. These feeds are supplemented with Zn, Fe, Cu, Mn, Se, I and P, however, they may also contain other trace elements supplied from common feed ingredients. Elevated levels of Zn, Cu, Cd and Mn have been found in sediments under sea cages and solid wastes generated by land-based fish farms. These minerals have a wide range of effects on benthic organisms particularly in their reproduction, recruitment success and survival. The impacts of these minerals are generally restricted to the immediate vicinity of aquaculture operations and are significantly influenced by water current flow and other environmental factors.

In their paper, Römkens *et al.* showed that Cu and Zn loads to the soil in the Netherlands are among the highest in Europe. Farm data and model calculations for dairy operations indicate that net average accumulation rates for both zinc and copper range from less than 50 to more than 500 gram ha^{-1} yr^{-1} depending on farm and soil type. The majority of both Cu and Zn added to soil originate from feeds and feed additives. If present loads are to be continued, they stated that adverse effects on soil quality will become evident within 100 years for Cu. Zinc accumulation is also ongoing, but ecological threshold levels are not exceeded within 250 years. Leaching from soil, however, already has resulted in a large emission to ground- and surface waters. The impact on surface water quality for both Cu and Zn in low-lying areas is substantial and reductions of emissions of Cu and Zn are crucial to reach upcoming EU water quality standards.

The paper by Öborn *et al.* showed a comparison of the mineral balances of organic dairy farming to conventional farming, as well as possible adaptive approaches to management. The farm-gate balances were established for 14 elements (Ca, Cd, Cu, Fe, K, Mg, Mn, Mo, N, Na, P, S, Se and Zn). The farm importation of all nutrients was small in both systems, being higher in the conventional system. Some elements, i.e. Cu, Fe, Mo, Mn, Se and Zn were added as minerals in the feed. However, due to the low bioavailability of most of the added minerals, a large proportion ends up in manure. The importation of the minerals Fe, Mn, and Zn largely exceeded the outputs resulting in soil accumulation. The surplus was twice as high in the conventional as compared to the organic system due to different feeding strategies. Since Fe and Mn are major soil constituents this is a minor problem, while the long term accumulation of Zn and potentially also Cu is of more concern. When the field balance was calculated, it showed that the Zn balance was positive in all the years, and higher in the conventional system compared with the organic system (390-460 and 230-320 g ha^{-1}, respectively). For Cd, Cu and Zn the atmospheric deposition and leaching were relatively more important input and output flows as compared to the major nutrients (N, P, K) where the agronomic flows were dominating.

Three papers were presented on treating manure and its effect on concentration of heavy metals in the resulting products.

Novak *et al.* reported the large expansion of the pig population in North Carolina in the United States of America over the last 25 years from 2.5 to 10 million head. In addition, pig production changed from small size (< 300 animal units) to larger size (> 1000 animal units) operations. The majority (59%) of North Carolina's pig population is concentrated in only four counties that collectively comprise only 6% of North Carolina's total land area. This has raised concern about manure disposal and the soil's long-term ability to assimilate nutrients. Novak *et al.* (2004) reported situations where Cu and Zn loading rates from lagoon effluent exceeded crop removal rates that resulted in Cu and Zn accumulation in soil. In that study, 10 years of lagoon effluent applications to a sandy North Carolina soil caused surface soils (0 to 15 cm depth) mean Cu and Zn concentrations to increase from 0.5 and 1.7, respectively, to 3.6 and 7.9 mg/kg. This 10 year accumulation of Zn is approaching the threshold limit (12 to 20 mg/kg) of some Zn-sensitive crops. The authors showed that filtration of pig lagoon effluent using constructed wetlands can result in Cu and Zn concentration reductions prior to land application.

The paper by Nicholson and Chambers described some commonly used methods of manure storage and treatment, and explored how these may influence heavy metal inputs to agricultural soils. Livestock manure may contain heavy metals from a number of sources including dietary supplements, veterinary medicines, soil ingestion and via the disposal of footbaths into manure stores. Most manure storage and treatment technologies will not remove heavy metals from manures *per se*, but these practices can have important implications for heavy metal inputs to agricultural soils. Substantial losses of N often occur during the storage of solid manures and liquid slurries, with the heavy metals becoming more concentrated per unit N remaining compared with 'fresh' manure. This, may lead to higher metal loading rates at the field-level where manures are applied on an N basis. Several technologies for treatment of animal manures, including anaerobic digestion for biogas production, were described together with their impact on the partitioning of heavy metals between the solid and liquid fractions. At a given N (or P) application rate, this will increase the metal addition rate from the solid fraction but decrease it from the liquid fraction compared to an unseparated slurry.

Other ways of disposal/utilizing poultry litter than the spreading of poultry litter on the land as fertilizer was presented by Van Ryssen. These are the feeding of litter to ruminants, incinerating the product to produce energy, pellet it for use as a fertilizer, etc. Disposing of the poultry litter is becoming an increasing problem because of the expansion and concentration of poultry production units in some areas and the fact that the potential for environmental pollution and threats to human health exists or is perceived to exist in all these methods of disposal. The feeding of poultry litter to livestock is a very controversial topic. Problems associated with the feeding of poultry litter have been thoroughly investigated especially in the USA, though this practice is now also prohibited in the USA. The chemical composition and especially the concentration of elements can vary tremendously in poultry litter. Further practices and occurrences that could have drastic effects on the mineral element composition of poultry litter are the dietary inclusion of certain drugs and other components containing high levels of trace elements. The impact of feeding poultry litter to ruminants, the effects on animal health, possible negative interactions among trace minerals (high Fe content) on mineral absorbability were further outlined.

Based on several dietary scenarios both for growing-finishing pigs and for breeding sows included their piglets to about 25 kg, Dourmad *et al.* calculated the concentration of Cu and Zn in manure, their annual application per ha, and the time needed for Cu and Zn to reach 50 and 150 mg/kg DM soil, respectively. These levels correspond to half of the maximum allowed concentration of Cu and Zn in sludge-treated soils in France. Using Cu as a growth promoter is currently authorized in the EU for pigs up to 12 weeks of age, with diets containing a maximum of 170 mg Cu/kg. After 12 weeks of age, the maximal level of incorporation is 25 mg/kg. Compared to the former EU regulation (before 2004), these newly set legal limits result in a drastic reduction of Cu in manure, by almost 60%. Nevertheless, supplies of Cu remain higher than usual published requirements (less than 10 mg/kg) and average retention efficiency is still less than 1%.

Supplementing weaning diets for piglets with 1500 to 3000 mg Zn/kg as ZnO was also reported to stimulate their growth. This practice is still allowed in many countries but not in the EU. In fact, in 2003 the maximal level of Zn incorporation in all pig diets was reduced to 150 mg/kg, compared to 250 mg/kg before. The level of 150 mg Zn/kg is much closer to the published requirement which varies between 100 and 50 mg/kg according to growing stage and the studies concerned. Compared to a situation in which weanling pigs are fed a diet with 2500 mg Zn/kg from 8 to 15 kg body weight and 250 mg/kg thereafter, the present EU regulation results in 53% reduction of Zn excretion.

With the present EU regulation Cu and Zn contents in DM (about 350 and 1250 mg/kg DM, respectively) are below the maximal concentration allowed in sewage sludge in France (1000 and 3000 mg/kg DM, respectively), but they exceed the concentrations allowed for organic fertilizers (300 and 600 mg/kg DM, respectively). With the hypothesis that 170 kg N/ha are spread each year, it will take 80 to 160 years for the soil to reach the maximal limit fixed. This is much longer than with the previous regulation (25 to 50 years).

The paper by Hill described the reduction of trace element excretion in pigs, which so far has received limited attention in the USA. An over-supplementation pattern of trace elements has developed in the United States because marginal or deficient dietary concentrations of minerals were well known to reduce productivity, and mineral costs were low compared to the cost of other nutrients. Like pigs in all phases of production, many birds in the broiler and turkey industries are fed dietary concentrations beyond those recommended by the National Research Council (NRC, 1998). In the USA, pig producers have utilized the performance benefits of pharmacological Zn and Cu especially in nursery diets for the last 10 to 15 years. However, due to concerns about the faecal excretion of these nutrients and other minerals, scientists have begun to look for alternative sources and synergistic interactions to maintain enhanced growth and reduce threats to the environment. Several examples were presented using pharmacological Zn levels in nursery diets and organic vs. inorganic mineral sources on performance, health and excretion of trace minerals.

Young piglets normally have a low feed intake and have to adapt to a lot of changes at weaning. For instance, the change from milk to solid feed imposes an urgent need for quick intestinal adaptations to the new feed components due to the shift from milk-borne to plant-borne nutrients. This drastic shift of nutrient supply may cause several problems such as growth retardation, bacterial infections and diarrhoea. In the paper by Poulsen and Carlson some underlying mechanisms were described to explain that Cu and Zn have a preventive effect on post-weaning diarrhoea when piglets are fed high dietary concentrations of Cu and Zn. Although only a few studies have focused on the effect of Cu in the gastrointestinal tract of pigs, the action of Cu is generally attributed to its antimicrobial activity. Also high dietary levels of ZnO reduce the bacterial activity in the digesta of the gastrointestinal tract in newly-weaned piglets compared with piglets fed low levels of ZnO. It was stated that the influence of high levels of ZnO on the gastrointestinal microflora resembled more or less the working mechanisms suggested for some growth-promoting antibiotics.

Feeding high dietary concentrations of ZnO to piglets showed that the increased performance and reduced frequency of diarrhoea was accompanied by an increase in plasma Zn concentration. These results suggest that high dietary Zn has a physiological effect in the young weaned piglet and that Zn exerts its effect via a general improvement in the Zn status of the piglets. The positive effect of high dietary Zn may also be due to reducing the intestinal mucosal susceptibility to secretagogues that activate chloride secretion but also other physiological mechanisms may be involved.

In their paper, Spears and Hansen discussed various experimental approaches and criteria for assessing bioavailability of essential trace minerals. Bioavailability is the proportion of an ingested trace element that is absorbed and utilized. Relative bioavailability of trace mineral sources have been determined using animals fed either diets below or diets considerably above their dietary requirement for the particular mineral. In theory, bioavailability studies should be conducted in animals receiving the mineral in question at dietary levels below their requirement for a specific function. This is because of homeostatic control mechanisms that attempt to maintain tissue trace mineral concentrations within a fairly narrow range.

A number of studies have estimated relative bioavailability of trace mineral sources by feeding the mineral at concentrations well above the animal's requirement and measuring tissue accumulation of the element. This is done because it is difficult to formulate diets deficient in certain trace minerals using practical feed ingredients. Estimating bioavailability in animals supplemented with high concentrations of trace minerals offers the advantage of using practical diets and greater statistical sensitivity. Depending on the trace mineral of interest, criteria that have to be used to estimate bioavailability include growth, enzyme activity, mineral absorption and retention, blood or tissue mineral concentrations and prevention of a certain deficiency sign. Relative bioavailability estimates of trace minerals from different sources can vary depending on the criterion used to assess bioavailability and the level of antagonists present in the experimental diets. Criteria used to determine bioavailability of Zn, Cu, Mn, Fe, Se and Co have been critically reviewed in their presentation.

Intrinsic trace element concentrations in feedstuffs used in diets for pigs are mainly suboptimal for Zn, Mn, Cu, Se and I. For supplementation inorganic (oxides, sulfates, chlorides, carbonates) and/or organic sources which can be chelates, metal amino acid complexes, metal proteinates or metal polysaccharide complexes are used. In order to reduce excretion of the trace minerals, those sources should be used that have a high absorbability. Bioavailability, which is characterized not only by the apparent absorbability but also by retention in different organs, mineral-specific enzyme activities (metalloenzymes), health and performance, varies relative to sulfates between 0 to 185% (Ammermann *et al.*, 1995). These authors concluded that oxides are less available than sulfates, and inorganic compounds are less available than organically-bond minerals. Männer performed an experiment to investigate the effect on mineral bioavailability (apparent absorbability, performance, haemoglobin and plasma concentrations) of three trace mineral sources (sulfates, chelates, glycinates) for Fe, Mn, Zn and Cu in weaned piglets. The three supplemented sources for Fe, Mn, Zn and Cu were included into the basal diet to obtain 90% of NRC recommendations.

From this study it was found that the average absorbabilities of mineral sulfates were not better than those naturally found in the unsupplemented control diet. The overall mineral absorbability of the chelates relative to the sulfates was improved by 19%, while the glycinates were best with an overall improvement of the apparent absorbability by 31% when compared to sulfates. His conclusion was that based on the method and response criteria he used, the organically-bound minerals studied had a higher absorbability in pigs than the inorganically-bound Zn. These organically-bound trace minerals may be of use to further decrease the excretion in faeces and manure.

Recommendations for essential trace elements, as well as, their nutritive values are still based on a gross basis and not on an absorbable basis. The reason for this and the limitations for further improvements were discussed by Windish and Ettle. Absorption and/or excretion of most of the essential trace elements are regulated by the so called homeostatic regulation which acts over a wide range of dietary supply, and results in elimination of trace elements from the body, as soon as intake exceeds metabolic requirement. This principle is an important part of current trace mineral recommendations as they usually include considerable safety margins to secure sufficient supply at unfavourable nutritional conditions and low bioavailability of trace mineral sources.

Calculation of gross requirement is done by a conversion factor usually set rather low in order to maintain security in adequate trace mineral supply even under adverse feeding conditions. Another step towards improving supply systems is quantification of bioavailability of dietary essential trace element sources. The final aim of such a concept is to tabulate feedstuffs for their capacity to provide metabolically available trace mineral quantities.

Essential trace elements are known to have various interactions with other dietary components. These interactions significantly affect the practical application of feed tables, as values from single feedstuffs are then not more additive in describing the entire feed. An example of this case is found with strong chelators such as phytate and oxalate, which may severely depress bioavailability of essential trace metals. It is also known that some chelators such as ascorbic acid or citrate enhance absorption of trace elements (Fe). In general, interactions increase the gross requirement of the respective trace element.

Organic trace mineral sources are often considered to be less affected by such interactions than inorganic sources. Especially in essential trace metals, however, hypotheses on the mode of action are still controversial, stressing a more pronounced availability of the (inorganic) trace mineral at the site of absorption, or absorption pathways apart from regular transport systems of the respective trace element. Organic trace mineral formulations seem to bring essential trace elements more effectively into the body, thus improving the security of sufficient trace mineral supply and providing an option to reduce safety margins.

Interactions between essential trace elements and other dietary factors severely limit the possibility of properly establishing the animals' requirement as well as the capacity of the feed components for provision of available trace elements. Especially the latter forces nutritionists to maintain safety margins when expressing recommendations.

Lall reviewed the current state of knowledge on trace element nutrition of fish and crustaceans with an emphasis on Fe, Zn, Cu and Se. Unlike terrestrial animals, fish and crustaceans have unique physiological mechanisms to absorb and retain minerals from their diets and water. Although the mineral uptake by the fish gill from water has been an area of active research, there have been few studies on the absorption of major trace elements from the gut. An excessive intake of minerals from either diet or gill uptake can cause toxicity and therefore a fine balance between mineral deficiency and surplus is vital for aquatic organisms to maintain their homeostasis either through increased absorption or excretion. Major gaps exist in the knowledge of mineral requirements and physiological functions in fish and crustaceans, as well as, bioavailability of trace elements from mineral supplements and feed ingredients. This makes it difficult to permit reliable prediction of the development of deficiency and toxicity states within the given states of culture in fresh, brackish and sea water environmental conditions.

Quantitative dietary requirements have been reported for six trace minerals (Zn, Fe, Cu, Mn, I and Se) for certain fish species, however, none for shrimps and other crustaceans. In diets containing high levels of plant protein, trace element supplementation is necessary to improve growth and bone mineralization of carnivorous salmonid and marine fishes.

The paper by Girard dealt with trace mineral status and immunity. The intestinal mucosa is the largest interface of the animal with the outside world which has two functions: it is a barrier to noxious outside stimuli and is a site of active exchange. An active immune system is crucial in protecting the animal against all kinds of infections and antigens. In case of anorexia or unsanitary conditions, absorption of trace minerals might not be sufficient. Even a moderate immune response increases the concentration of some trace minerals in blood and might mask some nutritional deficiencies. His paper evaluated how methods for reducing trace mineral concentrations in diets might affect the immune response in animals.

Another aspect of trace minerals, namely the interference with scrapie and bovine spongiform encephalopathy (BSE), was presented by Brown. These diseases, like many neurodegenerative disorders, have been linked to metal metabolism because the involved proteins would bind to metal. The normal protein is a Cu-binding protein and Brown's group has recently verified that it is a Cu-binding protein of high affinity. Previously, it has been suggested that the protein can also bind Mn and this will initiate its conversion to a protease-resistant form with an altered conformation. These results suggest that binding of Mn to prion protein (PrP) will result in an alteration in its biochemical properties, altering its conformation and increasing its resistance to proteases.

It was suggested that difference in the dietary intake of Mn and Cu could result in spontaneous prion disease. However, there is little evidence that dietary intake of Cu and Mn could initiate the disease in the absence of other factors. Nevertheless, there is now substantial evidence that prion diseases cause alteration in the levels of Mn in the brain and blood of infected animals before the onset of clinical signs of the disease. The origin of the increased Mn in the brains and blood of TSE infected animals is still unknown, and other possibilities were discussed in his paper.

Concluding remarks and conclusion

The livestock sector and aquaculture emerge as one of the top two or three most significant contributors to the most serious environmental problems, at every scale from local to global. Global production of meat and milk is projected to double from 2000 to 2050. The environmental impact per unit of livestock production must be cut by one half, just to avoid the level of damage worsening beyond its present level. As a first step the EU set maximal limits on concentrations of several essential trace minerals and heavy metals in feeds for livestock and fish. However, it is suggested that these levels should be revised with regard to achieving sustainability of the environment.

Several sophisticated and integrated models to estimate trace and heavy metal balances are available at national and farm levels. These raise awareness and clearly indicate that there is a net surplus of trace minerals per ha of land, and of Cu and Zn in particular. The largest input of Cu and Zn depends on animal density and animal type, feeding strategy and country (feeds, fertilizers, sewage sludge, atmospheric deposition). It has been shown that there is a large difference between farm balances even in the same region.

Applying manure to soils may lead to exceeding accepted mineral concentrations such as Cu and Zn in the soil together with unacceptable high concentrations in ground- and surface waters in the long term. The latter may be the first limiting factor with regard to eco-toxicological aspects. Differentiation in soil type, organic matter content, precipitation surplus, water table level, etc., however, is necessary to estimate long term effects on the environment. Aquaculture faces the same environmental problems as land animals.

The concentration of heavy metals in manure may be largely affected by treatment, and becomes higher in relation to its nitrogen or organic matter content. Use of constructed wetlands may reduce Cu and Zn concentration of manure before land application. In some countries poultry litter is used as a feed raw material for ruminants.

Trace minerals such as Cu, Zn, Fe, Mn and Se are essential elements for animals. Thus an insufficient supply of one of these elements impairs an efficient animal production and may lead to compromised animal welfare. As the raw materials may contain not enough of these trace minerals, they are added to the diet, mainly as inorganic compounds. The amount added depends on the requirement of the specific animal and the bioavailability of the mineral compound.

Concentrations of Cu and Zn much higher than the physiological requirements may result in stimulatory effects on animal performance and health. Several modes of actions were given for the improved pig performance. Although these high dietary concentrations are effective and cheap, they are not allowed in the EU. The environmental implications of using such high concentrations of Cu and Zn in pig diets were shown. In the future, further reductions in Cu and Zn excretion should be possible, resulting in a more equilibrated balance between manure spreading and uptake by plants.

Bioavailability of a mineral is a key factor in mineral recommendations for animals and fish. However, many interactions exist and quantitative understanding of interactions between essential trace elements and other dietary components is therefore a bottleneck for further progress in development of more precise and secure systems in essential trace element nutrition. Several approaches are possible to assess bioavailability of trace minerals. Recent data indicates that some organic mineral sources are more available to the animal, and thus may have the potential to reduce environmental contamination and increase productivity. As the dietary excess of trace minerals contributes to higher feed costs and undesirable ecological impacts, the concept of plenty to meet the safety margins must be revised. It has been debated whether a factorial approach to assess trace mineral requirements will be possible in the near future.

Knowledge of trace minerals on bioavailability and requirements of fish and crustaceans is scarce. A further complicating factor in assessing allowances for animals is the effect on immunity and health. Finally it has been shown that trace mineral metabolism is clearly linked with prion diseases.

It can be concluded that pressure from society will lead to reduction of mineral load to the fields and it may be speculated that concentrations of trace minerals in ground- and surface waters will be the first limiting factor over concentrations in the soil. In the near future pharmacological levels of Cu and Zn in diets will possibly not be allowed. Bioavailability and requirements of trace minerals need to be more specified and diversified to be able to reduce safety margins.

References

Ammermann, C.B., D.H. Baker and A.J. Lewis, 1995. Bioavailability of nutrients for animals: Amino acids, minerals, vitamins. Academic Press, San Diego.

Eckel, H., U. Roth, H. Döhler, F. Nicholson and R. Unwin (eds.), 2005. Assessment and reduction of heavy metal input into agro-ecosystems. Final report of the EU-Concerted Action AROMIS. KTBL publication no. 432, Darmstadt, Germany.

Novak, J.M., D.W. Watts, and K.C. Stone. 2004. Copper and Zinc accumulation, profile distribution, and crop removal in Coastal Plain soils receiving long-term, intensive applications of swine manure. Trans. ASAE 47, 1513-1522.

NRC, 1998. Nutrient Requirements of Swine. Natl. Acad. Press, Washington, DC.

List of authors

A. Ahmadi, Animal Science Department, Bu-Ali Sina University, Hamedan, Iran

H. Aliarabi, Animal Science Department, Bu-Ali Sina University, Hamedan, Iran

A. Alric, Ecolab UMR 5245 CNRS/UPS/INPT ENSAT BP 32607, 31326 Castanet-Tolosan cedex, France

N. Andres, Institute of Research and Food Technology (IRTA), Passeig de Gracia, 44,3r., 08007 Barcelona, Spain

A. Arun Kumar, Project Directorate on Poultry, Rajendranagar, 500 030 Hyderabad, India

C.M. Atkin, ASRC, Harper Adams University College, Newport, Shropshire TF10 8NB, United Kingdom

H. Bengtsson, Department of Soil Sciences, Swedish University of Agricultural Sciences (SLU), Box 7014, 750 07 Uppsala, Sweden

A. Berk, Institute of Animal Nutrition, FAL Braunschweig, Bundesallee 50, 38116 Braunschweig, Germany

G. Bertin, Alltech France, EU Regulatory Department, Place Marie-Jeanne Bassot, 92300 Levallois-Perret, France

L.T.C. Bonten, Alterra, Wageningen UR, P.O. Box 47, 6700 AA Wageningen, The Netherlands

J.Y. Bottero, CEREGE UMR 6635 CNRS – Université Paul Cézanne Aix-Marseille III; IFR PMSE 112, Europôle Méditerranéen de l'Arbois, BP 80, 13545 Aix-en-Provence Cedex 04, France

D.R. Brown, Department of Biology and Biochemistry, University of Bath, Bath BA2 7AY, United Kingdom

D. Carlson, Faculty of Agricultural Sciences, University of Aarhus, Institute of Animal Health, Welfare and Nutrition, Blichers Allé 20, P.O. Box 50, 8830 Tjele, Denmark

V.K. Chalasani, Project Directorate on Poultry, Rajendranagar, 500 030 Hyderabad, India

B.J. Chambers, ADAS Gleadthorpe, Meden Vale, Mansfield, Nottinghamshire NG20 9PF, United Kingdom

A. Chhabra, Dairy Cattle Nutrition Division, National Dairy Research Institute, Karnal, India

B.R. Choi, Milae Resource Ltd. Seoul, Korea

H.B. Choi, School of Agricultural Biotechnology, Seoul National University, Korea

L.B. Correa, Faculdade de Zootecnia e Engenharia de Alimentos, University of São Paulo, Duque de Caxias Norte, 225, CEP 13635-900, Pirassununga-SP, Brazil

R.D. Criste, National Research and Development Institute for Biology and Animal Nutrition (IBNA), Calea Bucureşti, 1, jud. Ilfov., 077015 Baloteşti, Romania

J.A. Cunha, Faculdade de Zootecnia e Engenharia de Alimentos, University of São Paulo – CP 23, CEP 13600-000, Pirassununga, Brazil

W. de Vries, Alterra, Wageningen UR, P.O. Box 47, 6700 AA Wageningen, The Netherlands

G.R. Del Claro, Faculdade de Zootecnia e Engenharia de Alimentos, University of São Paulo, Duque de Caxias Norte, 225, CEP 13635-900, Pirassununga-SP, Brazil

E. Doelsch, CIRAD, Environmental Risks of Recycling Research Unit, Station de La Bretagne, BP 20, 97408 Saint-Denis Messagerie Cedex 9, La Réunion, France

H. Döhler, Association for Technology and Structures in Agriculture (KTBL), Bartningstrasse 49, 64289 Darmstadt, Germany

J.Y. Dourmad, INRA, UMR1079 Systèmes d'Élevage, Nutrition Animale et Humaine, 35590 Saint-Gilles, France

M. Dupas, IDENA, 21 rue du Moulin, 44880 Sautron, France

S. Durosoy, Pancosma S.A.,Voie des Traz 6, 1218 Geneva, Switzerland

H. Eckel, Association for Technology and Structures in Agriculture (KTBL), Bartningstrasse 49, 64289 Darmstadt, Germany

S.G. Edwards, Animal Science Research Centre, Harper Adams University College, Edgmond, Newport, Shropshire TF10 8NB, United Kingdom

J. Els, Virbac RSA (Pty) Ltd, 38 Landmarks ave, Samrand Business Park, Centurion, 0157 South Africa

T. Ettle, Department of Food Science and Technology, University of Natural Resources and Applied Life Sciences Vienna, Gregor Mendel-Strasse 33, 1180 Vienna, Austria

N. Felix, Fisheries College and Research Institute, Tamilnadu Veterinary and Animal Sciences University, 628008 Tuticorin, India

G. Flachowsky, Institute of Animal Nutrition, FAL Braunschweig, Bundesallee 50, 38116 Braunschweig, Germany

J. Fleckenstein, Institute of Plant Nutrition and Soil Science, FAL Braunschweig, Bundesallee 50, 38116 Braunschweig, Germany

P. Fleischer, Clinic of Diseases of Ruminants, Faculty of Veterinary Medicine, University of Veterinary and Pharmaceutical Sciences Brno, Palackeho 1-3, 612 42 Brno, Czech Republic

P.J. Gerber, Food and Agriculture Organisation of the United Nations (FAO), Livestock Environment and Development (LEAD) Initiative, 00100 Rome, Italy

V. Girard, Faculté de médecine vétérinaire, Université de Montreal, Montréal-Québec, Canada

J.E. Groenenberg, Alterra, Wageningen UR, P.O. Box 47, 6700 AA Wageningen, The Netherlands

M. Guiresse, Ecolab UMR 5245 CNRS/UPS/INPT ENSAT BP 32607, 31326 Castanet-Tolosan cedex, France

G.M Gustafson, Centre for Sustainable Agriculture (CUL), Box 7047, 750 07 Uppsala, Sweden

S.L. Hansen, Department of Animal Science, North Carolina State University, Raleigh, NC 27695-7621, USA

J.-L. Hazemann, Laboratoire de Cristallographie, CNRS, UPR 5031, BP 166, 38042 Grenoble Cedex 09, France

N. Helle, TeLA Technische Lebensmittel- und Umweltanalytik, Fischkai 1, 2757 Bremerhaven, Germany

G.M. Hill, Michigan State University, 2209 Anthony Hall, East Lansing, MI, 48824, USA

J. Holmqvist, Institute of Agricultural and Environmental Engineering (JTI), Box 7033, 750 07 Uppsala, Sweden

S.A. Hosseini Siyar, Animal Science Department, Bu-Ali Sina University, Hamedan, Iran

H.C. Humberto, Faculdade de Zootecnia e Engenharia de Alimentos, University of São Paulo – CP 23, CEP 13600-000, Pirassununga, Brazil

Y.D. Jang, School of Agricultural Biotechnology, Seoul National University, Korea

C. Jondreville, INRA, UMR1079 Systèmes d'Élevage, Nutrition Animale et Humaine, 35590 Saint-Gilles, France

A.W. Jongbloed, Animal Sciences Group of Wageningen UR, P.O. Box 65, 8200 AB Lelystad, The Netherlands

S. Jonsson, Dept. of Agricultural Research for Northern Sweden, SLU, Patrons Allé 10, 943 31 Öjebyn, Sweden

D.T. Juniper, University of Reading, School of Agriculture, Policy and Development, Earley Gate, RG6 6ARReading, Berks, United Kingdom

D. Kampf, Orffa Deutschland GmbH, Augustastraße 12, 46483 Wesel, Germany

A. Khatibjo, Animal Science, Tarbiat Modares University, Tehran, Iran

Y.Y. Kim, School of Agricultural Biotechnology, Seoul National University, Korea

J. Kogut, Animal Sciences Group of Wageningen UR, P.O. Box 65, 8200 AB Lelystad, The Netherlands

S.P. Lall, National Research Council, Institute for Marine Biosciences, 1411 Oxford Street, Halifax, Nova Scotia B3H 3Z1, Canada

S. Legros, CIRAD, Environmental Risks of Recycling Research Unit, Station de La Bretagne, BP 20, 97408 Saint-Denis Messagerie Cedex 9, La Réunion, France

K.E. Lloyd, Department of Animal Science, North Carolina State University, Raleigh, 27695 NC, USA

A.M. Mackenzie, Animal Science Research Centre, Harper Adams University College, Edgmond, Newport, Shropshire TF10 8NB, United Kindom

K. Männer, Free University Berlin, Faculty of Veterinarian Medicine, Institute of Animal Nutrition, Brümmerstraße 34, 14195 Berlin, Germany

C.E. Marcat, Ecolab UMR 5245 CNRS/UPS/INPT ENSAT BP 32607, 31326 Castanet-Tolosan cedex, France

F.R. Martins, Faculdade de Zootecnia e Engenharia de Alimentos, University of São Paulo, Duque de Caxias Norte, 225, CEP 13635-900, Pirassununga-SP, Brazil

A. Masion, CEREGE UMR 6635 CNRS – Université Paul Cézanne Aix-Marseille III; IFR PMSE 112, Europôle Méditerranéen de l'Arbois, BP 80, 13545 Aix-en-Provence Cedex 04, France

H. Menzi, Swiss College of Agriculture (SHL), 3052 Zollikofen, Switzerland

N. Michael, ABS Global, 1525 River Rd, DeForest, Wisconsin 53532, USA

F.O. Miller, Faculdade de Zootecnia e Engenharia de Alimentos, University of São Paulo – CP 23, CEP 13600-000, Pirassununga, Brazil

J.E. Milley, National Research Council, Institute for Marine Biosciences, 1411 Oxford Street, Halifax, Nova Scotia B3H 3Z1, Canada

L. Misurova, Clinic of Diseases of Ruminants, Faculty of Veterinary Medicine, University of Veterinary and Pharmaceutical Sciences Brno, Palackeho 1-3, 612 42 Brno, Czech Republic

K. Mitchell, Valley Veterinarians Inc 2861 South K St, Tulare, CA 93274, USA

A.-K. Modin-Edman, Institute of Agricultural and Environmental Engineering (JTI), Box 7033, 750 07 Uppsala, Sweden

S.W. Moolenaar, Nutrient Management Institute, P.O. Box 250, 6700 AG Wageningen, The Netherlands

D. Nagalakshmi, College of Veterinary Science, Sri Venkateswara Veterinary University, Rajendranagar, 500 030 Hyderabad, Andhra Pradesh, India

A. Neels, Université de Neuchâtel, Institut de Microtechnique, Rue A.-L. Breguet 2, 2000 Neuchâtel, Switzerland

F.A. Nicholson, ADAS Gleadthorpe, Meden Vale, Mansfield, Nottinghamshire NG20 9PF, United Kingdom

S.I. Nilsson, Dept. of Soil Sciences, Swedish University of Agricultural Sciences (SLU), Box 7014, 750 07 Uppsala Sweden

J.M. Novak, United States Department of Agriculture, Agriculture Research Service, Coastal Plains Soil, Water and Plant Research Center, 2611 West Lucas Street, Florence, SC 29501, USA

I. Öborn, Department of Soil Sciences, Swedish University of Agricultural Sciences (SLU), Box 7014, 750 07 Uppsala Sweden

S. Oguey, Pancosma S.A, Voie-des-Traz 6, 1218 Geneva, Switzerland

F.A. Paiva, Faculdade de Zootecnia e Engenharia de Alimentos, University of São Paulo, Duque de Caxias Norte, 225, CEP 13635-900, Pirassununga-SP, Brazil

A.K. Panda, Project Directorate on Poultry, Rajendranagar, 500 030 Hyderabad, India

S.E. Pattinson, Animal Science Research Centre, Harper Adams University College, Edgmond, Newport, Shropshire TF10 8NB, United Kindom

L. Pavlata, Clinic of Diseases of Ruminants, Faculty of Veterinary Medicine, University of Veterinary and Pharmaceutical Sciences Brno, Palackeho 1-3, 612 42 Brno, Czech Republic

A. Pechova, Clinic of Diseases of Ruminants, Faculty of Veterinary Medicine, University of Veterinary and Pharmaceutical Sciences Brno, Palackeho 1-3, 612 42 Brno, Czech Republic

R.H. Phipps, University of Reading, School of Agriculture, Policy and Development, Earley Gate, RG6 6ARReading, Berks, United Kingdom

E. Pinelli, Ecolab UMR 5245 CNRS/UPS/INPT ENSAT BP 32607, 31326 Castanet-Tolosan cedex, France

P. Pouech, Apesa Hélioparc, 64053 Pau Cedex, France

H.D. Poulsen, Faculty of Agricultural Sciences, University of Aarhus, Institute of Animal Health, Welfare and Nutrition, Blichers Allé 20, P.O. Box 50, 8830 Tjele, Denmark

O. Proux, Laboratoire de Géophysique Interne et de Tectonophysique, UMR CNRS Université Joseph Fourier, 1381 rue de la piscine, Domaine Universitaire, 38400 St Martin d'Hères, France

M.V.L.N. Raju, Project Directorate on Poultry, Rajendranagar, 500 030 Hyderabad, India

S.V. Rama Rao, Project Directorate on Poultry, Rajendranagar, 500 030 Hyderabad, India

Y.R. Reddy, College of Veterinary Science, Sri Venkateswara Veterinary University, Rajendranagar, 500 030 Hyderabad, Andhra Pradesh, India

P.F.A.M. Römkens, Alterra, Wageningen UR, P.O. Box 47, 6700 AA Wageningen, The Netherlands

J. Rose, CEREGE UMR 6635 CNRS – Université Paul Cézanne Aix-Marseille III; IFR PMSE 112, Europôle Méditerranéen de l'Arbois, BP 80, 13545 Aix-en-Provence Cedex 04, France

U. Roth, Association for Technology and Structures in Agriculture (KTBL), Bartningstrasse 49, 64289 Darmstadt, Germany

J.B.J. van Ryssen, Department of Animal and Wildlife Sciences, University of Pretoria, Pretoria, South Africa

H. Saint-Macary, CIRAD, Environmental Risks of Recycling Research Unit, Station de La Bretagne, BP 20, 97408 Saint-Denis Messagerie Cedex 9, La Réunion, France

E. Salomon, Chemical Engineering, Lund University, Box 124, 221 00 Lund, Sweden

A.A. Saki, Animal Science Department, Bu-Ali Sina University, Hamedan, Iran

A. Saran-Netto, Faculdade de Zootecnia e Engenharia de Alimentos, University of São Paulo, Duque de Caxias Norte, 225, CEP 13635-900, Pirassununga-SP, Brazil

P. Schlegel, Pancosma S.A., Voie des Traz 6, 1218 Geneva, Switzerland

U. Schultheiß, Association for Technology and Structures in Agriculture (KTBL), Bartningstrasse 49, 64289 Darmstadt, Germany

P. Senthilkumar, College of Veterinary Science, Sri Venkateswara Veterinary University, Rajendranagar, 500 030 Hyderabad, Andhra Pradesh, India

L.H.O. Silva, Faculdade de Zootecnia e Engenharia de Alimentos, University of São Paulo – CP 23, CEP 13600-000, Pirassununga, Brazil

L.A. Sinclair, ASRC, Harper Adams University College, Newport, Shropshire TF10 8NB, United Kingdom

S. Slosarkova, Clinic of Diseases of Ruminants, Faculty of Veterinary Medicine, University of Veterinary and Pharmaceutical Sciences Brno, Palackeho 1-3, 612 42 Brno, Czech Republic

W.A. Smith, MULTIMIN® USA Inc., 31910 Country Club drive, Porterville, CA 93257-9610, USA

J.W. Spears, Department of Animal Science, North Carolina State University, Raleigh, NC 27695-7621, USA

M. Spolders, Institute of Animal Nutrition, FAL Braunschweig, Bundesallee 50, 38116 Braunschweig, Germany

H. Steinfeld, Food and Agriculture Organisation of the United Nations (FAO), 00100 Rome, Italy

H. Stoeckli-Evans, Université de Neuchâtel, Institut de Microtechnique, Rue A.-L. Breguet 2, 2000 Neuchâtel, Switzerland

A. Storch, ABS Global, 1525 River Rd, DeForest, Wisconsin 53532, USA

K. Sudhakar, College of Veterinary Science, Sri Venkateswara Veterinary University, Rajendranagar, 500 030 Hyderabad, Andhra Pradesh, India

G.S. Sunder, Project Directorate on Poultry, Rajendranagar, 500 030 Hyderabad, India

H. Sverdrup, Institute of Agricultural and Environmental Engineering (JTI), Box 7033, 750 07 Uppsala, Sweden

H.V.L.N. Swami, Alltech, Bangalore, India.

A.A. Szogi, United States Department of Agriculture, Agriculture Research Service, Coastal Plains Soil, Water and Plant Research Center, 2611 West Lucas Street, Florence, SC 29501, USA

M.M. Tabatabaie, Animal Science Department, Bu-Ali Sina University, Hamedan, Iran

I. Taranu, National Research and Development Institute for Biology and Animal Nutrition (IBNA), Calea Bucureşti, 1, jud. Ilfov., 077015 Baloteşti, Romania

D. Torrallardona, Institute of Research and Food Technology (IRTA), Passeig de Gracia, 44,3r., 08007 Barcelona, Spain

A. Untea, National Research and Development Institute for Biology and Animal Nutrition (IBNA), Calea Bucureşti, 1, jud. Ilfov., 077015 Baloteşti, Romania

D.W. Watts, United States Department of Agriculture, Agriculture Research Service, Coastal Plains Soil, Water and Plant Research Center, 2611 West Lucas Street, Florence, SC 29501, USA

D. Wilde, Alltech (UK) Limited, Alltech House, Ryhall Road, Stamford, Lincolnshire PE9 1TZ, United Kingdom

R.G. Wilkinson, Animal Science Research Centre, Harper Adams University College, Edgmond, Newport, Shropshire TF10 8NB, United Kindom

C.L. Williams, Animal Science Research Centre, Harper Adams University College, Edgmond, Newport, Shropshire TF10 8NB, United Kingdom

W. Windisch, Department of Food Science and Technology, University of Natural Resources and Applied Life Sciences Vienna, Gregor Mendel-Strasse 33, 1180 Vienna, Austria

M.A. Zanetti, Faculdade de Zootecnia e Engenharia de Alimentos, University of São Paulo, Duque de Caxias Norte, 225, CEP 13635-900, Pirassununga-SP, Brazil

Keyword index

H

haemocyanin 206
haemoglobin 148, 166, 184
health 216
heart 167
homeostatic control 162, 168, 170, 178, 180, 188, 215, 325
homocysteine 169
hoof disinfectants 40
hydrogen sulfide 80
hydrolysation degree 247

I

immune
 – functions 215
 – non-specific system 216
 – response 261, 278, 327
immunity 215, 326
increasing populations 21
infrared spectroscopy 250
insulin-like growth factor-I 154
intensive production 28, 115, 123
interactions 192, 261, 284, 326, 328
intestinal
 – morphology 154
 – permeability 154
invertebrates 81
iodine 209, 307
iron 93, 148, 166, 180, 195, 204, 267, 273, 281
 – glycinate 246

K

kidney 272

L

lagoon 46
lambs 278, 299
land use 22
layer 258
leaching 35, 95, 116, 133
lead 34
liming material 47
litter 101
liver 165, 166, 169, 272, 274, 277, 284

livestock 21
 – environment interactions 30
 – farming 39
 – grazing 22
 – product demand 21
lung 167

M

macronutrients 91
manganese 93, 144, 167, 180, 207, 234, 235, 261, 293
 – glycinate 246
manure 38, 42, 45, 47, 63, 96, 132, 133, 139
 – storage 55, 322
 – treatment 55, 322
mastitis 220, 221, 290
meat 319
meta-analysis 284
metallothionein 146, 153
methane 23
methylmalonate 169
microflora 151
micronutrients 91
milk 319
 – amyloid A 290
mineral 33, 82, 206, 210
 – balance 127
 – chelates 211
 – feed 91
 – loading rate 59, 60
 – nonessential 82
 – supplements 115, 128, 133
 – uptake 81
model 64, 116, 118
molluscs 77
molybdenum 93, 166, 236, 281, 284
monogastrics 161
muscle 168
muscular dystrophy 209

N

net-pens 79
neuronal function 232
nickel 34
nitrogen 81